Programmable Logic Handbook

Other McGraw-Hill Books of Interest

Boswell ▪ Subcontracting Electronics
Boswell and Wickam ▪ Surface Mount Guidelines for Process Control, Quality, and Reliability
Chen ▪ Computer Engineering Handbook
Christiansen ▪ Electronics Engineers' Handbook, 4/e
Coombs ▪ Electronic Instrument Handbook, 2/e
Coombs ▪ Printed Circuits Handbook, 4/e
Ginsberg ▪ Printed Circuits Design
Juran and Gryna ▪ Juran's Quality Control Handbook
Jurgen ▪ Automotive Electronics Handbook
Jurgen ▪ Digital Consumer Electronics Handbook
Manko ▪ Solders and Soldering, 3/e
Rao ▪ Multilevel Interconnect Technology
Sze ▪ VLSI Technology
Van Zant ▪ Microchip Fabrication, 3/e

Programmable Logic Handbook

PLDs, CPLDs, and FPGAs

Ashok K. Sharma

McGraw-Hill

New York San Francisco Washington, D.C. Auckland Bogotá
Caracas Lisbon London Madrid Mexico City Milan
Montreal New Delhi San Juan Singapore
Sydney Tokyo Toronto

Library of Congress Cataloging-in-Publication Data

Sharma, Ashok K.
Programmable logic handbook / Ashok K. Sharma.
p. cm.
Includes index.
ISBN 0-07-057852-4
1. Programmable logic devices. 2. Field programmable gate arrays.
I. Title
TK7872.L64S48 1998
621.39'5—dc21 98-10726
CIP

McGraw-Hill

A Division of The McGraw-Hill Companies

1 2 3 4 5 6 7 8 9 0 FGR/FGR 9 0 3 2 1 0 9 8

ISBN 0-07-057852-4

The sponsoring editor for this book was Stephen S. Chapman and the production supervisor was Tina Cameron. It was set in New Century Schoolbook by J. K. Eckert & Company, Inc.

Printed and bound by Quebecor/Fairfield.

This book is printed on recycled, acid-free paper containing a minimum of 50% recycled, de-inked fiber.

Contents

Preface

In generic terms, *programmable logic device* (PLD) refers to any type of integrated circuit that can be configured in the field by the end user for a logic design implementation. A major attribute of this category of devices is their user configurability, which differentiates them from more complex *metallization programmable gate arrays* (MPGAs) and custom ASIC approaches that tend to involve nonrecurring engineering (NRE) costs and longer design cycles.

This book discusses programmable logic devices including *complex PLDs* (CPLDs) and *field programmable gate arrays* (FPGAs) that have gained a tremendous popularity and grown into a multibillion-dollar market world wide. The growth of the PLD industry has been driven by applications as diverse as telecommunication, networking, and video games, as well as advanced military and space satellite systems and higher-performance custom computing applications. Currently, there are dozens of vendors of PLDs, CPLDs, and FPGAs, offering hundreds of product families along with their design support and software tools. A major goal of writing this book is to provide the PLD users—design and application engineers—with a useful resource guide on the latest available technologies, architectures, device selection trade-offs, package selection issues, and design and development tools. Chapter 1 provides an introduction to these various issues. Chapters 2 and 3 review product families, general architectures, programming technologies, and design and development tools from most major PLD vendors.

The PLD design is based on simple combinatorial and registered logic functions that are the basic building blocks for libraries supplied by the vendors. The design methodology involves design entry, design implementation, and design verification. For high-density CPLDs and FPGA designs, gate-level entry tools are cumbersome and therefore the use of hardware description languages such as VHDL and Verilog, and logic synthesis tools are becoming more popular. These PLD design issues are discussed in Chap. 4, which also includes special topics such as the *finite state machine* (FSM), *digital signal processing* (DSP), and PCI bus interfacing designs. This chapter also provides rules and guidelines for high-performance PLD based designs.

The current trend in programmable logic is toward higher densities as the manufacturers migrate to submicron 0.5 µm and finer geometry processes. This has resulted in devices with a wide variety of additional features such as JTAG interface, PCI bus compliance, and mixed-voltage (5 V/3.3 V) applications. The high-density, in-system programmable (ISP) devices offer significant advantages over ASICs to board-level designers by providing them with the flexibility of changing product features by device reconfiguration.

A major consideration in the PLD selection process is the evaluation of true logic capacity and performance of the target PLD and then matching it with application requirements. Programmable Electronics Performance Corporation (PREP) provides a comparison standard for the relative capacity of CPLDs and FPGAs by using certain benchmark circuits that are representative of typical applications. These published benchmark test results for various vendors' product families can be used as one of the device selection criteria. These performance issues are discussed in Chap. 5, which also provides an introduction to the general failures modes and mechanisms associated with CMOS technologies as well as certain reliability issues specific to PLDs, such as integrity of antifuse programming elements and FPGA metastability. This chapter also discusses various screening and quality conformance inspection (QCI) test requirements.

There have been many exciting developments in the field of reconfigurable logic using SRAM-based FPGAs. The dynamic reconfiguration has the potential of providing the designers with a capability for unlimited device reconfigurations *on the fly*, enabling them to build *virtual hardware*. A major application area has been the development of custom computing machines, which are high-performance FPGA-based reconfigurable logic systems, targeted toward computation-intensive applications. Chapter 6 reviews several of these commercially developed systems, such as configurable hardware algorithm mappable preprocessor (CHAMP), programmable active memories (PAM), SPLASH attached processor systems, Virtual Computer, and so on.

PLD use in military and space applications is growing because of their attractive attributes such as low cost, quick turnaround time, and technological advancements. A major hurdle in wider acceptance and usage of PLDs has been the concerns about their radiation tolerance in space applications. Chapter 7 provides an introduction to the space radiation environment, radiation effects, including total dose and single event phenomenon (SEP), radiation hardness assurance (RHA) testing with some actual test data, and radiation effects mitigation techniques.

Chapter 8 reviews latest PLD technology developments and future trends. The market for high-density PLDs continues to grow. There is a growing trend in use of intellectual property (IP) cores: predesigned and preverified functional blocks that are usually designed for ASICs but are becoming more popular now for CPLD and FPGA designers. These IP cores can enable very high-performance and high-density designs. Major PLD vendors such as Actel and Altera have announced availability of several cores and megafunctions that can be integrated into their product families. A new development has been the concept of system programmable gate arrays (e.g., the ES family of SPGAs by Actel) which, on a single chip, combine the performance advantages of gate-array or standard cell architecture and programmable logic design flexibility. Motorola has introduced analog FPGAs that can be programmed to provide analog functions such as programmable gain stages, adders, subtracters, filters, and sample-and-hold circuits.

The book includes a detailed glossary of PLD acronyms. An appendix is provided as a PLD vendor and tool resource guide, divided into two sections. Section I is a listing of PLD vendors, their e-mail/web addresses, product families, architectures, basic cell types, maximum circuit density, and so on. Section II is a listing of CPLD/FPGA design and development tools suppliers, software products for design entry, schematic, simulation, and synthesis tools for Windows platforms.

Ashok Sharma
Silver Spring, MD

Acknowledgments

I would like to thank various PLD manufacturers and suppliers for providing technical reviews and product specification sheets, including figures and permission to reprint them. Special acknowledgments are due to following people and their organizations:

Mike Sarpa, Actel Corporation; Michel Courtoy, Aptix Corporation; Mike Ross, Atmel Corporation; Jeff Wills, Altera Corporation; Don Parkman, Cypress Semiconductor; Sean Long, I-Cube; Ed Barnett, ICT, Inc.; David Gillooly, IMP Inc.; Steve Stark, Lattice Semiconductor; Robert Kuchta, Lucent Technologies; Fred Zlotnick and Connie Schultejans, Motorola Semiconductor; Cindy Huang, Philips Semiconductors; Gregory Judge, Quicklogic Corporation; Myles Standish and Teresa Farris, UTMC Microelectronic System Inc.; Stephan Cooper, Vantis Corp.; Steve Casselman, Virtual Computer Corp.; and David Varnholt, Xilinx Corp. Thanks are also due to William Hagen of IEEE Press; Steven Svoboda of Cadence Corp. and others for their informative articles in magazines such as ISD, Computer Design and EDN.

I would like to thank several people at Goddard Space Flight Center management for their support, including George Kramer, Ann Darrin, Robert LeBair, Robert Kichak and others. The author also acknowledges valuable support of Jeff Eckert (J.K. Eckert & Co.), executive editor Steve Chapman of McGraw Hill, and their supporting staff members.

Chapter

1

Introduction to Field Programmable Logic Devices

1.1 Evolution of Programmable Logic

In general, *programmable logic devices* (PLDs) is a generic term referring to any type of integrated circuits that can be configured by the end user for a particular design implementation. Since PLDs are programmed "in the field" by the end user, they are also called *field programmable logic devices* (FPLDs). The programmable devices have played a major role in the design of digital hardware by providing flexibility of user configuration for a wide variety of applications. One of the most commonly used programmable logic device has been one-time programmable read-only memory (PROM), consisting of an array of memory cells that can be programmed with the user-defined bit patterns of zeros and ones. These have been available in two different versions: (a) mask programmable devices, which are usually programmed by the manufacturer using custom mask for final interconnections, and (b) field programmable devices, which are user configurable. A major advantage of the PLDs are that they are less expensive at low volumes compared to the mask programmable devices.

Figure 1.1*a* shows the block diagram of PROM architecture. A more appropriate device that was developed as a follow-up to the PROMs, specifically for logic implementation, was the programmable logic array (PLA). The combination of PLA structure with PROM technology resulted in first programmable logic device called the *field programmable logic array* (FPLA), first introduced in the mid-1970s.[1] The FPLA has a fixed number of inputs, outputs, and product terms. However, both the inputs to the AND gates (the AND array) and the inputs to the OR gates (the OR array) are programmable. Figure 1.1*b* shows

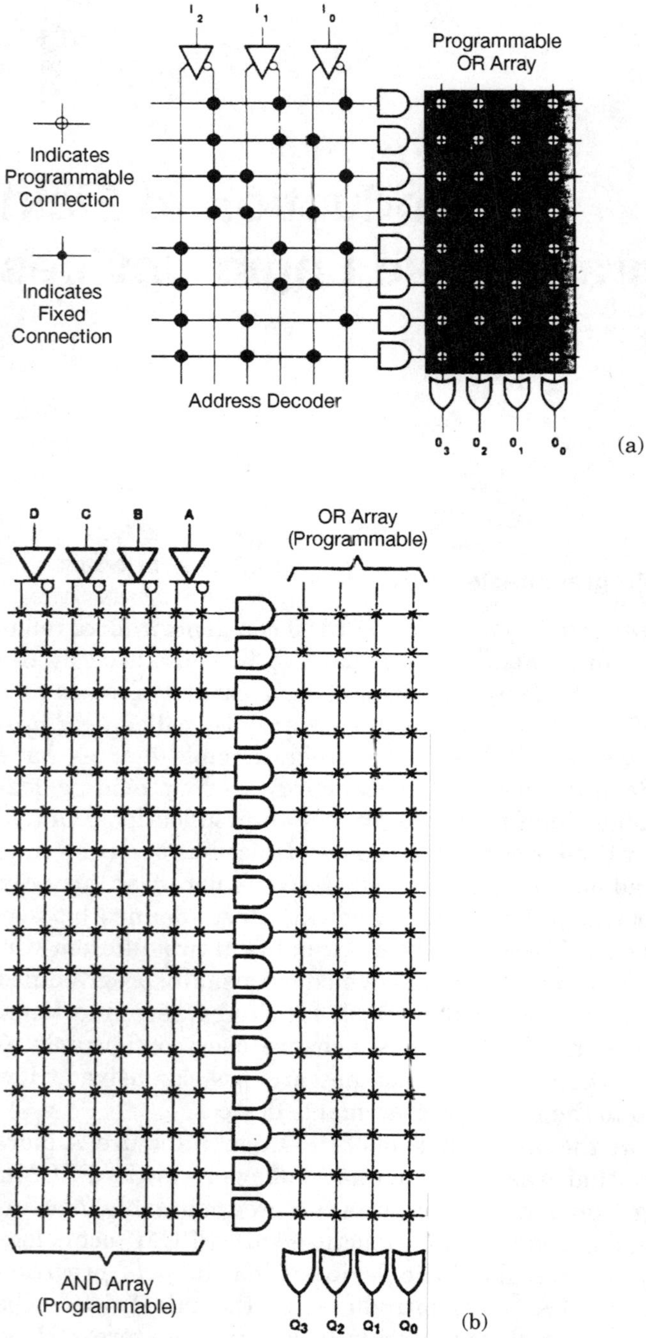

Figure 1.1 Block diagrams of (a) PROM architecture and (b) FPLA architecture.

a block diagram of the FPLA. The FPLAs had limited success, since they were quite slow and difficult to use. The program mode was a fuse map with no resemblance to the schematics or Boolean equations typically used in logic design.

In the late 1970s, programmable logic became widely accepted with the development of programmable array logic (PAL). The PAL architecture combines a programmable AND array with a fixed OR array, such that each output is the sum of a specific set of product terms. Also, development tools were created at the same time that allowed design logic input in Boolean equation form, thus simplifying the design process. Early PALs allowed only the programming of the AND array. These were soon followed by a large family of programmable devices from various vendors offering many improvements in basic PAL architecture.

The PAL devices are currently available with various types of options regarding number of inputs/outputs, the size of OR-gate, and flip-flops connected to the OR-gate outputs for sequential logic circuit implementation. The PALs can provide fairly high-speed performance characteristics for simple logic designs. In some device types, the simple AND/OR structure is replaced by either NAND/NAND or NOR/NOR structure to provide higher design flexibility.

1.2 Simple Programmable Logic Devices

Although, in the last few years, a number of low-density PLD architectural enhancements have appeared, simple PAL architectures in 20- to 24-pin packages (e.g., 16V8/20V8, 22V10) have become de facto industry standards. In general, low-density PLDs, including PLAs and PALs, encompass monolithic block-based structures with 20- to 44-pin packages and density ranging from 100 to several hundred gates. In this book, these devices will be referred to as *simple PLDs* (or SPLDs). The basic AND-OR structure makes SPLDs a favorite choice for implementing the logic equations in Boolean sum-of-products form. Some recently introduced features may include programmable I/O pins, flexible output-enable and bidirectional I/Os, programmable output polarity, flexible register configurations, and flexible clocking schemes.

The PLDs have been used to replace small- and medium-scale integrated (SSI/MSI) "glue-logic" chips to provide high packaging density. A single PLD can be functionally equivalent to tens of SSI ICs with 200 to 500 equivalent gate range. The secondary benefits of PLDs include reduced power, faster turn-around time, higher performance because of reduced interconnection length, and higher reliability. Low-end PLDs are made in both bipolar and CMOS processes. Bipolar fuse link technology, which was originally developed for the PROMs, has been the technology of choice for high-performance, low-density PAL devices. Simple PLDs in CMOS technology are available in EPROM-based (UV-erasable) as well as EEPROM (electrically-erasable) versions. These technologies will be discussed in more detail in Chap. 2.

The SPLDs, with their simple architecture, still offer a performance edge. As the manufacturers migrate down to 0.5 μm CMOS process, the leading edge devices have reached logic delays as low as 3.5 ns at operating frequencies up to 200 MHz. Although, the higher density, complex programmable logic devices (CPLDs) are closing the gap in terms of performance, SPLDs continue to offer the best performance available in the market. The SPLDs are easier to use, design in, and program, as compared to CPLDs because of their industry standard architectures.[2] The end products that benefit most from these SPLD characteristics are computer networking components and other telecommunication switching equipment where high overall performance is required.

The trend in programmable logic evolution toward higher densities, coupled with more integration and complex designs, means that SPLDs will eventually settle into a few niche markets. These will include simple, embedded devices requiring low-cost, low-power designs such as cellular telephones, video game machines, and hand-held web browsers. They are also being used in applications such as disk drives, communication network hubs, bridgers and routers.[3] As a result, some of the vendors with a strong presence in SPLD design and manufacturing are working to innovate their product features and extend their life cycles. Advanced Micro Devices (AMD) has made a commitment to its entire product line by sectioning its PLD business into a separate subsidiary, called Vantis, which is actively developing SPLD products in industry standard, high-speed, and low-power categories.

The innovations in universal SPLDs include offerings with user programmable output logic macrocells that can be configured with outputs for combinatorial or registered operation, and active-low or active-high polarity. For designs that require a nonstandard mix of inputs and outputs, or a variable number of combinational/registered/latched inputs and/or outputs for a given application, universal devices allow the designers to reduce inventory costs by stocking just one or two types of PLDs.

A promising new area for SPLDs is in low-voltage and low-power application developments for the portable systems market, such as cellular phones and notebook computers. The SPLDs have followed the curve down to 3.3 V, making them useful in wide range of products. Currently, many 20- and 24-pin SPLD designs are available in 5 V/3.3 V versions, as well as a variety of speed and power grades to provide a better range of trade-offs between cost and power consumption. Several versions of low-power SPLDs are being offered in the market. For example, Philips Semiconductors offers the CoolRunner® family of PLDs available for 3.3 V/5 V applications based on extended programmable logic architecture (XPLA®) and fast zero power (FZP®) to reduce static power consumption to "zero-power" and "quarter-power" modes (see Chap. 2.3.6).

Lattice Semiconductor has introduced 3.3 V, high-speed SPLDs, such as GAL20LV8D and GAL20LV10D, featuring propagation delays of just 3.5 ns and 4 ns, respectively. The PCI bus designs are a good market for high-speed SPLDs. ICT Inc., which builds medium-density PLDs called *programmable electrically erasable logic* (PEEL®) arrays with additional input lines, product terms, and macrocell functions, has introduced zero-power devices.

1.3 High-Density Programmable Logic Devices

One of the main disadvantages of low-density SPLDs is their architectural limitation, which means fully committed logic structures and fixed logic allocation. In comparison, high-capacity or high-density PLDs (HCPLDs/HDPLDs), also called *complex PLDs* (CPLDs) and *field programmable gate arrays* (FPGAs), attempt to solve this silicon inefficiency problem with more flexible block structure (which permits better logic allocation of macrocells) and through more flexible interconnects as the routing resources. In this book, the term *programmable logic devices* (PLDs) refers to simple as well as high-density programmable logic devices including CPLDs and FPGAs.

In 1985, a new programmable logic architecture was developed by Xilinx Corporation and termed the *logic cell array* (LCA). It served as a prototype for the following generation of FPGAs. The new architecture consisted of an array of independent logic cells surrounded by a periphery of I/O cells and included the routing source as a programmable interconnect, which allowed arbitrary interconnection of the logic cells. In this first LCA, each logic cell consisted of a combinatorial function generation and a flip-flop. Each I/O cell could be programmed for an input (registered or direct), an output, or a bidirectional pin. The programmable interconnect provided for clock distribution, routing for high-speed and low-skew signals, as well as local routing.

The two major components of high-density programmable logic architecture are logic elements and the interconnect structure. The logic elements are referred to as the macrocells, logic cells, configurable logic blocks, and so on. The interconnect structure in programmable logic devices determines how these logic elements are connected together across the device. The two terms most commonly used to describe high-density PLDs are *complex programmable logic devices* (CPLDs) and *field programmable gate arrays* (FPGAs). The distinction between the two is blurred, although a few characteristics for each stand out. CPLDs have long, fixed on-chip routing resources, which are usually interconnected via a switching matrix. They typically resemble arrays of multiple 22V10 having a large product-term base interconnected via the switch matrix. A notable exception to this is Altera Corporation's FLEX 8000 and FLEX 10K families, which use a RAM-based look-up table (LUT) as the basic logic element.

In SPLDs, all inputs and feedback connect to all product terms of all macrocells, and each SPLD macrocell has its own devoted product terms. The CPLD vendors have implemented architectural innovations to deal with the geometric nature of the die size increase as a function of the complexity. This has been accomplished through concepts such as product-term steering and product-term sharing between the macrocells. In CPLDs, the term *complex* refers to pin counts and number of internal logic macrocells available. As the device complexity increases, it has become typically impossible to offer chips that have an output pin available for every product term of the inputs. In addition to the architectural enhancements to deal with this problem, many vendors are also promoting the use of software development tools that provide high internal connectivity for their devices.

The basic performance specification quoted by the PLD vendors is the total propagation delay (T_{pd}), which the raw speed of the silicon from input(s) to output function on a pin. For SPLDs and CPLDs up to 44 pins from nearly every vendor, T_{pd} is usually a specified constant value. However, above 44 pins, where input-pin-to-output-pin routes are internally redirected, T_{pd} may instead be determined by the complexity of the logic function implemented and the design tools used. The evolution of simple programmable logic devices have typically spawned CPLDs of 44 pins/32 macrocells available with logic delays of 5 ns and operating frequencies to 180 MHz, while higher capacity 100-pin/128-macrocell devices are available with 7.5 ns delay. Figure 1.2*a* shows the SPLD versus CPLD performance history and forecast.[4] Figure 1.2*b* shows Dataquest projection of CMOS programmable logic market growth rate at 23 percent annually through the year 2000, and the high-density segment showing a growth rate of 30 percent for the same period.

The roles for CPLD and FPGA markets were clearly defined until recently. The higher-priced CPLDs offered predictable timing specifications and better performance, which made them a better choice for low-volume, high-speed applications. In comparison, FPGAs, with their higher densities and lower prices, had carved out a market for medium-volume and mid-level performance applications. However, a

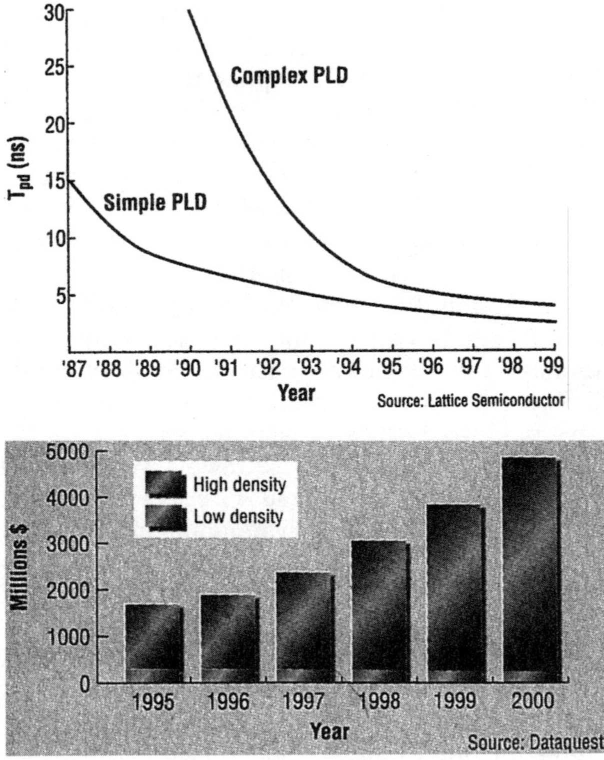

Figure 1.2 (a) Simple PLD vs. complex PLD performance history and forecast by Lattice Semiconductor[4] and (b) low-density vs. high-density CMOS programmable logic growth projections provided by Dataquest *(from Ref. 5).*

drop in CPLD prices, combined with features such as in-system programmability (ISP) and higher gate counts, have increased the demand for their applications. Nowadays, the distinction between the two different domains of programmable logic (i.e., CPLD and FPGA) applications has blurred. The growing market demand for CPLDs comes mainly from their use in such speed-sensitive areas as real-time video, DSP and image-processing applications, and communication networks—especially by the high speed demands of asynchronous transfer mode (ATM) and routers.[4]

The ISP feature provided by many CPLD families also makes them an attractive choice, since it enables the manufacturers to deliver “soft” devices to customers. The ISP allows the CPLD to be reprogrammed while it is still soldered on the board. The time to market can also favor CPLDs over FPGAs, since the CPLDs can be programmed and reprogrammed quickly, at the workbench, with very little nonrecurring engineering (NRE) costs. Both CPLDs and FPGAs can be programmed via flash or SRAM memory, but only CPLDs can be EPROM or EEPROM programmed. This implies that CPLDs are available with on-chip nonvolatile program storage that is also alterable (i.e., reprogrammable), and they are ready to run instantly on application of power. In contrast, the antifuse-based FPGAs are one-time programmable devices (OTPs), whereas for the SRAM-based FPGAs, the configuration program resides off-chip in an external device such as a serial EEPROM.

In high-density PLDs evolution, two important architectures have emerged: segmented-block-based PLDs and channel-array-based PLDs. The major differences between the two are the programmable block architecture and their interconnect structure. Each of these approaches has its characteristic strengths and weaknesses.

A segmented-block PLD basically consists of few large AND-OR blocks with an array of logic elements combined with I/O macrocells, which are interconnected by a centralized programmable interconnect scheme. Some examples of this approach include the Advanced Micro Devices/Vantis MACH family and Altera Corporation's MAX family of CPLDs. These devices are implemented in technologies based on CMOS EPROM or EEPROM. The CPLD architectures will be discussed in more detail in Chap. 2.

A global (nonsegmented) internal PLD architecture consists of a relatively fine-grained matrix of programmable logic blocks, which are surrounded by a border of uncommitted I/O blocks. These blocks are interconnected by distributed interconnect structures. These programmable devices are available in both one-time programmable (antifuse-based) and reprogrammable configurations. Some examples of reprogrammable channel-array-based FPGAs are Xilinx Corporation's logic cell array (LCA), Atmel's 6000 series reconfigurable logic, and Lucent Technologies' (formerly AT & T Microelectronics) ORCA series. These FPGA architectures will be discussed in more detail in Chap. 3.

In FPGA evolutionary development, various logic elements have been introduced to implement Boolean functions. One-time programmable FPGAs include Actel Corporation's ACT 1, ACT 2, and ACT 3 families, and QuickLogic Corporation's pASIC devices. SRAM technology-based devices use look-up tables (LUTs), whereas antifuse-based devices use multiplexer-based elements or sea-of-NAND gates blocks for logic implementation. A common thread in these various FPGA technologies has been the use of a segmented interconnect scheme as a signal routing resource for connecting the logic blocks.

The major strength of channel-array-based FPGAs is their register-rich architecture, many I/Os, complete decoupling of the logic blocks from I/O blocks, and programmable connectivity between the logic blocks and I/O blocks. The reprogrammable devices that provide interconnection between routing resources via an SRAM cell or an EEPROM bit are resistive (more than antifuse-based devices), which means increased routing delay because the data moves more slowly through the interconnection. Antifuse-based FPGAs such as Actel and QuickLogic devices, although typically faster than the SRAM-based FPGAs, are non-reprogrammable. The FPGAs have been used for highly pipelined applications such as datapath and slower, register-intensive logic designs. They are being increasingly used in high-speed custom computing applications.

In comparison, segmented block CPLDs are faster and more predictable, offering uniform, path-independent, and deterministic delays because of their continuous interconnect structure. The CPLD interconnect structures have continued to evolve as the vendors optimize them for speed and connectivity. The segmented block PLDs are more suitable in control applications such as implementation of complex finite state machines. Recently, new architectures have been developed (e.g., the Altera FLEX 10K series) to provide the high capacity and high register count (which are historically FPGA characteristics) but retain the fast, continuous interconnect scheme of CPLDs.

Figure 1.3*a* shows the block diagram of segmented architecture in which the macrocells communicate through a switch matrix.[4] Figure 1.3*b* shows the block diagram of nonsegmented approach, which eliminates the switch matrix for direct communication with the logic array. The nonsegmented approach increases the number of macrocells and

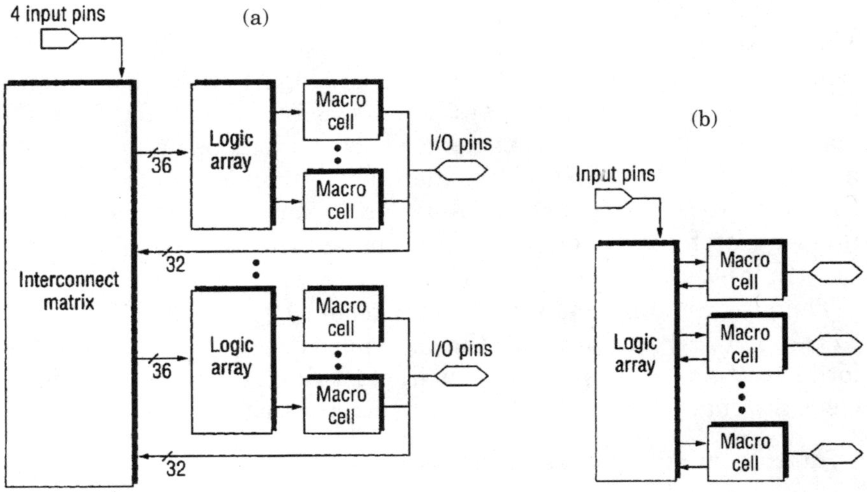

Figure 1.3 Block diagrams showing connectivity differences between (a) segmented and (b) nonsegmented CPLD architectures *(from Ref. 4)*.

product terms and enlarges the logic array. According to critics of non-segmented architecture, this means that, at higher densities, the device has to grow to accommodate more I/Os, and a larger device is likely to be more power hungry.

The debate continues over the technical merits of CPLDs versus FPGAs. There are basic architectural differences that cause CPLDs and FPGAs to have different degrees of flexibility in the same design environment. CPLDs, with their PAL-like architectures, have predictable speed characteristics and are less seriously affected by routing, compared to the FPGAs. Also, they have a switch-like matrix structure linking blocks of macrocells, which results in quite fast and predictable performance regardless of the gate count. The CPLD architecture has traditionally provided good solution for implementing control logic.

In comparison to CPLDs, with most of the channeled-array-based FPGA architectures, it is difficult to accurately predict the performance and logic capacity before design completion due to the routing and path-dependent delays. However, one of the great strengths of FPGAs is their truly flexible architecture, which lets the designers implement not only control logic but also the datapath and arithmetic and logic functions. The FPGAs are optimum choice for register-intensive designs. To keep FPGA timing reasonably predictable, designers usually limit the amount of logic placed within a device, which means lower logic resource utilization relative to the CPLDs.

The CPLD architectures are array based, which means that they have a relatively coarse logic block granularity, whereas the FPGAs are cell based, with finer logic block granularity. The array-based and cell-based devices are optimized for different types of logic functions. The large fan-in of CPLDs makes them ideal for control circuits that require many product-term inputs, whereas the low fan-in but register-rich FPGA architectures are usually more suitable for datapath applications. The state-machine-based control logic is an ideal implementation for the CPLD, where there is an abundance of wide input logic functions. An application illustrating suitability of the register intensive architecture of an FPGA is the implementation of a bidirectional FIFO.

The programming and interconnect technology used in CPLDs and FPGAs has a great impact on the granularity of the architecture, performance, and other application characteristics. There are three major elements characterizing an FPGA architecture, as follows:

- Internal building blocks (logic modules)
- Interconnection between logic modules
- Switching (or programming) elements

In an FPGA design implementation, each building block (or logic module) may be configured to perform a logic function whose complexity depends on the capacity of these building blocks. When necessary, two or more of these logic blocks may be connected to perform the required logic function. There are various types of interconnection schemes used, depending on the logic block architecture. As an example, in Actel architecture, the interconnection between the logic modules is provided by channels of wiring segments between the rows and columns of the logic module.

The routing architecture of an FPGA includes channels of wiring segments. The size of switching elements also introduces limitations in connecting these wire segments. A maximum routability is achieved if every horizontal wiring segment is connectable to every vertical segment. Therefore, a full crossbar switch connecting n horizontal wires to m vertical wires requires $n \times m$ switching elements. However, in practice, only a few of all possible connections between the segments are permitted.

The programming elements used include devices based on antifuse, SRAM cell, and EEPROM (or flash) memory cells. In the FPGA architectural trade-offs, the antifuse-based devices have significantly more routing resources than the SRAM-based FPGAs. Antifuse is a programming element that irreversibly changes from a high resistance to a low resistance state when a programming voltage is applied across its terminals. Two categories of antifuse elements are commonly used: dielectric and amorphous silicon. In the SRAM-based FPGAs, each switching element consists of a five-transistor RAM cell and a pass transistor called a *programmable interconnect point* (PIP). In the programmable interconnect, the PIPs control the wiring segments connections and are controlled by the configuration RAM cell. The wiring segments on each side of the PIP are connected or not connected, depending on the bit value stored in the RAM cell.

An architecture with more switching elements provides more options for routing, and possibly enhancing performance. However, the lower the resistance and capacitance of the switching element, the faster the interconnect speed. One of the main impacts of the size, resistance, and capacitance in the switching elements is their effect on the granularity of the logic blocks in the FPGAs. This affects the choice of selecting an ideal logic block to implement a given function. In SRAM-based FPGAs, more switching elements mean considerably more die area. Also, the SRAM switching elements are slow enough to require a larger logic block. In antifuse-based FPGAs (such as Actel's), since antifuse switches occupy negligible die area, a smaller logic modules is used. However, when needed, two or more of these logic blocks may be connected to each other.

In antifuse and SRAM-based FPGA architectures, the distinction is no longer as clear as earlier projected. Two leading antifuse-based FPGA manufacturers have recently announced plans to offer devices with mixed architectures. Actel plans to add programmable SRAM-based logic cells to accommodate in-circuit reconfiguration, while QuickLogic is using diffused, mixed-function blocks to complement its device's array of antifuse programmable logic. Another vendor, Motorola, has introduced a fine-grained SRAM-based architecture that employs a hierarchical interconnect structure that more closely resembles antifuse technology than the other SRAM-based devices.

The market for high-density programmable logic continues to grow. Some PLD vendors that were using 0.6 μm CMOS process to fabricate 25k-gate FPGAs containing 3.5 million transistors are already moving to a 0.35 μm process 100k-gate device containing 12 million transistors.[6] Lucent Technologies recently announced plans for 0.25 μm process FPGAs that will reach 225,000 gate densities. The development of higher-density devices has made it possible, and even necessary, to incorporate intellectual property (IP) cores into the PLD design. These prepackaged hard or soft cores are available for a wide variety of functions such as PCI bus controller, universal serial bus (USB) devices, network controllers, and even processor cores. Since many of these functions are synthesizable, they often can be modified for specific customer applications.

Major PLD vendors such as Xilinx, Altera, and others are beginning to integrate IP functions into their devices. For example, Xilinx's AllianceCORE and Altera's AMPP (Altera Megafunctions Partners Program) are offering dozens of IP cores for integration into the PLDs. The push for prepackaged IP functions and cores has been driven by cost and time to market of products that are rapidly growing in complexity. It has become difficult for the PLD vendors to design from scratch all of the needed complex logic functions in-house.

As FPGAs and CPLDs improve in density and architecture, they are becoming an alternative to gate arrays. Figure 1.4*a* shows that the projected market for high-density CPLDs (HCPLDs) is growing rapidly to take over many of the CMOS gate arrays lower end (<25,000 gates) applications. While some of the decline in gate array sales may be the result of customer migration to higher complexity gate arrays, a significant portion of the market is going to CPLDs and FPGAs. As HCPLDs technologies migrate to 0.5 to 0.25 μm feature sizes, the CPLDs and FPGAs are reaching thresholds in density, performance, and cost that makes them attractive alternatives to gate arrays. Figure 1.4*b* shows the curves for technology migration of minimum feature size and growth in usable gates/chip for high-density programmable logic.

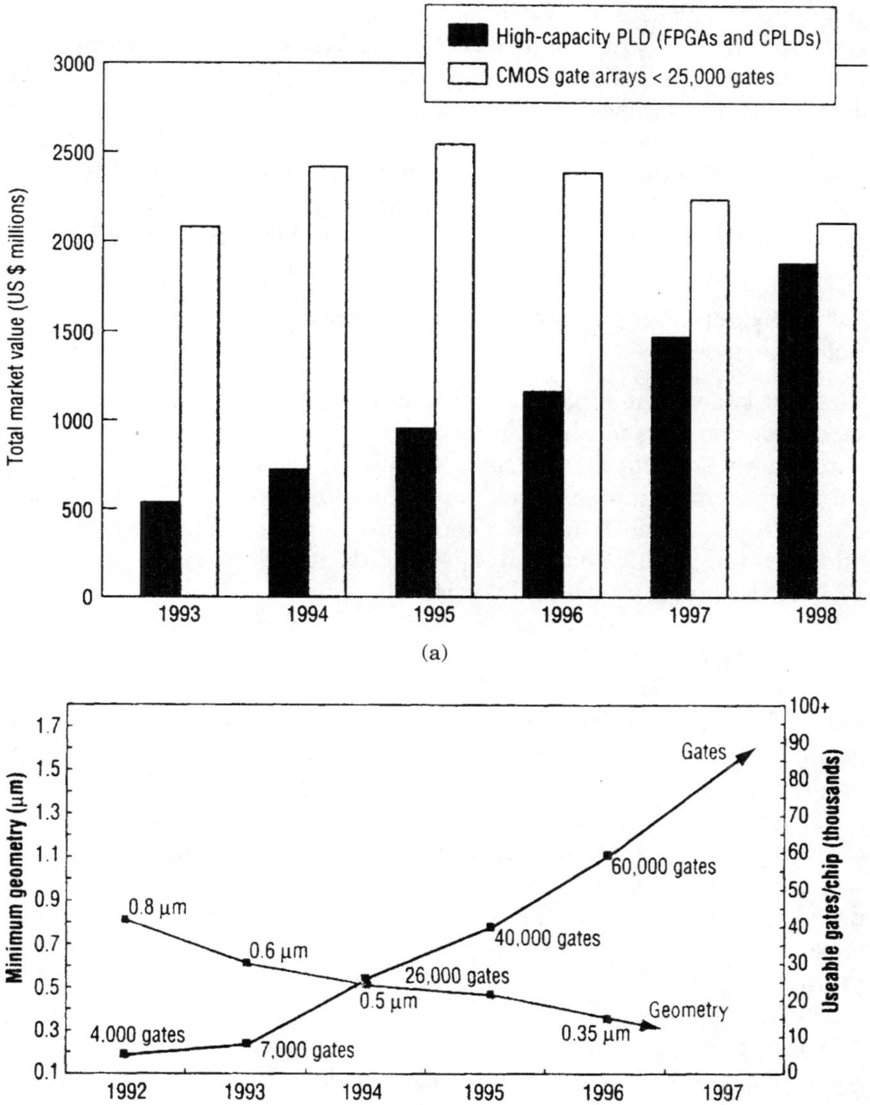

Figure 1.4 (a) Projected market growth for HCPLDs vs. CMOS gate arrays (<25,000 gates)[5] and (b) technology migration path for minimum feature size and growth in usable gat.

Some of the recent technological developments are blurring the line of distinction between high-density FPGAs and gate arrays. An effort to extend FPGA architectures to system-level functions has resulted in the emergence of a new class of devices called *system-programmable gate arrays* (SPGAs). They combine architectural attributes of the FPGAs and masked-programmed ASICs. SPGAs use a multitiered routing architecture designed to support a wide range of high-capacity devices by allowing multiple building blocks such as logic, memory, datapath, and processing to be integrated on a single die. This new class of programmable logic devices are targeted to provide the users both the flexibility and field programmability of an FPGA and access to complex cores, which was until now obtainable only in the ASICs.

1.4 In-System Programmability and Package Selection Issues

The first generation bipolar fuse-based PLDs required stand-alone device programmers for programming the fuses. As PLDs migrated through subsequent technologies such as UVEPROM and EEPROM, the designers continued to rely on stand-alone programming for loading the logic into each device. Therefore, some extra PLD sockets were added to the circuit boards in case the design changes required new PLDs to be programmed and inserted.[7]

The concept of in-system programming, commonly abbreviated to ISP, has developed from a niche to a standard feature on CPLDs to meet users' needs for simpler package handling, design prototyping, and field updates. The ISP logic traditionally refers to one generic group of PLDs, the nonvolatile reprogrammable devices, as opposed to one-time programmable (OTP) antifuse-based devices or infinitely reprogrammable SRAM-based FPGAs. A further enhancement of ISP concept is a move toward in-system reprogramming (ISR) capabilities, which allows the design changes to be made to a device already soldered down on the board. Thus, with ISP, a device can be programmed once it is placed on the board. However, with ISR, it can be reprogrammed for design changes without requiring board changes.

The move toward ISP has been accelerated by increasing device densities and the number of I/O pins, forcing devices into larger packages such as the 208-pin plastic quad flat-pack.[8] These packages have very delicate gull-wing leads that can be easily damaged from handling during programming, which can result in *fallout* as a result of bent leads. Currently, most CPLD manufacturers have begun offering ISP, not only on larger packages but on smaller packages also that are not so susceptible to bent-lead problems. A major benefit of ISP from the designers' point of view is the flexibility it provides by allowing them to hook up their boards directly to a workstation or PC for design and debug. Therefore, the use of ISP devices greatly reduces the design and debug cycle, as well as prototype development time. The ISR can also enable them to implement last-minute changes on the preassembled boards, thereby eliminating potential rework and speeding systems to the market.

Another potential benefit of ISP is to enable field upgrades of existing logic systems. In field tests, devices can be upgraded and tested out immediately. Communication systems are particularly hard to debug and test in a field environment. It is estimated that about 50 percent of high-density PLD applications are in the field of communications and telecommunications, and that is where the field upgrades are needed. These include private branch exchange (PBX) switching, wireless communication, local area network (LAN) equipment, and asynchronous transfer mode (ATM) networks.

The major drawback for SRAM-based ISP devices is their volatility—that is, the loss of logic configuration when the system power is turned-off. This means that, each time the system is powered up, the programming pattern must be loaded from an external memory device such as an EPROM. This causes two major problems: (a) SRAM-based devices require a power-up time delay for loading the data from the EPROM memory device, and (b) additional EPROM devices are required that use up the valuable board space.

The EEPROM technology is best suited for ISP, since it requires none of the support circuitry that is needed for SRAM-based devices, thus saving critical board space. After selecting a device technology, the designers need to consider programming requirements. Some ISP devices require special 12 to 14 V supply voltages for device programming. Many ISP devices in the SRAM and EEPROM technologies offer 5 V logic supply voltage for programming and reprogramming.

After the selection of device technology and programming requirements, another consideration is the simplicity of the programming interface, whether serial or parallel. ISP devices with serial interfaces usually require fewer interface signals, which makes them more reliable and cost effective, as well as easier to design into the system. The serial interface ISP devices are available in both SRAM and EEPROM technologies. Parallel programming options require extra layout resources for running interface signals across the board.

To reap the potential benefits of an effective ISR, there are three main aspects of an ISP solution that must be addressed:

- The use of standard programming interface protocol
- Short programming times
- A device architecture that efficiently accommodates the design changes

A standardized protocol makes ISR more practical. In-system programming or reprogramming is performed by placing the devices in a serial chain and then using some scheme for cycling the data through all of them.

The joint task action group (JTAG) test access port (TAP) has been widely accepted as a way of connecting devices in a chain as IEEE 1149.1 boundary scan architecture has become an important testability standard. Many available boards already have a JTAG interface, and some standards, such as PCI, include the JTAG pin in their interface. Although each vendor's devices may program a little differently, it is possible to link them all, since the JTAG scheme allows a way of bypassing the devices in a chain. In the absence of a common interface, the devices from different vendors must be placed in separate chains, complicating connections and testing. If the JTAG test access port is present and the CPLDs have JTAG compatible ISP, then no extra interface is needed. The CPLDs can be programmed anywhere in the JTAG chain.

It is always preferable to use devices that have short programming time specifications. These can range from a few seconds to several minutes. The programming time, when using a cable on a PC, is determined by the speed of the PC, which can be very fast. The parallel programming of multiple devices can reduce overall board programming times considerably.

The element that has the greatest long-term impact on the effectiveness of programming solution is the architecture, which determines the difference between in-system one-time programmability and the true in-system reprogrammability (ISR). The ability to change the contents of a device while it is soldered onto a board does not provide any benefit if the device is not useful to the system after reprogramming. This can happen in following cases:

1. The device architecture is not routable enough to accommodate the design changes in the same pinout as the original design, or
2. The timing model is so complicated that the design changes affect the critical timing paths to an extent that they fail to meet the application requirements unless pinouts are changed.

The *pin locking* (or *pin fixing*) on a device refers to a practice of pin assignment to specific I/O functions, which does not allow them to be later reassigned in the design process.[9] If the pin assignments are made without regard to the logic operations being performed by the device, then pin fixing can become a problem. Some device architectures provide great design flexibility. The *routability* describes the overall impact of a number of individual portions of an architecture, especially the ability to route signals from pins to the desired blocks and macrocells and the ability to allocate product terms in a flexible manner. Some device architectures intentionally limit routability to keep down the device cost, which raises the possibility that the pinouts may have to be changed, even if simple logic changes are made.

The ability of a design to route is critical, but equally important is the ability of an architecture to maintain its timing characteristics in the face of design changes. Many devices have complicated timing models where the timing of a path is highly dependent on its route. The simplest possible timing model ensures that any signal from any source will reach any other pin in a given time period. In addition to an architecture that is well suited to ISR, the accompanying tools should be good enough to retarget the design into an existing pinout.

In summary, there are several methods that a designer can use to implement the ISP, as follows:[10]

- The manufacturer's software is used via a cable hooked to the PC. The software bypasses nonprogrammable JTAG devices while programming the CPLDs.
- The manufacturer's software creates a tester-specific file to be integrated into the board test program, or creates a generic serial vector format file (SVF), which can typically be translated into a tester-specific file through software supplied by the tester manufacturer.
- Embedded reconfiguration is accomplished using a microcontroller. In this approach, the designer can embed JTAG ISP capability on a microcontroller board using programming information from the CPLD vendor. Thus, instead of programming via a cable or on a tester, a user can transmit updated configuration data to on-board memory and then start the microcontroller program to update the JTAG-ISP CPLD.

The growing density of complex and high-performance PLDs is a challenge to the vendors to provide packages with optimum thermal, electrical, and mechanical characteristics while remaining small and adaptable to automated board-level assembly operations. Recent innovations in the PLD industry have led to the development of plastic quad flatpacks (PQFPs), thin plastic quad flatpacks (TQFPs), and ball-grid array (BGA) packages. These packages use variety of thicknesses and heat sink methods, and employ finer lead pitches to improve performance. Figure 1.5 shows some of these packaging options, which are a function of both the gate count and pin count.[11]

There are two basic package styles: through-hole and surface mount. The dual in-line packages (DIPs) and pin-grid arrays (PGAs) are popular through-hole technologies, while leadless chip carriers (LCCs) and quad flatpacks (QFPs) represent surface mount approaches. The through-hole board mounting requires package inser-

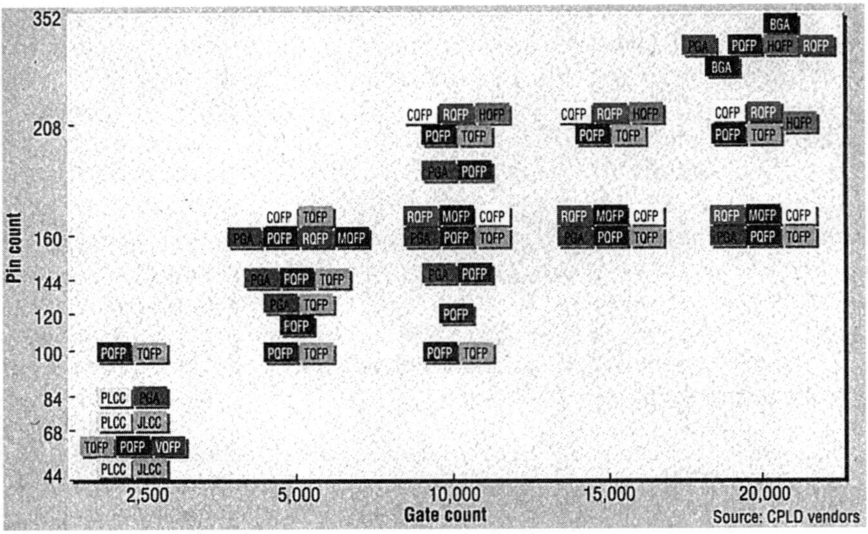

Figure 1.5 Various PLD packaging options that are functions of both device gate count and pin count *(from Ref. 11).*

tion and solder plating, and these ceramic packages occupy more board space than the surface mount technologies (SMTs). However, these packages are mechanically stronger during handling processes and better suited for military and space applications.

Surface mount technology, which has become quite popular for commercial and industrial applications, allows devices to be mounted on both sides of the printed circuit board. In addition to the cost savings derived from the higher board density, surface mount devices are generally more suitable for use with automatic placement equipment. These packages help designers to reduce board space, minimize parasitic capacitances and inductances, and shorten the design-to-production cycle. A disadvantage of flatpacks is their use of fine-pitched leads that can be damaged during component testing and handling.

The appropriate PLD package selection depends on the targeted application and requirements of the design environment. The package of choice must have low thermal resistance as well as minimum capacitance and inductance for an optimum power and performance tradeoffs. The thermal characteristics of packages exercise a major influence over the device performance and reliability. As the device density and power dissipation rises, the PLDs require packages with lower thermal resistance (θ_{JA}). The testability requirements of high pin count packages demand fine-pitched and reliable test sockets, test boards, and package trays for automatic test equipment. Also, minimizing the damage from handling becomes more challenging for the designers using variety of QFPs, including thin quad flatpacks (TQFPs) and very thin quad flatpacks (VQFPs).

A relatively new introduction has been the ball-grid array (BGA) packages used for high-density PLDs that combine the use of plastic material with metal heat sink and employ solder balls instead of QFP's fragile fine-pitch leads. The BGA packages are supplied in smaller sizes and thinner shapes than comparable QFPs. As complex PLDs grow in density, power, and performance, their packages are becoming increasingly complex and susceptible to damage from excessive handling. To minimize the damage from handling, PLD vendors are now offering JTAG boundary scan standard (IEEE 1149.1) testing as well as ISP through JTAG-compliant ports on many devices.

1.5 Design and Development Methodologies

The basic design methodology for PLDs consists of three interrelated steps: design entry, design implementation, and design verification. Typically, generic tools are used for design entry and simulation, but architecture-specific tools are needed for design implementation. Schematic entry is quite suitable for glue-logic designs but is less effective for behavioral design approach. For high-density FPGA designs, gate-level entry tools are often cumbersome and, therefore, the use of logic synthesis and high-level description languages (HDLs) is becoming more popular. The behavioral design entry methods including Boolean equations and state machine descriptions are supported by many EDA vendors.[12] However, this approach requires FPGA "fitters," which are device-specific tools for logic optimization to fit the target FPGA architecture.

A high-level language design can be ported to different architectures, which provides several advantages to the system designers, such as

1. Technology selection can be postponed until later in the development cycle, when requirements are better defined.
2. Design migrations (e.g., FPGA to gate-array) to reduce cost during the product life cycle are simplified.
3. Portions of the design can be easily reused in the future products, even if those products use different technologies.

Two types of HDLs are commonly used for programmable logic. Single module-level HDLs are employed to describe small modules within a schematic-based hierarchical design. Some examples of this are widely supported PAL Assembler (PALASM), Data I/O's Advanced

Boolean Expression Language (ABEL), and Logical Devices' Universal Compiler for Programmable Language (CUPL) design languages—all of which were developed to support the small PLDs of the late 1970s. The system-level HDLs, which support more complex constructs and higher levels of abstraction, were developed in the late 1980s. The most popular of these high-level languages are Very High-Speed Integrated Circuits (VHSIC) HDL (VHDL) and Verilog HDL. System-level HDLs are modular, since the multiple modules constructed using the HDL can be linked into large designs.

In general, logic synthesis is the process of generating an optimized gate-level netlist for a logic design from a higher-level structural description. Logic synthesis tools are the key elements of top-down design methodologies. In theory, a design description in HDL is technology transparent, relying on synthesis compilers for mapping the logic into whatever device architecture is being used. In reality, technology-independent design with HDLs is difficult, since different approaches and coding styles are required for different FPGA architectures. However, as FPGA technology continues to develop, it will become easier to generate device-independent HDL code for FPGA designs.

Most FPGA development systems support the use of hierarchical design techniques in which the top level drawings are descriptions defining the major functional blocks, and lower level descriptions define the logic within each block. The hierarchical approach has several advantages such as ease of debugging, partitioning of large, complex designs, and storage of commonly used functions in a library as macros for repeated use. Most libraries provided by the FPGA vendors contain enough information to support board-level simulation. Alternatively, the FPGA behavioral models (e.g., SmartModel library from Logical Modeling) can be used with a variety of third-party simulators to support system-level verification.

The companies such as Cadence Design Systems, Mentor Graphics, Synopsys, and Viewlogic are offering universal computer aided engineering/electronic design automation (CAE/EDA) tools that enable PLD designs using a variety of device independent methodologies. The design logic can be entered through schematic entry tools or specified using high-level description languages. The design is processed and multiple solutions are generated, based on the user specified criteria such as types of devices to consider, maximum power consumption, and so forth. The resulting solutions can be combinations of CPLDs and/or FPGAs that meet the specified requirements.

Some PLD users do not need the power or want the cost of high-end EDA tools. For these users, proprietary PLD tools have adequate capability to meet their needs, e.g., AMD/Vantis-only versions of popular design packages such as Micro Sim's Design Center, and Data I/O's

ABEL and Synario. These tools offer schematic capture, timing simulation, automatic device selection, and multiple device partitioning.

Therefore, the design application, goals, and resources available ultimately determine whether to use design tools supplied by an FPGA vendor, an EDA vendor, or a combination of both. Since most of the CPLD/FPGA family architectures vary considerably, device-specific synthesis tools may be needed for optimum results. Therefore, a combination of the design tools from third-party and PLD vendors may be the desired approach for a given application.

Figure 1.6 shows a typical HDL-based design flow methodology.[13] The third-party tools including design entry, simulation, synthesis, and timing analysis, constitute the front end of the design. For the design densities exceeding 10,000 gates, it is usually recommended that the design entry mode should primarily be in either VHDL or Verilog, although some portions of the design can be entered by more traditional means, such as the schematics or state diagrams. The PLD vendors supply libraries representing the targeted devices that allow EDA vendors to tune their front-end tools to a specific device family and its macrocell architecture. Also, the PLD vendors usually develop and support back-end tools, which include placement and routing, or fitting of the design onto the targeted architecture. Some vendor tool

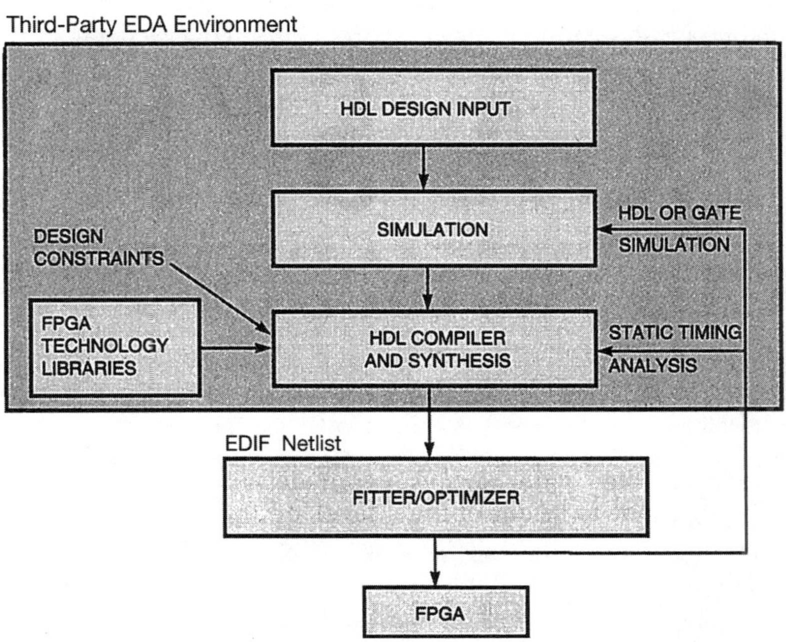

Figure 1.6 Typical design flow methodology *(from Ref. 13).*

suites may include a floor planner for more accurate timing analysis before the final place and route.

A program called a *device fitter* manipulates a designer's generic logic description until the best possible fit with a particular PLD or FPGA is obtained. Thus, the fitter's input may be a set of sum-of-products equations generated by a design entry tool such as schematic capture or behavioral description software. As PLD architecture continues to evolve, the demands on fitter software have grown to include the following:

- Two-level logic synthesis
- Multilevel logic synthesis
- Library and technology mapping
- Design partitioning
- Signal placement and routing

The two-level synthesis embodies a variety of techniques for logic optimization, many of which are extensions of algorithms employed for the simpler PLDs. Multilevel synthesis is the process of sorting a two-level design into three or more logic levels with a goal to reduce the amount of device resources required to implement the logic function within the delay constraints. Increasing the number of logic levels often reduces the required device resources but increases delay in the logic path; conversely, reducing the levels of logic speeds up the design, often at the cost of additional device resources. Therefore, the fitter must take into consideration the target architecture, the design logic, and the delay and resource utilization constraints.

Many FPGA designers are adopting a top-down methodology consisting of HDL design and behavioral simulation similar to that for ASIC designs. The change in methodology is being driven by an ever increasing device complexity and evolution of PC-based design tools.[14] A major hurdle was the wide variety of FPGA technologies and architectures, ranging from Actel's antifuse and multiplexer-based devices to Xilinx's SRAM-based look-up tables for design logic implementation. Therefore, the efficiency of using synthesis for FPGAs had been quite limited, as compared to the performance that could be obtained by using schematic entry and preoptimized macro functions. However, recently, both the high-end Unix-based ASIC tool vendors and PC tool suppliers have managed to bring higher levels of efficiency for the HDL synthesis into FPGAs, which for some designs matches that obtainable from a schematic entry approach. Tools are now available that allow designers path migration to the HDL-based designs without sacrificing large amounts of silicon.

A major difference between the FPGA designers and ASIC designers is their verification process. While the ASIC designer uses extensive simulation to assure first-pass silicon success, the PLD designer is able to perform in-circuit final verification with little impact on project schedule. The ASIC design means high cost and fabrication cycle time of weeks (and often months), compared to the PLD designs, which are relatively low cost and offer a quick turn-around time.

The increased productivity advantages of logic synthesis are more visible as the designs get longer and more complex (e.g., 10,000 to 100,000 gates). However, for some of these designs, logic synthesis may produce larger and slower circuits than the good gate-level implementations. Synthesis is most effective when the designer achieves some expertise with the HDLs and logic synthesis tools. Some designers find it difficult to transition from a visually oriented, schematic-based design to an abstract, textual format language such as VHDL. Also, the synthesis tool efficiency depends on the design description, i.e., a detailed description may overly constraint the synthesis tool, whereas a very abstract design description may yield inaccurate results.

Many FPGA designs are entered at the register transfer level (RTL) to be technology independent. In some tools, local optimization to improve area and delay are not considered during the HDL entry. In a VHDL compiler, several high-level, technology-independent optimization steps are performed for area improvement. An example of such high level optimization technique is resource sharing, which allows the VHDL compiler to look for opportunities to share arithmetic functions to reduce area, while trying to meet the timing requirements.

After high-level optimization, the remaining portions of the design are converted from Boolean descriptions into FPGA modules using technology-independent optimizations. In this process, the Boolean logic is structured to meet the design's timing constraints. Then, the sequential elements are mapped into complex flip-flops. A Boolean matching technique is used to find the maximum combinational logic, which can be mapped into a flip-flop. The circuits created by logic synthesis must be analyzed and compared to the design constraints. Static timing analysis is performed to find and display the critical delay path. The interconnect delays are estimated using a statistical model, which estimates the net delay based on the net's fanout. For mapping to the SRAM-based architectures, estimating prelayout timing can be quite a challenge.

Once the design has been verified to be correct, it is transferred to physical design tools. A structural netlist is generated to describe the logic netlist, and timing constraints are passed using standard or proprietary formats. The place and route tools use the timing constraints specified by the logic synthesis to calculate most critical nets and prioritize those for improvements. Once the design is placed and routed, post-routing delays may be extracted from the circuit. These post-layout and post-routing delays can be fed back into the synthesis tools to perform static timing analysis. The post-routed delay information, along with the structural netlist of the design, can also be read into the simulator to perform full timing simulation.

Many PLD vendors and EDA tool suppliers are moving toward the adoption of libraries of parameterized modules (LPMs), which allows silicon vendors to supply predefined, tightly crafted logic functions for use with the synthesis compilers. The design migrations are aided by standards such as LPM, which define a set of generic, parameterized logic functions. Each module describes a frequently used, high-level function (e.g., a counter, decoder, or ROM). The parameters associated with each module are used to fully specify the function for a given application. The major PLD vendors supply "LPM compilers" that are capable of generating optimized implementation of these functions for the targeted architecture (e.g., Xilinx X-BLOX). Thus, the synthesis tools need to synthesize only those functions into the LPM descriptions (as opposed to gate-level implementation), which allows for architecture-specific optimization.

1.6 PLD versus Custom ASIC/Gate Array Design Issues

The system designer seeking a technology that offers higher logic integration has the choice of standard off-the-shelf PLDs or an ASIC approach, which includes full custom design, traditional gate arrays, or masked programmable gate arrays (MPGAs). The selection of a suitable approach depends on factors such as functional fit, total cost effectiveness, and design cycle time. ASICs, both full custom and semicustom (gate array/standard cell), have several important features, which makes them good candidates for high-density logic designs, such as:[15]

- Lower cost per function for high-volume designs
- Customized design configuration and device operation, providing a better functional fit for a particular application
- Advanced CAE tools including high-level language support and logic synthesis to increase front-end design productivity
- Design security, as a result of customization

The major disadvantages of silicon customization in ASIC approach are time and nonrecurring engineering costs. In comparison, the PLDs provide many of the expected time-cost benefits of a standard component, such as off-the shelf availability, which provides quick reaction to the design changes or production needs. This translates to shorter design cycles and lower costs. The disadvantages of PLDs are:

- FPGAs have on-chip programming overhead circuitry to manage programming elements such as antifuses or SRAM/EPROM cells. As a result, the FPGAs are larger and slower than the equivalent MPGAs.
- The connection paths in an FPGA are slowed by the programming circuitry. Since more area is required for the same amount of logic as compared to an ASIC approach, the interconnect paths are longer, adding higher resistance and capacitance, and further slowing down the resulting circuitry. A typical FPGA design may be two or three times slower than the equivalent ASIC design.
- The area penalty for field programmability is significant, since some FPGA designs may be typically 10 times larger than the same gate capacity on the equivalent MPGAs.

Table 1.1 lists the selection criteria for various logic options such as standard components, PLDs, gate arrays, standard cells, and full-custom approach for relative comparison.[16] The PLD advances in high densities above the 100k-gate level, as well as performance improvements, are making them attractive candidates for gate array applications. Another approach is taking FPGA/CPLD designs to volume production through the "FPGA conversion" process, which provides many competitive advantages. The usual rationale for conversion to masked arrays is to save money on cost-per-gate basis, when the production volume for a product reaches a certain crossover point. However, ultimately, not only the high volume should justify conversion, but the design should have reached a maturity level where no further enhancements or addition of new features is anticipated in the near future.

A major consideration in design conversions is whether the converted masked array design will work successfully. FPGA design conversion involves the translation of a netlist to gate array vendor's design environment. The architecture of a masked gate array usually differs significantly from the original FPGA architecture. As a result, timing, functionality, and pin layout compatibility may be compromised.

The development of synthesis and optimization software for FPGAs is making the design migration path to ASICs somewhat easier. The companies offering this type of software include Synopsys Inc. and Exemplar Logic. Synopsys' FPGA compiler offers synthesis that is technology independent from VHDL or Verilog design descriptions, or a

TABLE 1.1 Listing of Selection Criteria for Various Logic Options Such as Standard Components, PLDs, Gate Arrays, Standard Cells and Full-Custom Approach, for Relative Comparison[16]

Criterion	Standard components	PLDs	Gate arrays	Standard cells	Full custom
Time to market	Short to medium	Short	Medium	Medium	Long
Development lead time	Immediate	Immediate	Weeks to months	Weeks to months	Years
Development cost	None	Low	Medium to high	Medium to high	Very high
Availability	High	High	Medium	Medium	Low
Available sources	Many	Many	Few	Few	Few
Volume dependence	Low	Low	High	High	High
Application support	Much	Much	Some	Some	None
Architectural flexibility	Low	Medium to high	High	Higher	Highest
Design change ease	Medium	Medium to high	High	Higher	Highest
Performance	Low	High	High	High	Very high
Density	Low	Medium	Very high	Very high	Very high
Solution efficiency	Low to medium	Low	High	High	Very high
Design change cost	Low	Medium	High	High	Very high

logic-level description. It means that the designs specified in HDL can be directly translated into one or more of the ASIC technologies offered by various vendors. The FPGA compiler is supported by leading FPGA vendors including Actel, Altera, Lucent Technologies, Xilinx, and so on. Actel supports migration for its antifuse-based ACT family through its industry alliance and commitment to a program called *technology-transparent design* (TTD). The TTD concept enables the designers to have access to design tools from a variety of leading EDA vendors and also provides a vendor-independent migration path to masked gate arrays.

In a typical FPGA-to-MPGA design conversion, the customer's netlist, in one of the industry standard FPGA formats such as EDIF , is provided as an input to the process. Typical development steps include design layout, inserting scan for testability, and developing test vectors through automatic test pattern generation (ATPG). After the conversion is completed, simulation results using the test vectors or a retargeted netlist are returned for customer review and design sign-off. After customer approval, the prototype development begins.

During the FPGA design conversions, routing efficiency is an important consideration for factors such as speed, die size, and packaging. Since MPGAs, as well as custom ASICs and traditional gate arrays, involve metal mask layers, chip routing can be very efficient, which allows not only higher speed but also a smaller die size. The extra speed attained from the efficient routing and a smaller die size makes MPGAs faster than most FPGAs currently available in the market. Another important factor is the drive capability. MPGAs, full custom ASICs, and traditional gate arrays typically feature "scalable" drive outputs, which allows the customers to select the output drive strength at the time of conversion. The pinout flexibility also has to be reviewed. In FPGA designs, many of the pinouts are predefined, providing designers less flexibility in pin assignments, whereas in an MPGA, ASIC, or gate array designs, each pin may be configured as an input, output, bidirectional, or power/ground pin.

For FPGA to ASIC conversion, some of the issues that have to be taken into consideration are as follows:

- Technology and architectural compatibility between FPGA design and target ASIC
- Conversion reliability of FPGA source files into ASIC source data
- Reliability checks for logic and timing errors and other circuit design hazards during the conversion process
- Verification of compliance with fanout rules, or placement and routing rules from one technology to another

It is essential to select proper software environment and logic synthesis tools to retarget the design descriptions specified by the FPGA source library and optimize them for the targeted ASIC technology. Once the design is retargeted, it is possible to add enhancements to improve the routability and testability of the design for smaller die size and yield improvements. During the conversion process, there may be certain types of FPGA cells that cannot be optimized by synthesis software. It may be necessary to create a "mapping" library for mapping these specific cells, and this mapping library may consist of a set of schematics containing logic from one library that duplicates the function of a cell in another library. This can allow creation of schematics containing the target ASIC vendor's macros with the same names as the FPGA cells being mapped. During the entire conversion cycle, special care has to be exercised regarding the target technology ASIC design rules.

1.7 Programmable Logic Device Selection Criteria

The choice and selection of a proper PLD involves trade-offs in density and performance. The smallest PLDs are usually the fastest but may not be able to accommodate the design in a single device. The maximum rating of an industry standard 22V10 PLD can vary from vendor to vendor, but there are leading edge devices specified for 175 MHz operation, and roughly 4 ns delay from input to output. A CPLD is a better choice in more complex designs, since it is equivalent to having many smaller PLDs on one chip. For a 7.5 ns delay, the CPLDs can reach approximately 150 MHz operation. However, in many cases, the maximum performance specification of a device provided by the manufacturer may be applicable only to the simplest portions of the design. Therefore, predicting accurate design performance based on the manufacturer's data sheet specifications is not a trivial matter. As a rule of thumb, the smaller the device, the higher are the chances of better performance predictability.[17]

The CPLDs are somewhat less predictable, compared to simple PLDs, but vary from family to family. Some CPLDs have specifications for timing delays that depend on factors such as the number of product terms used and the fanout. The software compiler and tools ultimately make the routing and design implementation decisions. However, there are other CPLD families that have more straightforward timing specifications and are therefore easier to use.

FPGAs are the least predictable of PLDs, since it is difficult to provide accurate specifications for parameters such as the propagation delay and maximum frequency, which depend on the placement and routing. Therefore, an accurate timing prediction before the final design implementation is difficult to achieve. In general, an FPGA is slower than a CPLD or a simple PLD, depending on the application. However, in terms of capacity, some of the latest FPGA architectures provide densities that are comparable to the low-end gate arrays.

For a given design, the selection of a PLD based on density requirements depends on determining the true capacity of a device, i.e., the number of usable gates versus the number of available gates. The vendors for both the CPLDs and FPGAs data sheet specifications quote the number of gates in numerous ways, such as available gates, gate-array-gates, PLD equivalent gates, NAND-gate equivalents, and usable gates. Therefore, the gate count is a fairly useless specification except when comparing specific devices from the same manufacturer. In general, two approximate yardsticks for evaluating the density of a given PLD (but only as a starting point) are the number of registers and I/O pins. Both performance and power consumption are also important design considerations in selection of a target PLD. Typically, flash- or EEPROM-based PLDs require more power than SRAM- or antifuse-based devices.

A major consideration during an FPGA selection process is to recognize that each vendor's device may exhibit fundamental architectural differences, not only in the individual logic blocks but also the routing structure. An FPGA manufacturer's data sheets may have some critical parameters specified differently than the other vendors, making it hard to make a general capability and performance comparison. Also, the speed of an FPGA depends on the interconnection between the logic blocks and, thus, on the design layout. Each design, and even each iteration of the same design, can use very different routing paths. Therefore, the resulting interconnect impedance of the nets will determine the actual speed of the device. The final design performance depends not only on the target FPGA resource utilization, but also on the quality of place-and-route software.

The state machines are one of the most popular applications of programmable logic, and their implementation approach in CPLDs or the FPGAs differ because of the basic architectural differences. An architectural advantage that some CPLDs have over the FPGAs is the programmable T-type register. For some state machine designs, such devices can significantly reduce the amount of required logic, producing a fast and compact implementation. The counters implemented in the CPLDs using T-type registers usually require fewer product terms than those implemented in the CPLDs that use D-type registers.

For state machine applications that are targeted into an FPGA, one has to choose a state encoding technique that minimizes the logic between each state transition. An example of this type of state encoding is the "one-hot" state encoding, which uses one register per state and is discussed in Sec. 4.6.1. Therefore, the selection of a target CPLD or FPGA for a state machine application depends on the performance requirements and the logic complexity required for each state transition. The CPLDs are inherently faster than the FPGAs in state machine applications. However, if state transition logic requirements are a major concern for the state machine design application, rather than the performance, then an FPGA may be a more appropriate choice than a CPLD. Alternatively, one can explore both the FPGA and CPLD options after synthesis and place and route, and the device selection decision can be based on actual simulated performance results in the application.

In general, FPGA architectures are more suitable for register-intensive arithmetic applications than are the CPLD architectures. Most FPGAs have a much finer-grained architecture than that of the CPLDs, and the degree of granularity varies amidst FPGA families. The finer the granularity, more flexibility it offers in implementing the logic associated with arithmetic circuits. The FPGAs are well suited for implementing high-performance and custom computing applications that will be discussed in Chap. 6.

In addition to PLD selection, the design tools for description, synthesis, optimization, and implementation are also equally important. In selecting a tool, one can use a tool from the PLD vendor that is tailored specifically for the vendor's products or a tool from the third-party tool vendor. The PLD vendor's tools support a given architecture, so that they can perform architecture-specific synthesis at the front end and are usually more efficient than are the third-party tools. Some PLD vendor's tools support a wider range of product families including CPLDs and FPGAs. However, one may already have a design environment built around a third-party vendor's tools and prefer to utilize that instead of procuring and learning a new tool every time a different vendor's device is targeted. Historically, the CPLDs have been more amenable to third-party approaches that generally synthesize a product's logic into a standard format. In comparison, the FPGAs, with their widely varying FPGA architectures, require different synthesis and optimization strategies.

Therefore, the design methodologies, architectural differences, and interconnect technologies offered by each candidate CPLD and FPGA, as well as the design tools, are important considerations in the selection of an appropriate device for a given application. Some of the special features provided such as the JTAG boundary scan, on-chip RAM, and fast-carry logic may also be the influencing factors in device selection. In addition, some other criteria for device selection includes following:[18]

- Target CPLD/FPGA device cost
- Time to marketing goals
- Number of registers
- Number of I/O pins
- Vendor's macro library, including hard and soft macros
- Design tools/software cost and quality
- Special device features offered
- Power consumption
- Packaging

Another yardstick for PLD selection process are the Programmable Electronics Performance Corp. (PREP) benchmarks developed for comparing the relative device architecture and technology implementation results on the device performance and capacity. These benchmarks will be discussed in Chap. 5.

In summary, for PLD-based design and product developments, the following are the major issues and trade-offs that must be taken into consideration from a user's point of view:

1. Schedule
2. Product requirements such as performance, area, and cost
3. Compatibility with the existing tool flows[19]

If schedule (or time to market) is the major concern for product success, then the following are some of the key factors in ensuring that the PLDs do not become a bottleneck in the overall design process:

- Performance and capacity must be predictable. It means that when a design is partitioned, the various functions and performance requirements should be predefined for each block to ensure that the blocks can achieve those goals prior to implementation. This will prevent the schedule slip and delays.
- The vendor must support fully automatic layout (place and route) that achieves desired performance and capacity targets.
- Implementation and timing verification can represent a major portion of the design process. The selected FPGA vendor must have the necessary tools and macros to move quickly through this process.

The product requirements are the basic definitions of an end product and may include factors such as cost, performance, area, function, I/Os, and power consumption. This is where the trade-offs have to be made. If performance is the primary goal, then the following approach can be used to evaluate the potential solutions:

1. Determine the performance of each major function in the design, and keep track of capacity information, since high capacity can significantly affect the performance of many PLD architectures.
2. Perform a critical path analysis to determine whether required performance is met. If performance goals are not satisfied, then speed-graded versions or optional design implementations should be considered.
3. Determine from the PLD supplier whether the functions can be effectively mapped into the silicon at the desired performance and capacity goals.

Another important factor driving the move toward the ASIC and PLD-based designs is to reduce the product size and board area. Therefore, if minimizing board area is a major concern, then following should be considered:

1. Reduce package count when possible.
2. Use smaller outline packages when possible, such as TQFP and VQFP.
3. Use the smallest package that still meets minimum I/O requirements.

A major trade-off area is cost versus performance. The actual device cost can be one of the most difficult areas to quantify when comparing multiple PLD vendors, since each vendor has a unique architecture that may be better suited for some functions than the others, and each vendor has its own device performance specifications. Therefore, if cost is the primary concern, then the basic approach is the same as outlined for performance as the major goal, except that some trade-offs will have to be made. This includes review of optional design implementations or a relaxation of performance requirements. It may also involve selection of a smallest device that accommodates the major functions, and ensuring that it fits the cost goals.

References

1. William S. Carter, "The evolution of programmable logic," paper supplied by Xilinx Corporation.
2. John H. Mayer, "SPLDs search for a niche," *Computer Design,* Feb. 1997, pp. 28.
3. Peter Varhol, "Simple PLDs still a design alternative," *Computer Design,* April 1997, pp. 30–33.

4. Dan Tanner, "Performance, price cuts, ISP, heat up the CPLD market," *Computer Design,* Dec. 1996, pp. 91–95.
5. Peter Varhol, "CPLDs grow in density, intelligence," *Computer Design,* June 1997, pp. 117–120.
6. John H. Mayer, "Programmable logic knows no bounds," *Computer Design,* Feb. 1997, pp. 3–4.
7. Richard Mitchell, "In-system programming," *Computer Design/ASIC Design,* Dec. 1993.
8. Bryon Moyer, "In-system programming: It's a mixed bag," *EE Times,* April 21, 1997, Issue: 950.
9. Doug Conner, "In-system programmable logic simplifies prototyping to production," *EDN,* Sept. 26, 1996, pp. 37–46.
10. Kristen Ahrens Vest, "The promises of ISP," ISP roundtable presentation made on WWW *Pldsite.*
11. Mike Cho, "CPLD complexity raises package-selection issues," *Computer Design,* April 1997, pp. 14–18.
12. Bradley Fawcett, "FPGA development tools: Keeping pace with design complexity," paper supplied by Xilinx Corporation.
13. James Lipman, "High-density FPGA synthesis speeds system design," *EDN,* July 20, 1995, pp. 32–42.
14. "Technology/trends in FPGA design tools," *Electronic Products,* Nov. 1995.
15. Altera/Product Information Bulletin # 6.
16. Larry Waller, "Years of strong growth lie ahead for high-density programmable devices," *Integrated System Design,* Oct. 1997, pp. 28–36.
17. Richard Kapusta, "Options dot the programmable logic landscape," *EDN,* July 6, 1993, pp. 107–116.
18. Ann Watson Swager, "Choosing complex PLDs and FPGAs," *EDN,* Sept. 17, 1992, pp. 77–84.
19. Comparing FPGA Solutions: Actel Data Book 1996 Application Note, pp. 4–23/4–24.

Chapter

2

Simple and Complex Programmable Logic Devices

2.1 General Architecture

In general, low-density PLDs have an architecture consisting of PLA- or PAL-like AND-OR structure or NAND-NAND, NOR-NOR multi-level logic capability, combined with integrated array of logic and I/O macrocells and I/O pins.[1] PLDs range in density from 100 to 500 gates, supplied in 20- to 44-pin packages. Figure 2.1 shows the architecture of a simple PAL device consisting of a programmable AND plane plus a fixed OR two-level logic array. The four most important components of PAL architecture are: (1) the AND-plane, (2) the OR-plane, (3) the output storage elements, and (4) the I/O pins.

The AND-plane provides interconnection of the array inputs to the AND gate (product terms) for implementation of both the logic and control functions. The size of the AND array plane is determined by the total number of inputs (in both true and complement forms) and total number of product terms. The OR-plane determines the connectivity, i.e., the "logical ORing" of the product terms (PTs) from the AND gates to the outputs. The OR-plane architecture defines the number of OR outputs, the number of PTs per output, the distribution of PTs per output, and the coupling of ORed gate outputs with storage elements.

The architecture of output storage elements specifies the types and structure/organization of the outputs; flip-flops are used for storage elements, providing clocking and feedback flexibility. In a PAL device, the output(s) can be configured to have either sequential (registered) or combinatorial capability, as well as polarity control for active high

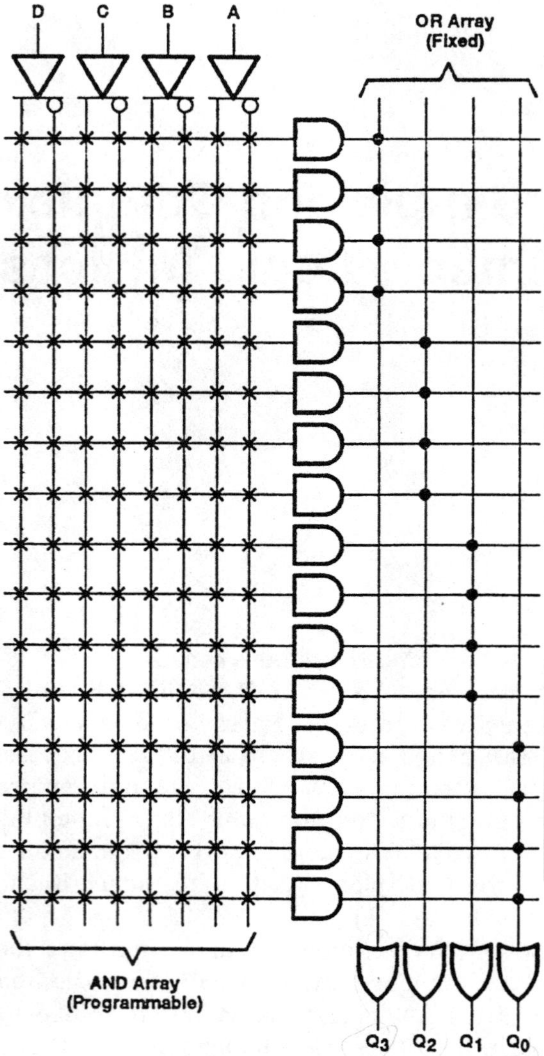

Figure 2.1 Simple PAL device architecture.

and active low. The flip-flops used for output storage elements can be edge-triggered D-type, J-K, S-R, or T flip-flops, or transparent latches. The architecture of the I/O pins specifies whether a pin may be configured as a dedicated input pin, a dedicated output pin, or programmable as a dynamically controllable bidirectional I/O pin.

A popular example of simple PLD architecture is the AMD/Vantis PAL family (such as PAL16R8 and 22V10 devices), which are available in bipolar TTL, programmable fuselink technology. Both of these devices are industry standards and widely second-sourced by various companies. The symbol "16R8" designates that the PAL has a maximum of 16 inputs (8 dedicated inputs and 8 inputs/outputs). The "R" refers to the type of outputs provided by the PAL and implies that each output is "registered" by a D flip-flop. Similarly, the "22V10" has a maximum of 22 inputs and 10 outputs. The symbol "V" means each output is "versatile" and can be configured in various ways (registered or nonregistered). Vantis also supplies universal PAL devices (the PALCE family) based on EECMOS process. These are discussed in Sec. 2.3.1. The PALCE family uses the macrocell concept, which was first introduced in PAL22V10 devices and is now an industry standard.

Another example of SPLDs is the generic array logic (GAL) family from Lattice Semiconductor (see Sec. 2.3.5). Lattice Semiconductor pioneered the use of a CMOS process and electrically reprogrammable cells based on EEPROM technology. Although the early CMOS PLDs such as GAL could not match the bipolar speeds, they offered reduced power consumption, which is a key consideration in many designs such as military and space applications. The faster PLDs available today are about 10 times speedier than the original PALs, offering propagation delays as short as 3.5 ns and with even shorter delays expected for enhanced versions. This has become possible by employing oxide-isolated submicron processes (e.g., 0.35 μm), coupled with the use of UVEPROM or EEPROM storage cells as the configuration elements. The original 16xx and 20xx families of SPLDs have grown considerably with enhancements of the 22V10 architecture, developed by AMD/Vantis and several other companies.

In general, a CPLD consists of few basic elements, i.e, logic blocks, a programmable interconnect scheme between the blocks, and I/O cells. The multiple logic blocks communicate with each other through a global interconnect. Each logic block in a CPLD is similar to a simple PLD, containing inputs, a product-term array, a product-term allocation function, macrocells, and I/O cells. The logic blocks execute logic equations that are simple sum of products expressions and stores the results into registers located in the macrocells. A central switching matrix interconnects and routes signals to and from the logic blocks. Three major resources available in all CPLDs are (a) the number of macrocells that the device assigns to each logic block, (b) the number of inputs going into each logic block from the global interconnect, and (c) the logic block product terms.

The number of macrocells refers to the basic registers available to implement logic functions such as the counters and state machines. Therefore, the number of macrocells, a measure of device complexity, is a limiting factor in design of certain logic functions. The second important resource is the number of available inputs going into the logic blocks from the matrix. For example, if each logic block of 16 macrocells can have up to 36 inputs, the logic for those macrocells can accommodate only up to 36 input signals. However, if the design requires additional input signals, then multiple logic blocks may be required. The third resource is the number and allocation of product terms. The total number of product terms that the device can OR together is a parameter that places an upper bound on the amount of logic that can fit in a CPLD. Some CPLDs can OR up to 16 or 20 product terms. In others, the number of product terms may vary from one location to another within the device.

There are CPLDs offering a pool of shared or redirectable terms, called *expanders,* that are used to increase the number of OR terms, at the expense of additional delay. If the logic design requires additional OR terms, multiple passes can be programmed through the array at the cost of speed. It should be noted that the CPLD product-term allocation scheme may be more important than the absolute number of terms. Some device architectures allow product terms to be assigned to individual macrocells through a product-term steering scheme and allocating the same product term to multiple macrocells through the product-term sharing. These product-term steering and sharing schemes increase the device efficiency by utilizing the available chip resources. CPLDs are sometimes also classified as programmable multilevel devices (PMDs), which implies a collection of simple PLDs with a programmable interconnect structure.

A CPLD performance depends on its interconnection approach for routing signals between the logic blocks and I/O pins. There are two different aspects of routability, as follows:

- *The number of signals that the global interconnect provides into each logic block.* If the maximum number of signals that can be routed to a logic block is smaller than the number required for implementation of a particular logic function, the design will have to be partitioned among multiple logic blocks.
- *Routability efficiency, i.e., the probability of finding an interconnect that routes all the required signals into the appropriate logic blocks.* As the number of signals that must be routed into a logic block approaches the maximum, it becomes more difficult to find a way to interconnect the signals. When the input location of these pins is fixed during the design phases, it becomes harder to route them to appropriate logic blocks.

In CPLDs, two approaches are commonly used for interconnections, as follows:

1. A crosspoint switch configuration, which is completely routable, can be used so that any combination of simultaneous signals can be routed to or from any logic block through the global interconnect. This interconnect scheme sacrifices speed and die size. An example of this approach is Cypress Semiconductor's multiple array matrix, MAX340 EPLD family (see Sec. 2.3.3).
2. A multiplexer-based approach used by many vendors distributes all signals through the multiplexers instead of a crosspoint switch system to reduce interconnect size and for speed enhancement. The implementation of multiplexer-based interconnects and routability efficiency varies from one vendor to another. An example of this approach is Cypress Semiconductor's Flash370 CPLD family. Figure 2.2 shows the CPLD logic blocks and I/O pins communicating with each other through the programmable interconnect.

The CPLDs, with their wide input structures and the sum-of-products architectures, are preferred choices for implementing fast state machines, high-speed control logic, and fast decode applications. Some CPLD manufacturers provide additional features such as in-system programmability (ISP), which is the ability to program and reprogram a device while it is in the system. This feature minimizes device handling and thereby helps eliminate the possibility of bent leads in fine-pitch packages. It also facilitates board-level test capability for system-level fault diagnosis and debug.

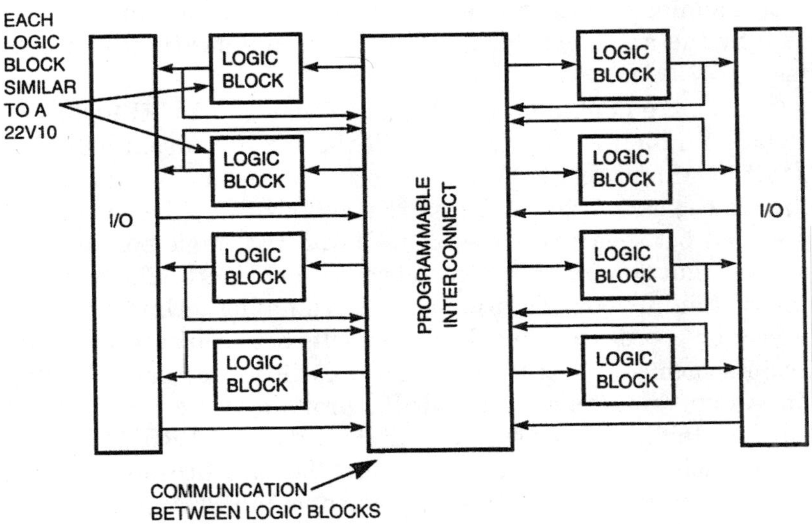

Figure 2.2 Multiplexer-based switch to provide programmable interconnect between logic blocks and I/O pins (*from Ref. 2, with permission of Cypress Semiconductor*).

2.2 Programming Technologies

The earliest PLDs developed were based on bipolar processes by adapting existing PROM technologies. The bipolar PROMs used lateral fuses as programming elements. Each fuse is simply a small section of metal alloy (e.g., nichrome or polysilicon) positioned between the two interconnect wires and in series with a bipolar transistor that supplies the current required to blow the fuse. In contrast to the lateral fuses, which normally connect two wires and can be blown to separate them, the vertical fuses are normally open circuits. The vertical fuses used in PROMs are not separate sections of the metal alloys, but each fuse is part of a bipolar transistor so that, in a normal (unblown) state, it is equivalent to two back-to-back diodes isolating the two wires. However, if sufficient current is forced through the transistor, an avalanche effect occurs, shorting out one of the diodes and connecting the two wires.

The first bipolar PLDs developed offered reasonably high-speed operation and relatively simple and reliable programming. Fusible-link PLDs (like PROMs) contain an array of programmable interconnect points, each consisting of a narrow bridge of metal that may be left intact or removed to enable or disable the interconnection. This bridge of metal is referred to as a *fuse* and can be made of variety of materials, ranging from nichrome to titanium-tungsten or platinum silicide. These fuses are "blown" using pulses of elevated voltages and certain amounts of fusing current. The various fusible-link PLDs have different programming requirements and algorithms, which are usually specified by the manufacturer in their device specifications and data sheets.

For CPLDs, the dominant switch technologies are UVEPROM and EEPROM, and for FPGAs, they are SRAM and antifuse. In CPLDs, the UVEPROM and EEPROM transistors are used as programmable switches in the same way as fuses are used in SPLDs, i.e., a transistor is placed between two wires to facilitate the implementation of wired-AND functions. The device selection for a particular application must take into consideration the programming technology used. This selection process also involves trade-offs between several important factors such as volatility versus nonvolatility, reprogrammability and in-system programmability (ISP), program time and erasure time, and testability.[3] This section will describe UVEPROM and EEPROM technologies as applicable to AMD/Vantis PLD families. The SRAM and antifuse technologies, as applicable to FPGAs, will be discussed in Sec. 3.2.

The electrically erasable technology used by the Vantis PLD family is a single-poly, double metal n-well CMOS process with 1.2 μm feature size and $L_{eff} = 0.8$ µm. The erasable CMOS technology is based on the concept of stored charge on a transistor with a "floating gate", i.e., the gate has no connection. Actually, the transistor has two gates: one that floats, and the second that acts as a control gate used to establish the field across the floating gate. In the programmed state, there is a net deficit of electrons in the floating gate, and the resulting positive charge turns the transistor ON. In the erased state, there are enough electrons on the control gate so that the negative charge turns the transistor OFF.

There are two basic ways of transferring the charge on to the floating gate: (a) hot electron injection, and (b) tunneling. Hot electron injection is used for UV-erasable devices (UVEPROMs), and tunneling is used for the EEPROMs' cell-based technology. Figure 2.3*a* is the schematic illustration of programming by hot-electron injection used for UVEPROMs.[4] A bias is set up between the source and drain of the transistor and between the control gate and the substrate. The channel is pinched off, causing a strong current flow. The two applied fields combine to form a field in the diagonal direction but, due to the oxide barrier, electrons cannot cross in that direction. However, some electrons acquire enough energy to cross the potential barrier in the shortest direction, i.e., from the channel to the floating gate. This is referred to as *hot electron injection* (the term *hot* referring to high electric fields). Once an electron crosses over to the floating gate, it is essentially trapped there.

During programming, large fields are set up so that significant number of electrons are injected into the floating gate. The erasing process requires exposure to ultraviolet light, which causes the electrons to cross back over the oxide barrier. Therefore, the device package must have a window to let the UV pass through, which makes the part expensive and is a distinctive drawback. Also, the windowed parts take about 90 minutes to erase. Some manufacturers supply these parts as one-time programmable (OTP) in plastic packages. These are not erasable and have no advantage over the fuse-based technology.

Electrically erasable devices use Fowler-Nordheim tunneling as the mechanism for getting charge onto the floating gate. This is defined roughly as the tunneling that occurs as a result of field placed across the barrier that the electrons tunnel through. For an electrically erasable cell, the tunnel oxide is about one-third the thickness of UV-erasable part. Therefore, the tunneling occurs at relatively low fields, even though the field used to cause Fowler-Nordheim tunneling is about five times the field required for UV-erasable parts. Electrical erasure has advantages over the UV erasure both from cost and reprogrammability perspectives, since no expensive ceramic packages with windows are required. This makes them cost-effective in high-volume production.

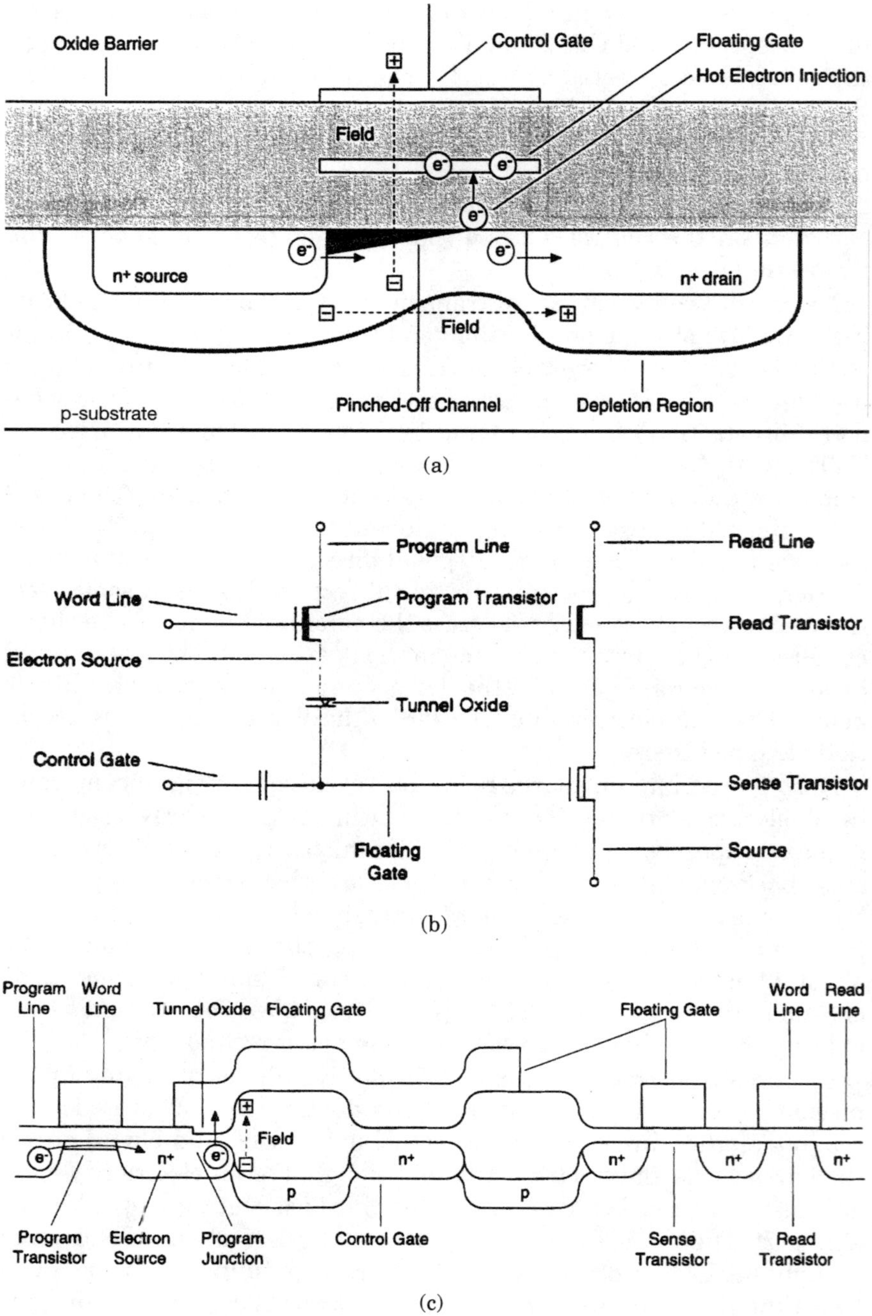

Figure 2.3 Schematics of (a) UV-erasable device programming by hot-electron injection (b) electrically erasable PLD programming cell, and (c) electrically erasable PLD programming cell cross section. *(Reprinted from Ref. 4 with permission of Vantis Corp.)*

Figure 2.3*b* shows the electrically erasable PLD programming cell schematic, and Fig. 2.3*c* shows the cross section. To improve device performance, the programming cell has been divided into the programming portion and datapath portion. The programming half-portion requires long-channel transistors capable of sustaining high electrical fields, whereas the data path portion requires the short-channel transistors that are fast. In electrically erasable technology, the programming and erasure are complementary procedures. A cell is considered to be programmed if there is a charge deficit on the floating gate, which provides a positive voltage; it is erased if there is excess charge on the floating gate, generating a negative voltage. This implies that programming a PLD only requires turning on those cells that are needed rather than turning off all of the cells that are not needed.

In electrically erasable devices, when programming or erasing the device, a voltage is applied between the program and control gate nodes so that the direction of the voltage determines whether the cell is erased or programmed. During erasure, the control gate is given a positive voltage while the program node is grounded. This draws electrons from the program transistor across the tunnel oxide to the floating gate, turning the read transistor OFF. For programming the cell, the program node voltage is elevated while the control gate is grounded, reversing the electron flow. This causes enough electrons to flow off the floating gate, leaving behind a net positive charge, which turns the transistor ON.

2.3 SPLD/CPLD Vendor Families and Development Tools

2.3.1 AMD/Vantis PAL and MACH families

Vantis is a design, development, and marketing subsidiary for all AMD PLD devices that include programmable array logic (PAL) and macro array CMOS high (MACH®) families offered in electrically erasable technology.[5] The functional requirements of a given design determine the device architecture most appropriate for that application. The functional criteria include such issues as the clocking scheme, macrocell flexibility, and the output drive. The combination of number of inputs/outputs required and number of product terms needed to implement a particular design determines the device selection.

For synchronous designs that are best suited for simple state machines, encoders/decoders, multiplexers, and other similar logic applications, D-type registered or combinatorial logic is needed. The choice for these applications includes the Vantis PALCE (programmable array logic CMOS electrically erasable) family device types such as PALCE16V8, PALCE20V8, and PALCE24V10, depending on the number of inputs required. For applications that require D, T, J-K, or S-R register capability, such as counters and large state machines, PALCE610 offers this flexibility. PALCE29M16 may be appropriate for applications that require embedded or buried registers. There are devices available for asynchronous applications as well, such as PALCE20RA10 and PALCE29MA16.

For both synchronous and asynchronous medium- and high-density design applications, Vantis offers MACH® families of devices that can allow multiple smaller PLD designs to be consolidated into one device.[6] MACH devices use PAL blocks that are interconnected using a switch matrix. The MACH families—consisting of MACH1/MACH2, MACH4, and the latest addition MACH5 devices—are differentiated by factors such as the number of macrocells and I/O pins available, macrocell structure, clocks, number of interconnects, and performance capabilities. Some of the major architectural attributes of PAL and MACH families are described in the following sections.

PALCE family. The programmable array logic CMOS electrically erasable (PALCE) family devices implement a standard Boolean logic transfer function consisting of the sum of products by using a programmable AND array driving a fixed OR array. As an example, Fig. 2.4 shows the logic block diagram, including output macrocells for PALCE22V10 family. The product terms are connected to the fixed OR array and with a varied distribution from 8 to 16 across the outputs. The device is produced with an electrically erasable cell "link" at each input to the AND gate array, and the connections may be selectively removed by applying appropriate voltages to implement the required design function. The product terms with all connections opened indicate a logical HIGH state, whereas the product terms connected to both true and complement of any single input assume the logical LOW state. The OR array sums up the products and feeds them to the output macrocell, each of which can be programmed as registered or combinatorial, and active high or active low. The output configuration choice is made according to the user's design specifications and is determined by two bits controlling the two multiplexers in each macrocell.

Another universal PAL device family member PALCE610 can accommodate logic functions with up to 20 inputs and 16 outputs. Eight product terms are brought to each macrocell for logic implementation. This device has 16 I/O macrocells that can be individually configured per the user's specifications as either registered or combinatorial. The registers can be configured as D, T, J-K, or S-R flip-flops. The state ma-

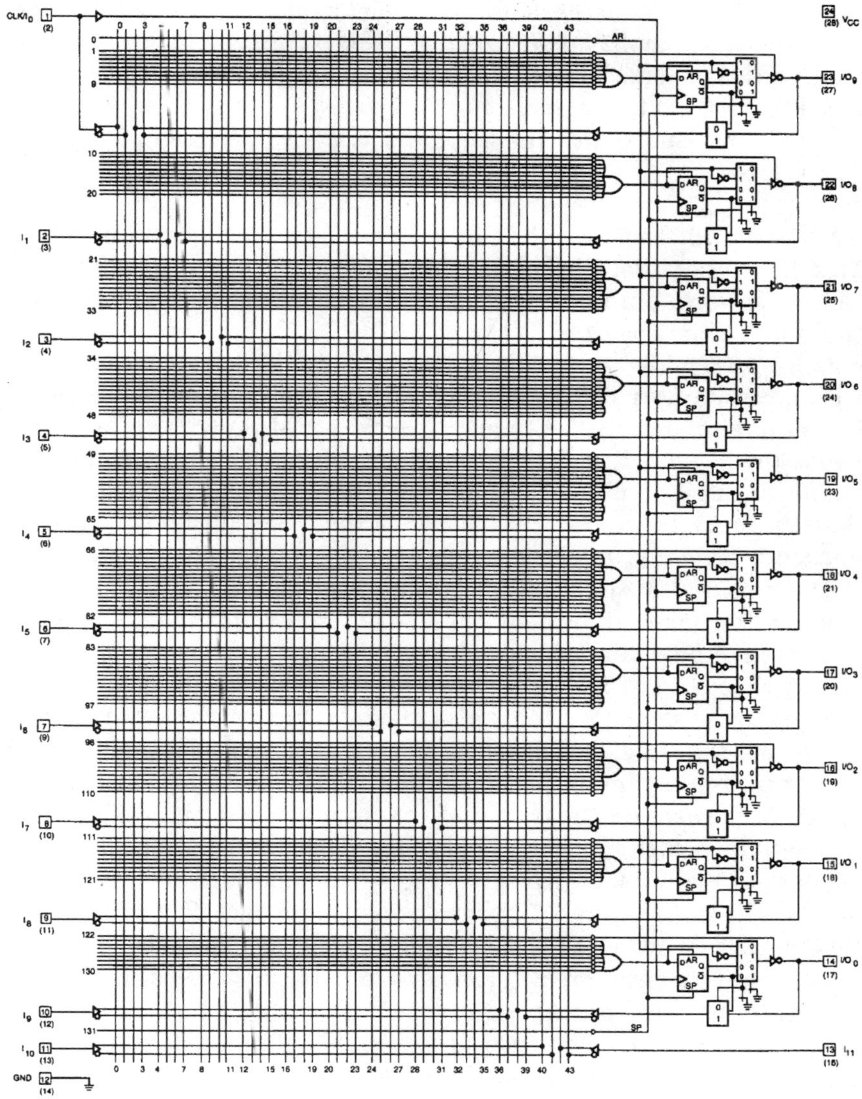

Figure 2.4 AMD PALCE22V10 logic block diagram including output microcells. *(Reprinted from Ref. 5 with permission of AMD.)*

chines that require use of embedded or buried registers can be better implemented using Vantis devices such as the PALCE29M16. The buried register operation is quite useful when a state machine uses bits that do not need to be brought outside the chip. This allows the pin associated with the macrocell to be available as an input. The macrocells can be configured as latches or registers. Also, like PALCE22V10, there is varied product-term distribution among the macrocells. There is a preload capability using a global product term to define the preload condition. This preload capability allows arbitrary states to be loaded directly into the register, which makes it unnecessary to cycle through long, complex vector sequences to get the device into a particular state.

MACH1/MACH2 families. The first- and second-generation Vantis MACH families offer PLDs in densities ranging from 32 to 128 macrocells, worst-case pin-to-pin delays specified of 5 ns, and clock-to-output times of 3.5 ns. The MACH 1 and MACH 2 families consist of 5 base options (10 devices) providing various density-I/O combinations, each with a counterpart equivalent that includes JTAG-compatible in-system programming (ISP). A major attraction of MACH 1/MACH 2 families is the *speedlocking* feature, a patented architecture that allows a stable, fixed pin-to-pin delay, independent of logic paths and design refits for up to 16 product terms per output. Speedlocking® provides guaranteed timing, regardless of logic or routing resource utilization.

The MACH 1/MACH 2 families consist of multiple, optimized PAL blocks interconnected by a switch matrix. This switch matrix takes all dedicated inputs and signals from the input switch matrices and routes them to the PAL blocks, as needed. The feedback signals returning to the same PAL block also must go through the switch matrix to ensure that the PAL blocks communicate with each other with guaranteed fixed timing. These PAL blocks can be viewed as independent PAL devices on the chip that communicate with each other only through the switch matrix. Each PAL block consists of following major elements: (1) product-term array, (2) a logic allocator, (3) macrocells, and (4) I/O cells. Figure 2.5*a* provides a general overview and description of basic components of MACH 1/MACH 2 families' architecture.

The product-term array generates the basic logic, although the number of product terms per macrocell is variable. The logic allocator distributes the product terms to the macrocells per design requirements. The macrocells configures the signal, and the I/O cell delivers the final signal to the output pin. The logic allocator is a block within which different product terms are allocated to the appropriate macrocells in groups of four product terms called *product term clusters*. The availability and distribution of product terms is automatically considered by the software while placing and routing functions within the PAL block.

Two different types of macrocells are used: the output macrocells and buried macrocells. Only the MACH 2 family uses buried macrocells, which effectively doubles the number of macrocells available without increasing the pin count. Both macrocell types can generate

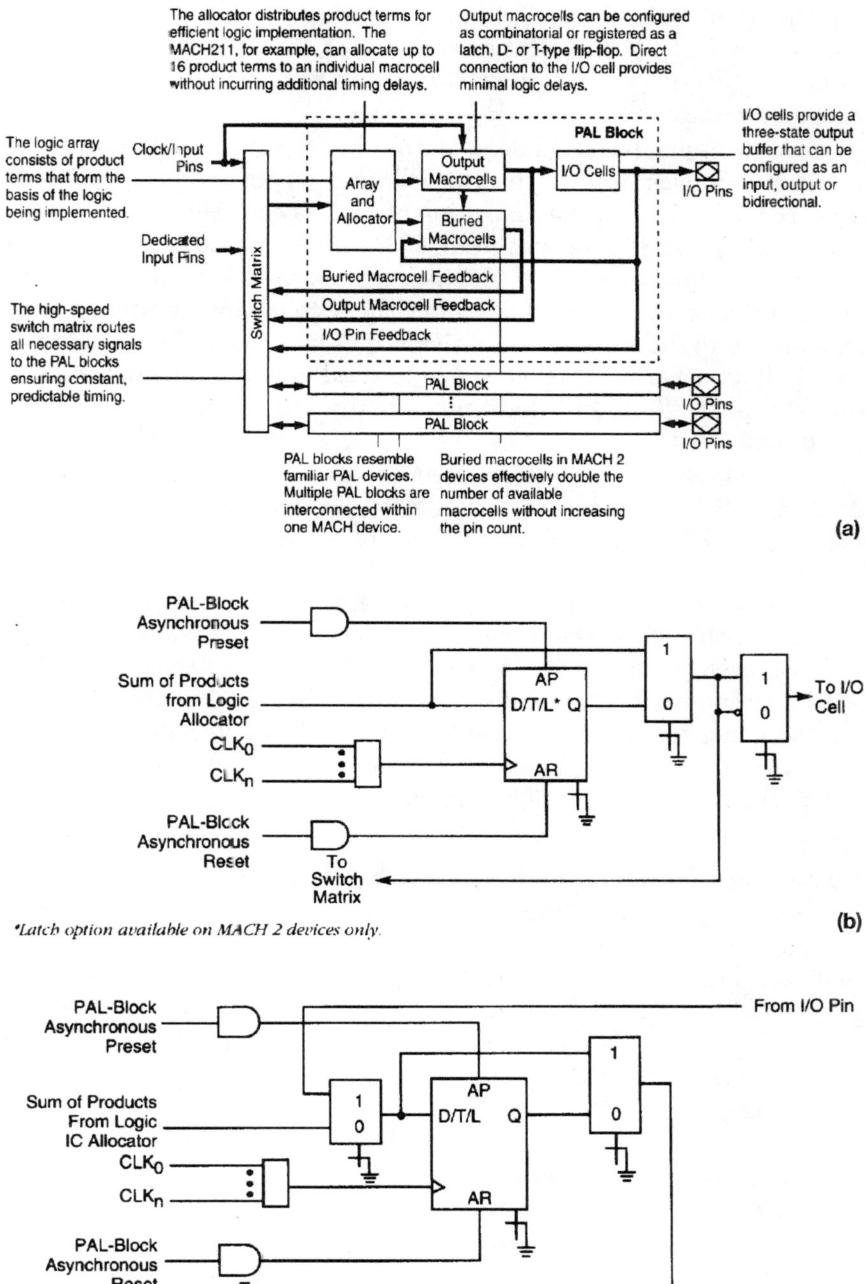

Figure 2.5 MACH 1/MACH 2 families: (a) general overview of architecture, (b) output macrocell, and (c) buried macrocell *(reprinted from Ref. 6 with permission of Vantis Corp.).*

registered or combinatorial outputs. For the MACH 2 family, a transparent low latched configuration is provided, which allows the register to be configured as a T-type or D-type flip-flop. Figure 2.5 shows the output macrocell and the buried macrocell.

The output macrocell sends its output back to the switch matrix, via internal feedback, and to the I/O cell. The feedback is always available, regardless of the configuration of the I/O cell. The buried macrocell does not send its output to an I/O cell but provides it only as an internal feedback signal to the switch matrix. This allows the designer to generate additional logic without requiring additional pins. In addition to the capabilities of the output macrocell, the buried macrocell allows the use of registered or latched inputs. The input register is a D-type flip-flop, and the input latch is a transparent low D-type latch.

An example of MACH 1/MACH 2 families architecture is MACH231SP, which has approximately 12 times the logic macrocell capacity of PALCE22V10. It consists of eight PAL blocks interconnected by a programmable switch matrix, which connects the PAL blocks to each other and all input pins. Each PAL block contains a 64 product-term logic array, a logic allocator, 8 output macrocells, 8 buried macrocells, and 8 I/O cells. The switch matrix feeds each PAL block with 32 inputs. The logic allocator in the MACH231SP takes the 64 logic product terms and allocates them to 16 macrocells, as needed. Each macrocell can be driven by up to 16 product terms. Figure 2.6 shows the block diagram of a MACH231SP device.

MACH 4 family. MACH 4 devices are fabricated on an advanced 0.35 µm EECMOS process, offering densities ranging from 32 to 256 macrocells. MACH 4 family features include multitiered central switch matrices, enhanced logic arrays, logic allocator with an XOR gate and multi-clocking scheme, synchronous/asynchronous logic and flexible set/reset options, and mixed 3.3 V/5 V design capabilities. All MACH 4 devices with the exception of M4-128N/64 (formerly the MACH435) are in compliance with IEEE 1149.1 standard and utilize the JTAG interface for in-system programming (ISP).

The MACH 4 family is an extension of the basic architecture of MACH 1/MACH 2 product families and consists of multiple optimized PAL blocks interconnected by a central switch matrix, which allows signal routing and communication between various PAL blocks. Each PAL block consists of a product-term array, a logic allocator, macrocells, an input switch matrix, an output switch matrix, I/O cells, and a clock generator. The logic allocator distributes the product terms to the macrocells, as per design requirements. The macrocell configures

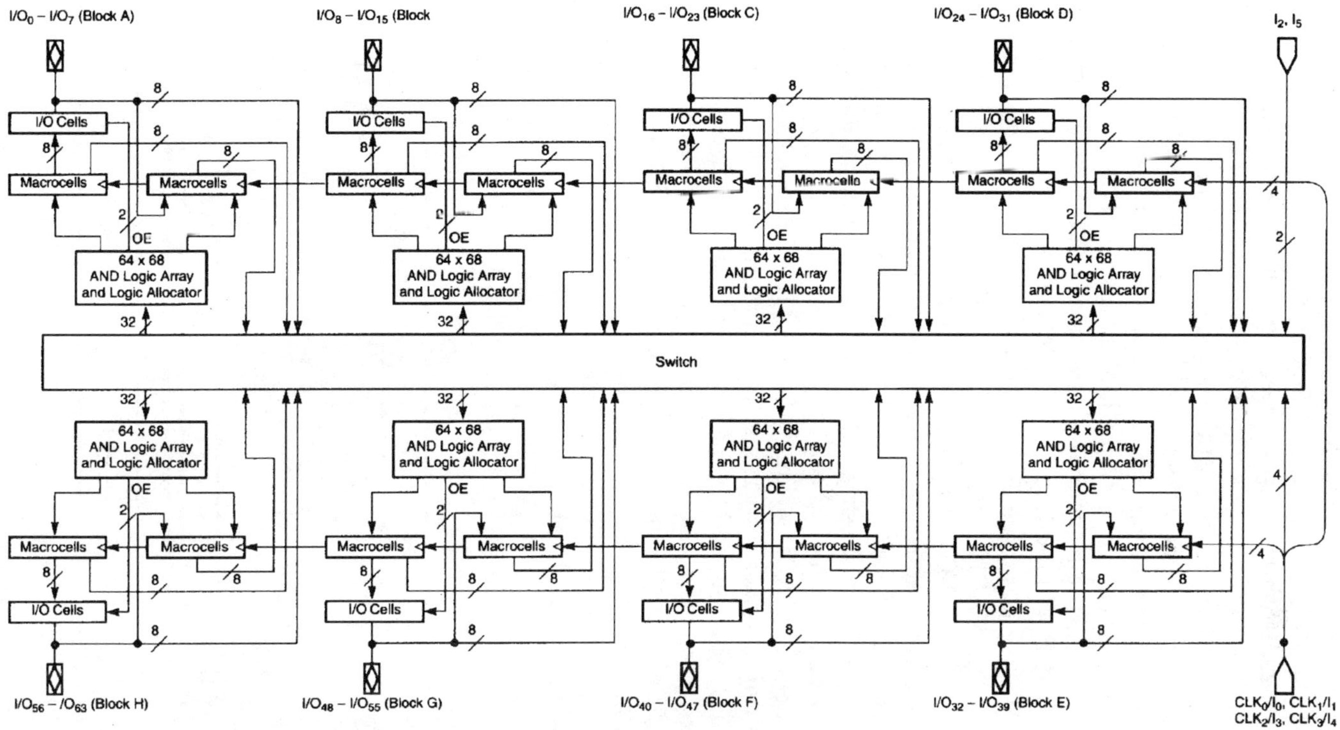

Figure 2.6 Block diagram of MACH231SP device *(reprinted from Ref. 6 with permission of Vantis Corp.).*

the signal largely by determining the storage characteristics. Macrocell signals are routed to I/O cells and the I/O pins by the output switch matrix.

The input switch matrix optimizes routing of the inputs and provides additional ways for these signals to enter the central switch matrix. The central switch matrix takes all dedicated inputs and signals from the input switch matrices and routes them to the PAL blocks as needed. The MACH 4 family speed-locking feature allows allocation of up to 20 product terms to an individual macrocell, with the assistance of an XOR gate, without incurring additional timing delay. Figure 2.7 shows the MACH 4 block diagram and PAL block structure.

In MACH 4 devices, each PAL block has twice as many macrocells as I/O cells. The MACH 4 output switch matrix allows for half of the macrocells to drive I/O cells within a PAL block. Each I/O cell can choose from eight macrocells, whereas each macrocell has a choice of four I/O cells. The I/O cell consists of a programmable output enable, a feedback path, and a flip-flop. An individual output enable product term is provided for each I/O cell.

An example of MACH 4 family is MACH 4-192 (M4-192) device, which has approximately six times the macrocell capability of the popular MACH111SP and additional functional features. It consists of 12 PAL blocks interconnected by a programmable central switch matrix and up to 20 product terms per macrocell can be assigned. Figure 2.8 shows the MACH 4-192 block diagram.

MACH 5 family. The MACH 5 family has many advanced features, such as in-system programmability, JTAG testability, advanced clocking options, power management system that allows incremental power

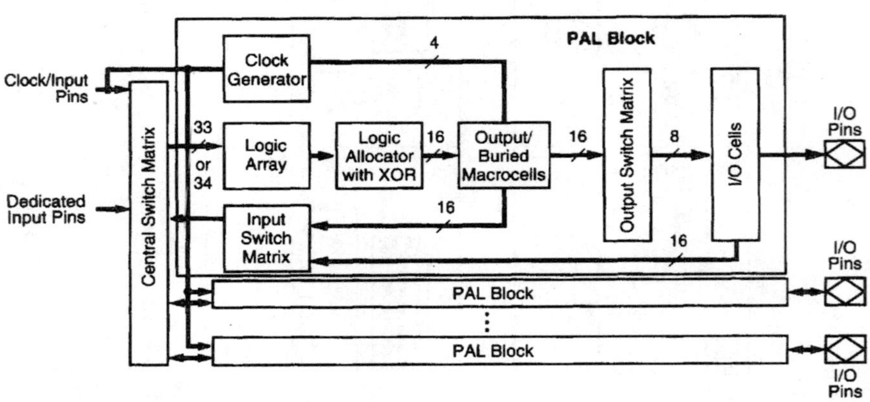

Figure 2.7 MACH 4 family block diagram and PAL block structure *(from Ref. 6 with permission of Vantis Corp.).*

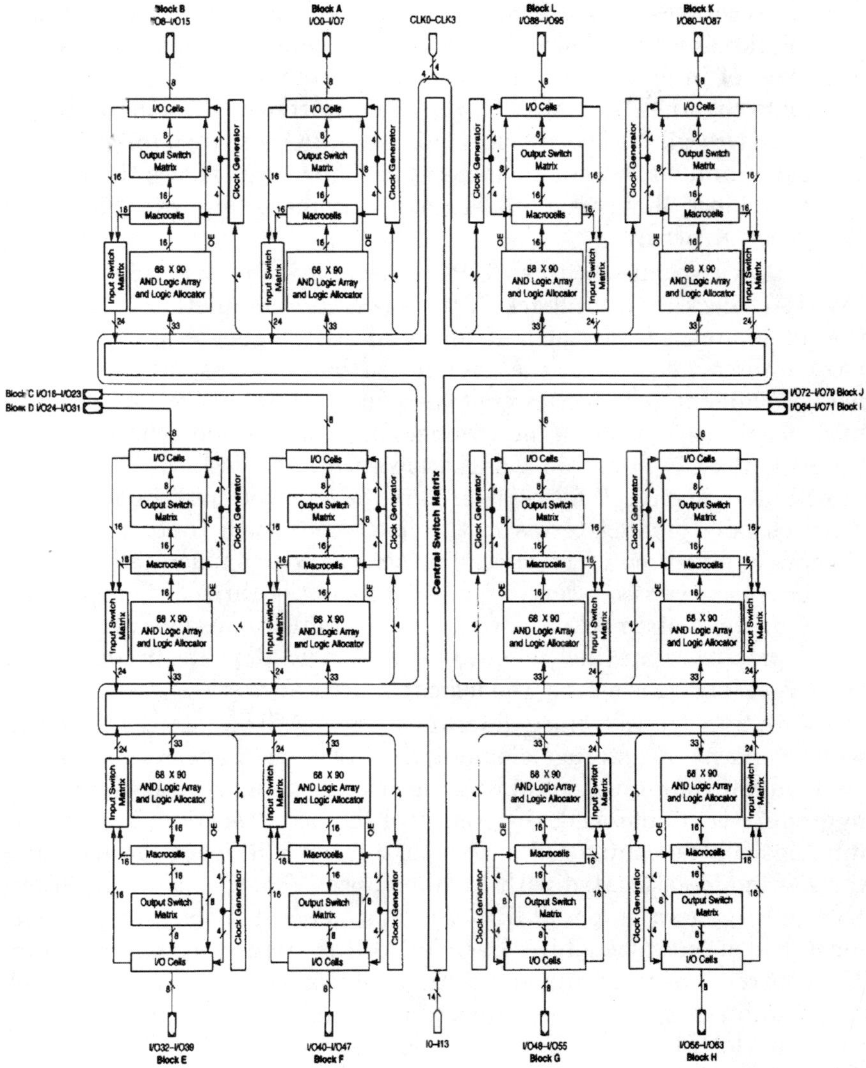

Figure 2.8 MACH 4-192 (M4-192) family block diagram *(from Ref. 6 with permission of Vantis Corp.).*

reduction and PCI Local Bus Specification compliance.[6] The MACH 5 devices are fabricated in Vantis EECMOS 0.5 and 0.35 μm L_{eff} technology, offering densities ranging from 128 to 512 macrocells, and are specified for pin-to-pin delays as fast as 7.5 ns. The routing resources include a logic switch matrix, which reassigns logic to the macrocells.

The MACH 5 architecture consists of PAL blocks connected by two levels of interconnect. Each group of four PAL blocks is given its own routing resources, called *block interconnect*. The combination of four PAL blocks and their block interconnect is called a *segment*. The second level of interconnect, called *segment interconnect,* ties all of the segments together. The MACH 5 PAL block consists of following elements: macrocell, logic array, logic allocator, I/O cell, registered control generator, and output enable generator. Figure 2.9*a* shows the MACH 5 family block diagram with segment numbers. Figure 2.9*b* shows PAL block structure.

The I/Os associated with each PAL block have a path directly back to that PAL block local feedback. If the I/O is used in another PAL block, the interconnect feeder acts as an input switch matrix and assigns a block interconnect line to that signal. The block and segment interconnect provide interconnections between any two signals in a device. The block feeder assigns block interconnect lines and local feedback lines to the PAL block inputs. Two inputs from each PAL block can also be fed directly to a macrocell for registered input applications. The logic allocator assigns product terms to the macrocells such that up to eight clusters of four product terms can be steered to one macrocell, and the product terms can be steered in a basic cluster of three or four product terms. Instead of an output switch matrix, which reassigns macrocells to the pins to retain the pinout, the wide logic allocator produces the same result by reassigning the logic to macrocells.

Figure 2.9*c* shows a block diagram of the MACH 5 family macrocell, which consists of a storage element, a control (clock and set or reset or latch enable) bus, and the routing resources. The macrocell can be configured for combinatorial, registered, or latched operation. The D-type flip-flops can be configured as T-type, J-K, or S-R operation by using the XOR gate associated with each macrocell. There is one output enable (OE) generator per PAL block that generates the two product-term driven output enables. Each I/O cell is simply an output buffer. The control generator provides four configurable clock lines and three configurable set/reset lines to each macrocell in a PAL block. Any of the four clock lines and any of three set/reset lines can be independently selected by any flip-flop within a block. The clock lines provide synchronous global (pin) clocks, asynchronous product-term clocks, sum-term clocks, and latch enables.

Vantis design software and development tools. Vantis provides software design support for MACH devices through its own development system and device fitters integrated into third-party CAE tools. MACHXL® software is a complete development system for a PC, supporting Vantis' MACH devices. It supports design entry with Boolean equations, truth tables, and high-level HDLs syntax. The MACHXL development system also includes functional simulation and static

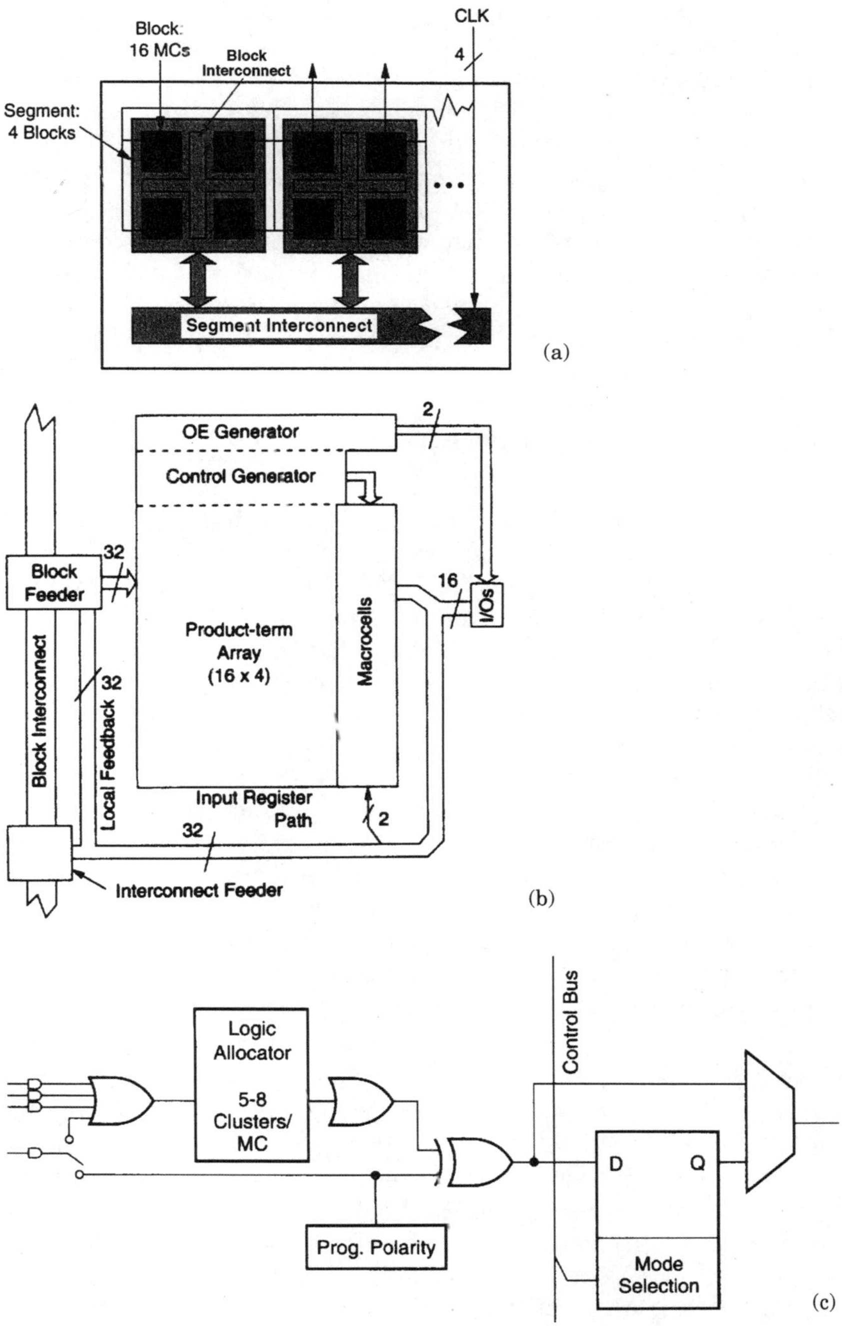

Figure 2.9 MACH 5 family device: (a) block diagram with segment numbers, (b) PAL block structure, and (c) macrocell diagram *(from Ref. 6 with permission of Vantis Corp.)*.

timing analysis. The fitter technology included in the software is incorporated into third-party tools from leading CAE vendors such as Synario, Viewlogic, Mentor Graphics, Cadence, and MINC.

Interface bits and MACHXL configurations are also available to support design entry and verification with other vendors such as Synopsys, OrCAD, Exemplar, and others. These MACHXL configurations and interfaces accept EDIF 2.0.0 netlists, generate JEDEC files for MACH devices, and create industry standard SDF, VITAL-compliant HDL and Verilog output files for design simulation. In-system programming support for MACH devices is offered through MACHPRO® software that enables device programmability through the JTAG compliant port and PC interface.

2.3.2 Altera simple and complex PLDs

Classic PLDs. Altera's Classic family of general-purpose PLDs is based on the CMOS UVEPROM process, featuring device densities ranging from 300 usable gates for EP610/EP6101 devices to 900 usable gates for EP1810 configuration.[7] The basic architecture of the Classic PLD family includes following elements: (a) macrocells, (b) programmable registers, (c) output enable/clock select, and (d) feedback selection. The macrocells can be individually configured for both the sequential and combinatorial logic operation. Eight product terms form a programmable-AND array that feeds an OR gate for combinatorial logic implementation. An additional product term is used for asynchronous clear control of the input register; another product term implements either an output enable or a logic-array-generated clock. For the implementation of registered functions, each macrocell register can be individually programmed for D, T, JK, or SR operation. Also, macrocell registers can be individually clocked either by a global clock or by any input or feedback path to the AND array.

MAX 5000 EPLD family. Altera's multiple array matrix (MAX) 5000 family provides higher logic density than the Classic PLD family, offering devices with 32 to 192 macrocells (approximately 600 to 3,750 usable gates). MAX 5000 EPLDs have macrocells that are combined into groups called *logic array blocks* (LABs). Each macrocell has a programmable-AND/fixed-OR array and a configurable register that provides D, T, JK, or SR operation with independent programmable clock, clear, and preset functions. For the implementation of complex logic functions, each macrocell can be supplemented with sharable expander product terms to provide more than 32 terms per macrocell. The MAX 5000 family is based on 0.8 μm CMOS EPROM process and is specified for pin-to-pin logic delays as fast as 15 ns. An enhanced version of this family MAX 5000A is based on 0.65 μm CMOS EPROM process and is specified for pin-to-pin delays as fast as 10 ns. Both MAX 5000 and 5000A devices are fully pin-, function-, and programming-file compatible.

The MAX 5000 architecture is based on linking multiple, flexible logic array blocks (LABs) via programmable interconnect array (PIA), a global bus that is fed by all I/O pins and macrocells. The architecture also includes 8 to 20 dedicated inputs, each of which can be used as a high-speed, general-purpose input.

The MAX 5000 EPLD family devices contain from 1 to 12 LABs, each of which consists of a macrocell array and an expander product-term array. The number of macrocells and expanders in the arrays varies with each device type. Figure 2.10*a* shows the block diagram of MAX 5000 LAB. The macrocells are primary resource for logic implementation. However, additional logic capacity is available from expanders, which can be used to supplement the capabilities of any macrocell. Each LAB has 32 shared expanders (except for EPM5032, which has 64). The expander product-term array consists of a group of unallocated, inverted product terms that can be used and shared by all macrocells in the LAB to create combinatorial and registered logic. These flexible macrocells and sharable expanders facilitate variable product-term designs and provide more flexibility than the fixed-product-term architecture.

All macrocell outputs are globally routed within a LAB via the LAB interconnect. The output of the macrocells also feed the I/O control block, which consists of a group of programmable tristate buffers and I/O pins. In the device types with multiple LABs, all LABs are connected by the programmable interconnect array (PIA). The PIA, which is fed by all macrocell and I/O pin feedbacks, routes only the signals required for implementing logic in a LAB.

The MAX 5000 macrocell consists of a programmable logic array and an independently configurable register. The register can be programmed to emulate D, T, JK, or SR flip-flop operations, as a flow-through latch, or bypassed for combinatorial operation. The combinatorial logic is implemented in the programmable logic array in which three product terms that are ORed together feed one input to an XOR gate. The second input to this XOR gate is used for complex XOR logic functions and for De Morgan's inversion. The XOR gate output feeds the programmable register or bypasses it for combinatorial operation. Figure 2.10*b* shows the block diagram of a MAX 5000 macrocell.

Each MAX 5000 LAB supports either global or array clocking. The global clocking is provided by a dedicated clock signal (CLK), which offers fast clock-to-output delay times. Since each LAB has one global

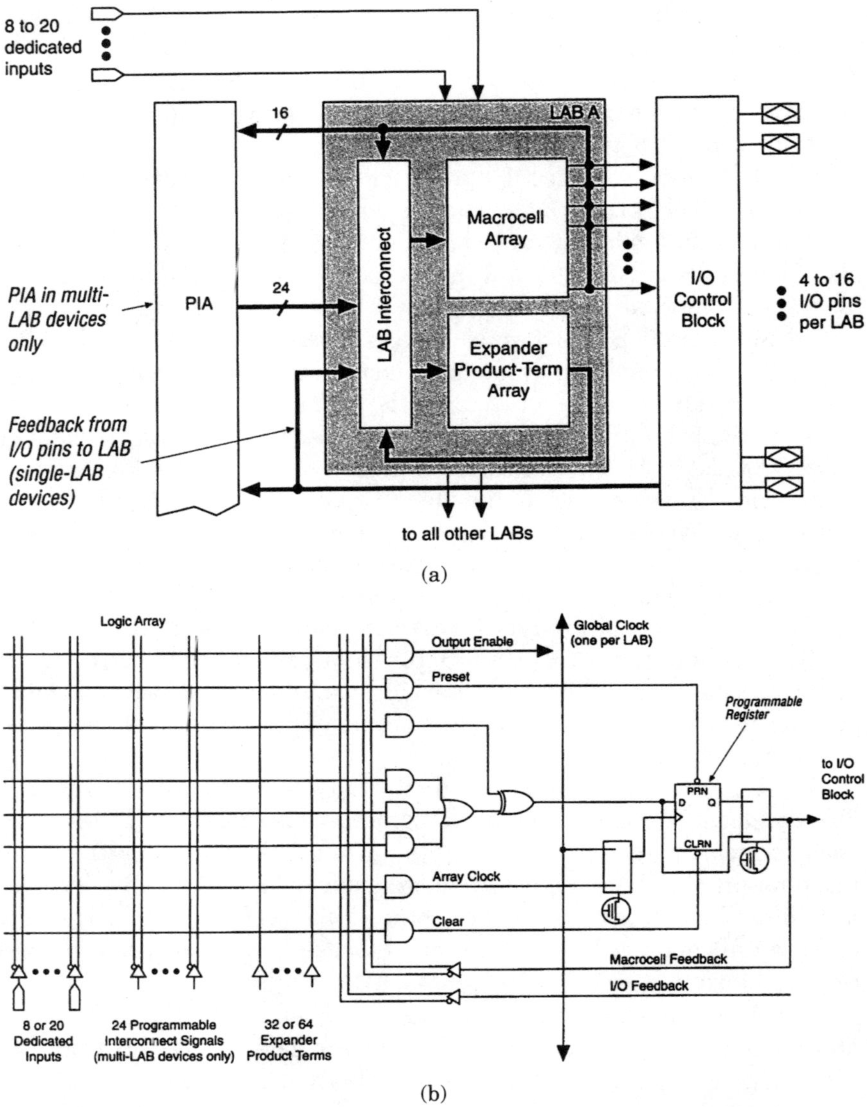

Figure 2.10 MAX 5000 block diagram of (a) LAB architecture and (b) macrocell *(from Ref. 7 with permission of Altera Corp.).*

clock, all flip-flop clocks within the LAB can be positive-edge triggered from the CLK pin. In the array clocking mode, each flip-flop is clocked by a product term. Any input pin or internal logic can be used as a clock source. Array clocking allows each flip-flop to be configured for positive- or negative-edge triggered operation, providing the macrocell with increased flexibility.

Each LAB has an I/O control block, which allows each I/O pin to be individually configured for input, output, or bidirectional operation. This I/O control block is fed by the macrocell array. A dedicated macrocell product term controls a tristate buffer, which drives the I/O pad.

MAX 7000 EPLD family. Altera's second-generation MAX 7000 family is fabricated with EEPROM-based CMOS technology, with specified pin-to-pin delays as fast as 5 ns. The MAX 7000 devices contain from 32 to 256 macrocells combined into groups of 16 macrocells called *logic array blocks* (LABs). Each macrocell has a programmable-AND/fixed-OR array and a configurable register with independently programmable clock, clock enable, clear, and preset functions. As in the case of MAX 5000 family, to build complex logic functions, each macrocell can be supplemented with both sharable expander product terms and high-speed parallel expander product terms to provide up to 32 product terms per macrocell.

The MAX 7000 basic architecture elements are similar to MAX 5000E except for some improvements and include LABs, macrocells, expander product terms (sharable and parallel), programmable interconnect array (PIA), and I/O control blocks. The MAX 7000 architecture provides four dedicated inputs that can be used as general-purpose inputs or as high-speed, global control signals (clock, clear, and two output enable signals) for each macrocell and I/O pin. The MAX 7000 macrocell can be individually configured for both the sequential and combinatorial operation. The macrocell consists of three functional blocks: the logic array, the product-term select matrix, and the programmable register.

Altera's MAX7000S family features ISP capabilities, and devices with 128 or more macrocells have built-in JTAG boundary-scan test circuitry. Another enhanced version of MAX 7000 family called MAX 7000E features: additional global clocking and output enable controls, more interconnect resources, faster input registers, and a programmable slew rate. The MAX 7000S devices are pin-, function-, and programming file compatible with all MAX 7000 family members, including the MAX 7000 E devices. Figure 2.11 shows the block diagram of MAX 7000S/E architecture. The multiple LABs are linked by the PIA. Each LAB is fed by the following signals: (a) global controls that are used for secondary register functions, (b) 36 signals from the PIA that are used for general logic inputs, and (c) direct input paths from I/O pins to the registers that are used for fast setup times.

In MAX 7000E devices, two global clock signals are available. All I/O pins have a fast input path to a macrocell register. The dedicated path allows a signal to bypass the PIA and combinatorial logic, for being clocked to an input D flip-flop with a fast (3 ns) input setup time. For

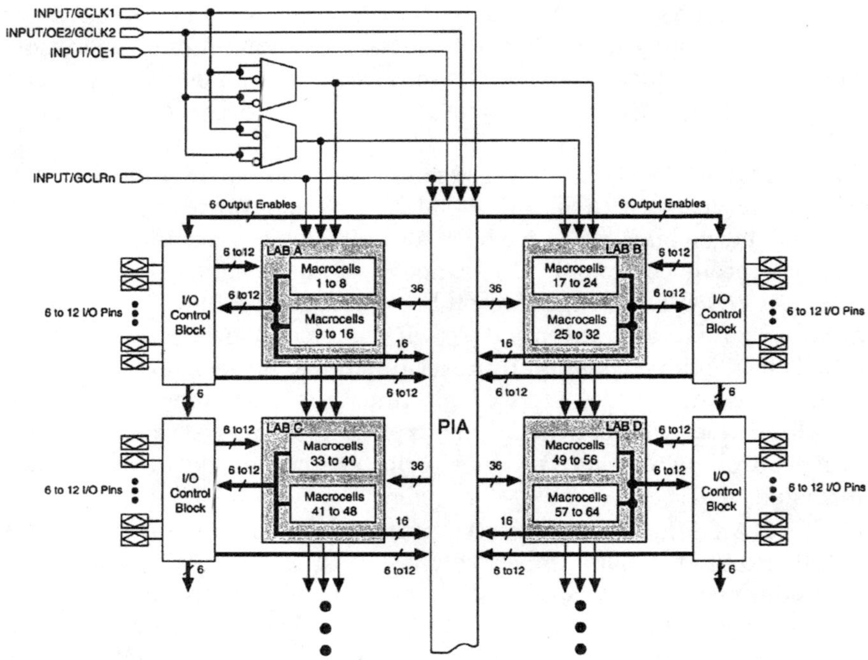

Figure 2.11 Block diagram of MAX 7000S/E architecture *(from Ref. 7 with permission of Altera Corp.)*

complex logic functions, the architecture offers both sharable and parallel expander product terms directly to any macrocell in the same LAB. These parallel expanders allow up to 20 product terms to directly feed the macrocell or logic, with 5 product terms provided by the macrocell and 15 parallel expanders provided by the neighboring macrocells in the LAB.

In MAX 7000 architecture, the PIA is a global bus and programmable path that routes all logic between the LABs. All dedicated inputs, I/O pins, and macrocells outputs feed the PIA, which makes them available throughout the device. Only the signals required by each LAB are actually routed from the PIA into the LAB. The I/O control block allows each I/O pin to be individually configured by one of the output enable signals or directly connected to GND or VCC. The architecture provides dual I/O feedback such that macrocell and pin feedback are independent. When an I/O pin is configured as an input, the associated macrocell can be used for the buried logic.

MAX 9000 EPLD family. The MAX 9000 family of EPLDs are in-system programmable devices based on Altera's third-generation MAX architecture, which extends the logic array block (LAB) interconnect approach of the MAX 7000 architecture to provide higher densities and comparable performance. These are fabricated in 0.65 μm CMOS EEPROM technology, which provides densities ranging from 320 to 560 macrocells (approximately 6,000 to 12,000 usable gates) and specified pin-to-pin delays of 15 to 20 ns. The MAX 9000 EPLDs have macrocells that are combined into groups of 16 macrocells called *LABs*.

Each macrocell has a programmable-AND/fixed-OR array and a configurable register with independently programmable clock, clock enable, clear, and preset functions. For higher flexibility and logic utilization improvement, each macrocell offers a dual-output structure that allows the register and product terms to be used independently. For complex logic functions, each macrocell can be supplemented with both sharable and high-speed parallel expander product terms (as in MAX 7000 family) to provide up to 32 product terms per macrocell. The MAX 9000 family provides programmable speed/power optimization, which can allow speed-critical portions of the design to run at high speed/full power and noncritical portions at reduced speed/low power.

The MAX 9000 architecture consists of the following elements: LABs, macrocells, expander product terms, fast track interconnect, dedicated inputs and I/O cells. Figure 2.12 shows the block diagram MAX 9000 (a) architecture and (b) LAB. The MAX 9000 LABs consist of 16-macrocell arrays that are fed by the LAB local array. The multiple LABs are linked together via the FastTrack Interconnect, which is a series of continuous channels that run the entire length and width of the device. The I/O pins are supported by the control blocks (IOCs) located at the end of each row (horizontal) and column (vertical) path of the FastTrack Interconnect. Each LAB is fed by 33 inputs from the row interconnect and 16 feedback signals from the macrocells within the LAB. All of these signals are available within the LAB in their true and inverted forms.

In addition, 16 shared expander product terms are available in their inverted form, for a total of 114 signals that feed each product term in the LAB. Each of these LABs is also fed by two global clocks and global clear signals that can be used for register control signals in all 16 macrocells. The LAB drives the row and column interconnect directly.

The MAX 9000 macrocell consists of three functional blocks: (a) the product terms, (b) the product term select matrix, and (c) the programmable register. The macrocell can be individually configured for both sequential and combinatorial logic operations. The combinational logic is implemented in the local array, which provides five product terms per macrocell. The product-term select matrix allocates these product terms for use as either the primary logic inputs (to the OR and XOR gates) to implement combinatorial functions, or as secondary inputs to

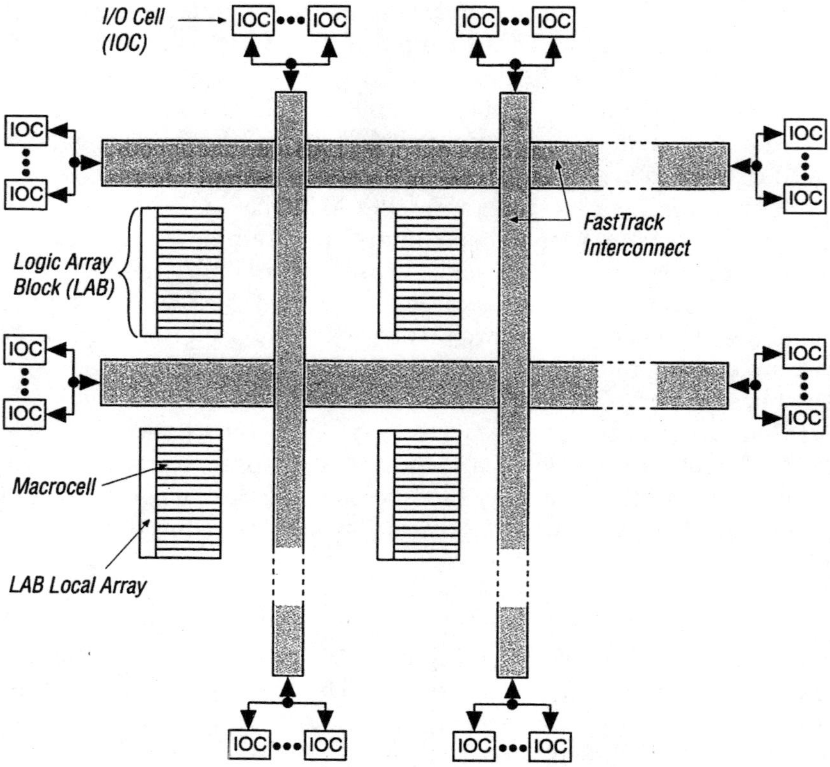

Figure 2.12a Block diagram of MAX 9000 EPLD architecture (from Ref. 7 with permission of Altera Corp.).

the macrocell's register clear, preset, clock, and clock enable control functions. For registered functions, each macrocell register can be individually programmed for D, T, JK, or SR operation with the programmable clock control.

In the MAX 9000 architecture, FastTrack Interconnect, providing connection between macrocells and device I/O pins, has a fairly predictable performance compared to the segmented routing structure used in FPGA designs. The LABs within the MAX 9000 device are arranged in a matrix of row and columns. Each row of LABs has a dedicated row interconnect that routes signals both into and out of the LABs in that row. The row interconnect can then drive the I/O pins or feed other LABs in the device. Each row interconnect has a total of 96 channels.

Each macrocell in the LAB can drive one of three separate column interconnect channels. The column channels run vertically across the entire device and are shared by the macrocells in the same column. A row interconnect channel can be fed by the output of the macrocell

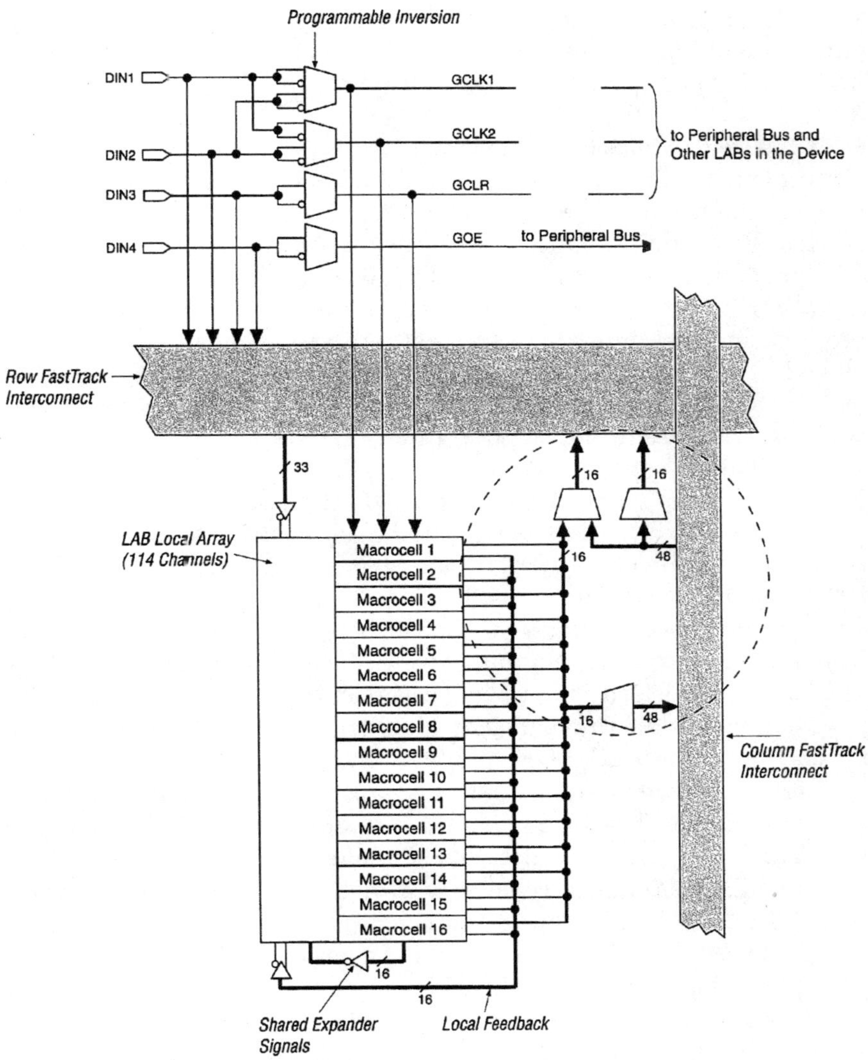

Figure 2.12b Block diagram of MAX 9000 EPLD LAB *(from Ref. 7 with permission of Altera Corp.)*. The dotted circle indicates that details of LAB connections to row and column interconnect are available in Ref. 7.

through a 4-to-1 multiplexer that the macrocell shares with three column channels. If the multiplexer is used for a macrocell-to-row connection, the three column signals can access another row channel via an additional 3-to-1 multiplexer. In any LAB, the multiplexers provides all 48 column channels with access to 32 row channels.

In addition to general-purpose I/O pins, MAX 9000 devices have four dedicated input pins, typically used for global clock, clear, and output enable control signals. These dedicated inputs can also be used as general-purpose data inputs because they can feed the row FastTrack Interconnect. The I/O pins can be used as input, output, or bidirectional pins.

FLEX 6000 and FLEX 8000 PLD families. Altera's Flexible Logic Element Matrix (FLEX) family combines the benefits of fine-grained architecture and high register count, which are characteristics of FPGAs for efficient logic implementation, with the high speed and predictable interconnect delays of EPLDs. The FLEX 6000 devices are fabricated on a 0.5 μm, triple-layer metal process. The logic and interconnections are implemented through SRAM based, compact four-input lookup tables (LUTs) and programmable registers. The logic array contains logic array blocks (LABs), each consisting of 10 logical elements (LEs) that communicate through a local interconnect structure.

A flexible and efficient architecture called OptiFLEX™ is used for LAB interleaving that gives any LE the flexibility to access the local interconnect of its own LAB as well as the adjacent LABs. Each LE contains a four-input LUT, a programmable register, and dedicated paths for carry and cascade functions. The logic array is routed through FastTrack Interconnect, a series of fast, continuous row and column channels that run the entire length and width of the device. Each FastTrack row and column feeds multiple I/O elements (IOEs), which provides programmable slew rate and individual tristate output enable control for each pin. The FLEX6000 family also supports FastFLEX™ I/O, a feature that provides a direct path from LE to the I/O pin for fast clock-to-output timing. The FLEX 6000 family has many common features with FLEX 8000 family, as described in more detail below.

The FLEX 8000 family consists of device types ranging in densities from 26 to 162 LABs (approximately 2,500 to 16,000 usable gates). The FLEX 8000 architecture consists of a large matrix of compact, fine grained building blocks called logic elements (LE), which are grouped into sets of eight to create logic array blocks (LABs) that are arranged in rows and columns. Each FLEX 8000 LAB is an independent structure with common inputs, interconnections and control signals. The LABs provide a coarse-grained structure for high performance and easy routing. The signal interconnections within the FLEX 8000 devices, and to and from the device pins are provided by the FastTrack interconnect. The I/O pins are supported by the I/O elements (IOEs) located at the ends of rows and columns in a layout similar to that shown in Figure 2.12(a). Each IOE contains a bidirectional I/O buffer and a flip-flop that can be used as either an input or output register.

Logic elements (LEs) are the smallest unit of logic in the FLEX 8000 architecture and contain a four-input lookup table (LUT), a programmable flip-flop, a carry chain, and a cascade chain. The LUT is a function generator for computation of any function of four variables. The programmable flip-flop in LE can be configured for D, T, JK, or SR operation. The clock, clear, and preset control signals on the flip-flop can be driven by the dedicated inputs, general-purpose I/O pins, or any internal logic. For purely combinatorial functions, the flip-flop is bypassed, and the LUT output goes directly to the output of LE.

The carry chains and cascade chains connect adjacent LEs without using local interconnect path. The carry chain supports high-speed counters and adders, whereas the cascade chain implements wide-input functions with minimum delay. These carry and cascade chains connect all LEs in a LAB and all LABs in the same row. Figure 2.13 shows block diagram of (2.13*a*) LE and (2.13*b*) LAB architecture.

A logic array block (LAB) consists of eight LEs, their associated carry and cascade chains, and the LAB local interconnect. Each LAB provides four control signals that can be used in all eight LEs. Two of these signals can be used as clocks, and the other two for clear/preset control. The multiple LABs are arranged into a matrix of column and rows. Each row of LABs has a dedicated row interconnect that routes signals both into and out of the LABs in the row. The row interconnect can then drive I/O pins or feed other LABs in the device. Each LE in a LAB can drive up to two separate column interconnect channels. Therefore, all 16 available column channels can be driven by the LAB. Each column of LABs has a dedicated column interconnect that routes signals out of the LAB into the column. The column interconnect can then drive the I/O pin or feed into the row interconnect to route the signals to other LABs in the device.

The FLEX 8000 logic elements can operate in one of four modes, each of which utilizes LE resources differently. In each mode, seven of the ten available inputs to the LE (the four data inputs from the LAB local interconnect, the feedback from the programmable register, and the carry-in and cascade-in from the previous LE) are directed to different destinations for implementing the desired logic function. The three remaining inputs to the LE provide clock, clear, and preset controls for the register. The four LE operating modes are briefly described below.

1. *Normal mode.* The normal mode is suitable for general logic applications and wide decode functions that can take advantage of a cascade chain. In this mode, four data inputs from the LAB local interconnect and the carry-in are the inputs to a four-input LUT. The LUT output can be combined with the cascade-in signal to form a cascade chain through the cascade-out signal.

2. *Arithmetic mode.* The arithmetic mode offers two three-input LUTs that are best suited for implementing the adders, accumula-

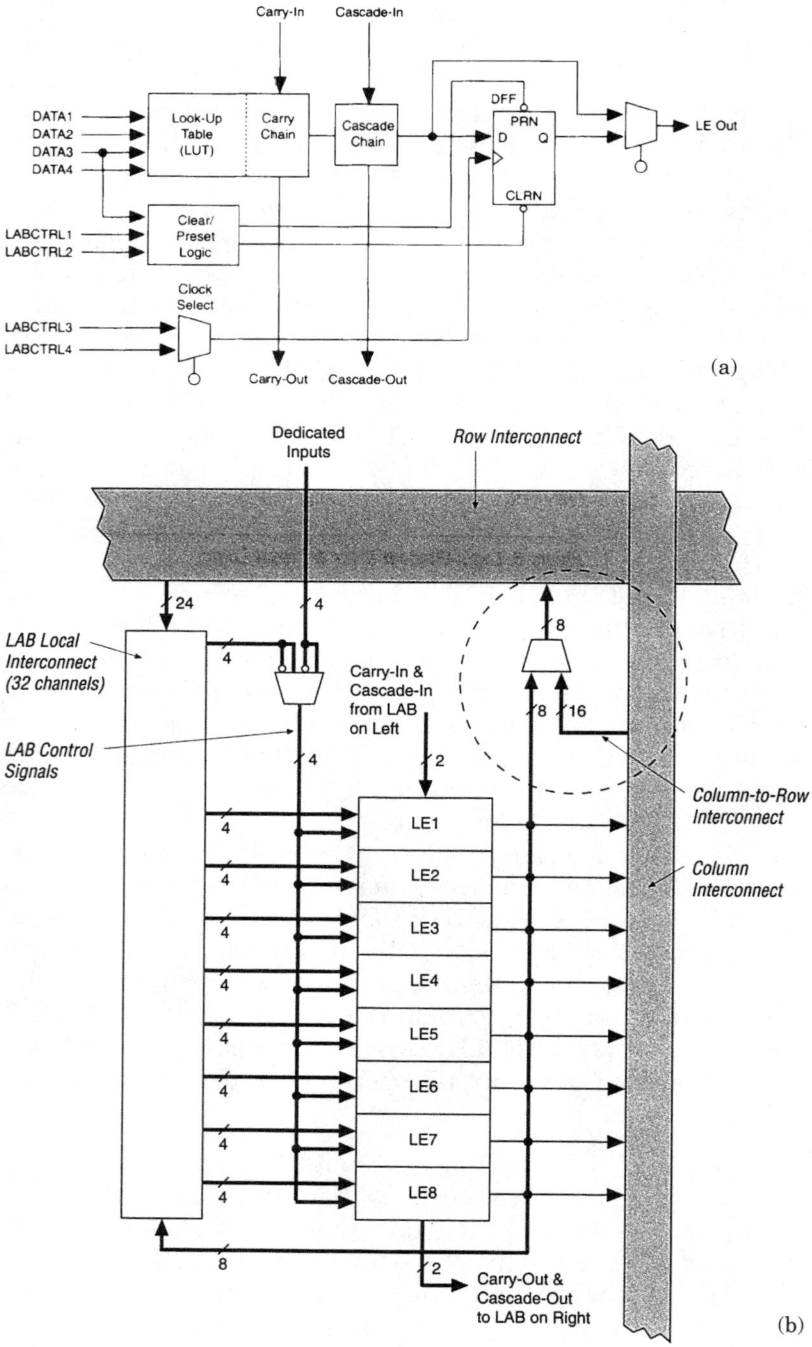

Figure 2.13 FLEX 8000 family block diagrams: (a) architecture and (b) logic element (LE) *(from Ref. 8 with permission of Altera Corp.).*

tors, and comparators. One LUT provides a three-bit function, and the other generates a carry bit.

3. *Up/down counter mode.* The up/down counter mode offers counter enable, synchronous up/down control, and data loading options. These control signals are generated by the data inputs from the LAB local interconnect, the carry-in signal, and output feedback from the programmable register. Two three-input LUTs are used—one for generating the counter data and the other for fast carry bit. A 2-to-1 multiplexer provides synchronous loading.
4. *Clearable counter mode.* This mode is similar to the up/down counter mode but supports a synchronous clear instead of the up/down control. The clear function is substituted for the cascade-in signal in the up/down counter mode.

FLEX 10K embedded PLD family. Altera's FLEX 10K family, the latest offering, is a family of embedded array programmable logic devices based on configurable CMOS SRAM elements, with device densities ranging from 10,000 to 100,000 gates (for both logic and RAM). The newest member of Altera's FLEX 3.3V 10KA family is built on a 0.35 μm, four-layer metal CMOS process that contains 4,992 logic elements and 24,576 bits of on-chip RAM. The FLEX 10K architecture is similar to that of embedded gate arrays, which are used for logic implementation in a conventional "sea-of-gates" approach. Two major components used to build logic are (a) logic array block (LAB) and (b) embedded array block (EAB). The LABs are used to implement logic functions such as eight-bit counters, address decoders, and small state machines. The EABs are used to build the memory blocks and megafunctions such as microprocessors, multipliers, and DSPs. Figure 2.14 shows the block diagrams of FLEX 10K architecture and EAB.

The FLEX 10K LAB and logic element (LE) structure is similar to that for FLEX 8000 family. It consists of eight logic elements (LEs), which are connected through a fully populated interconnect matrix, allowing any LE in a LAB to be connected to any other LE in the same LAB. The LE in this architecture consists of a four-input LUT, which is a function generator to compute any function of the four variables. In addition, each LE contains a programmable flip-flop with a synchronous enable, a carry chain, and a cascade chain. The LE has two outputs, one of which drives the local, and the other drives either the row or column FastTrack Interconnect. The two outputs can be controlled independently in a feature called *register packing* to improve the LE utilization.

The FLEX 10K architecture provides two types of dedicated high-speed data paths to connect adjacent LEs without using local intercon-

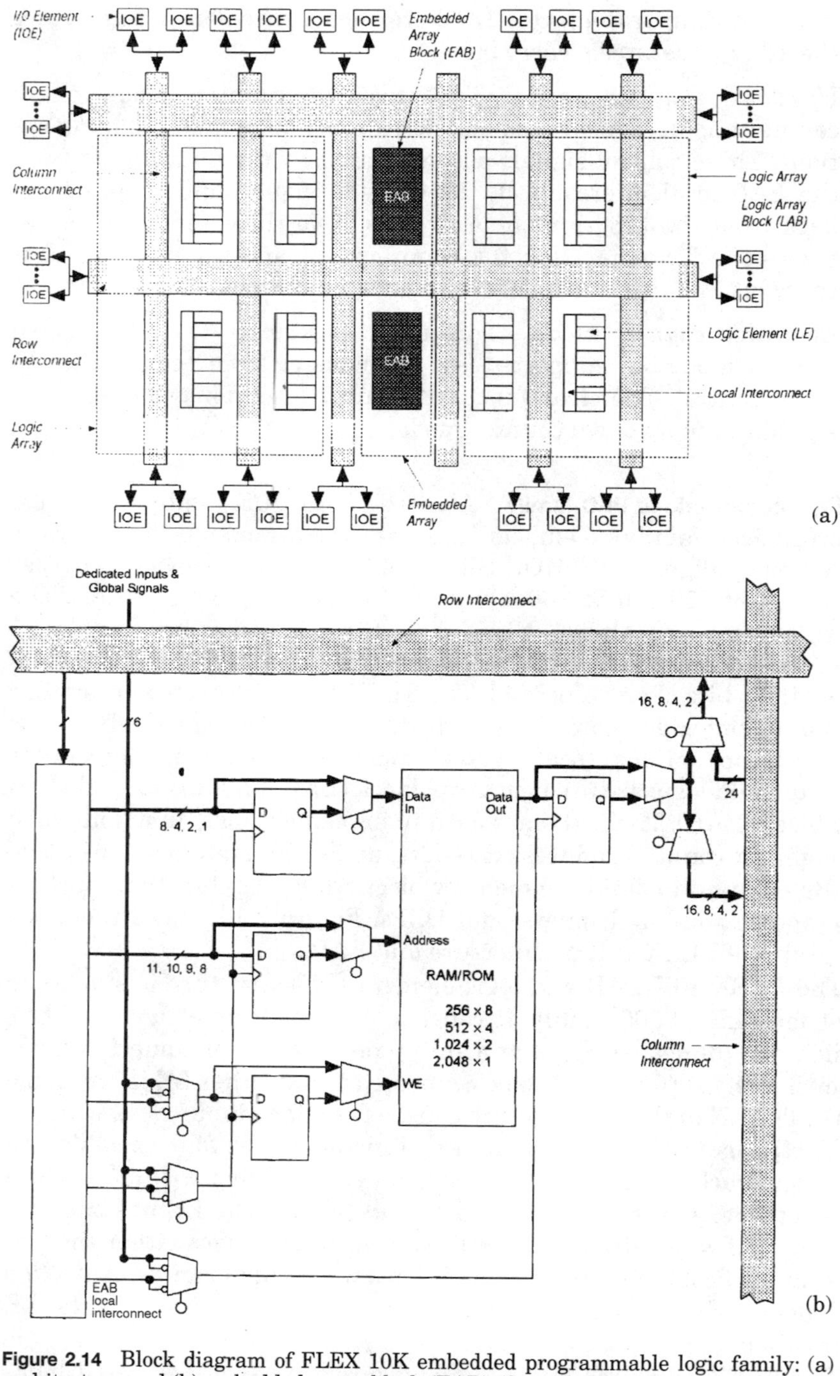

Figure 2.14 Block diagram of FLEX 10K embedded programmable logic family: (a) architecture and (b) embedded array block (EAB) *(from Ref. 7 with permission of Altera Corp.).*

nect paths, as follows: carry chains and cascade chains. These carry and cascade chains connect all LEs in a LAB and all LABs in the same row. The carry and cascade chains larger than eight bits are automatically implemented by linking the LABs together. The cascade chains help implement functions that have a very wide fan-in. Adjacent LUTs can be used to compute portions of functions in parallel; the cascade chain serially connects the intermediate values. As in FLEX 8000 family, the FLEX 10K LE can operate in one of the following four modes: (1) normal mode, (2) arithmetic mode, (3) up/down counter mode, and (4) clearable counter mode.

The eight LE blocks, called LABs, can be used to create medium-sized blocks of logic such as eight-bit counters, address decoders, and state machines, or they can be combined across LABs to create larger logic blocks. Each LAB represents about 96 usable gates of logic. The LABs are arranged into rows and columns. The FLEX 10K FastTrack Interconnect, similar to that for FLEX 8000 architecture, provides connection between LEs and the device I/O pins. Each row of LABs is served by a dedicated row interconnect that can drive the I/O pins and feed other LABs in the device. The column interconnect routes signals between the rows and can drive I/O pins. Each column of LABs is served by a dedicated column interconnect. The column interconnect can then drive the I/O pins or drive another row's interconnect for routing the signals to other LABs in the device.

The embedded array consists of a series of EABs, each of which can provide 2,048 bits that can be used to create RAM, ROM, FIFO, or dual-port RAM. For logic implementation, each EAB can contribute 100 to 300 gates toward complex logic functions such as multipliers, microcontrollers, and DSPs. The logic functions are implemented by programming the EAB during configuration with a read-only pattern, creating a large LUT. In this LUT, combinatorial functions are implemented by looking up the results (rather than computation) and is faster than using the algorithms implemented in general logic. For example, a single EAB can fit a 4×4 multiplier with eight inputs and eight outputs. These EABs can be used independently, or multiple EABs can be combined to implement larger functions.

The EAB is a flexible RAM with inputs and output registers that have programmable options for varying the memory width and depth. The EABs have an advantage over FPGAs, which implement blocks of on-board RAM as arrays of small, distributed RAMs. As the size of FPGA RAM increases, the delays associated with these RAMs becomes less predictable. In contrast, the EABs can be used to implement large, dedicated blocks of RAMs that reduce these timing and routing concerns. The EABs can be used to configure synchronous RAMs, which are easier to implement than asynchronous RAMs of typical FPGAs. When used as a RAM block, it can be configured in sizes of 256 × 8, 512 × 4, 1K × 2, or 2K × 1. Multiple EABs from different rows can be cascaded to build wider and/or deeper RAM. To support this multiconfigurable capability, a programmable decoding scheme is used for addressing and steering of byte-wide output data to the correct format.

The EAB provides flexible options for driving and controlling the clock signals. Therefore, different clocks can be used for the EAB inputs and outputs. Each EAB is fed by a row interconnect and can drive out to row and column interconnects. Each EAB output can drive either of the two row channels and either of two column channels.

Altera design software and development tools. Altera provides MAX + PLUS II software that fully supports all its FLEX PLDs, MAX EPLDs, and Classic PLD families. It also provides an architecture-independent design environment through Altera Hardware Description Language (AHDL) that includes automated design compilation, design partitioning, design rule checking, timing simulation and analysis, delay prediction for speed critical paths, automatic error location, device programming and verification, and a design archiver. MAX + PLUS II software imports VHDL, EDIF, Verilog HDL, and OrCAD netlist files directly and exports post-routing netlist files in the same formats.

The users can also describe their designs with the industry-standard library of parametrized modules (LPMs). The LPM standard offers scalable macrofunctions such as counters, adders, and multiplexers, and preserves high-level design implementation for optimal implementation (see Sec. 4.5). The MAX + PLUS II compiler automatically generates optimized, architecture-specific implementations of the LPMs primitives. The LPMs primitives can be entered with industry standard design entry tools or in schematic text designs created with MAX + PLUS II.

The MAX + PLUS II programmer uses programming files generated by the compiler to program Altera devices. It allows the designer to program, verify, examine, blank-check, and functionally test the device.

2.3.3 Cypress Semiconductor EPLD/CPLD families

Universal PAL devices and EPLDs. Cypress offers industry-standard 20- and 24-pin PLD device architectures, universal PAL devices, and 28-pin application-specific devices based on 0.65 μm CMOS EPROM technology and 0.65 μm CMOS flash reprogrammable technology.[8] The Cypress PALC20 family is made up of electrically programmable and UV-erasable devices, based on n-well CMOS process, and supplied in four functional configurations designated as 16R8, 16R6, 16R4, and 16L8.

The reprogrammable memory cells serve the same function as the fuses in most bipolar devices. Before programming, the AND gates or product terms are connected via the reprogrammable memory cell to both the true and complement inputs. The programming alters the transistor threshold of each cell so that no conduction can occur, which is equivalent to disconnecting the input from the gate or product term. The programmability feature of PLDs allows the users to customize the devices for implementation of their unique logic requirements. Cypress PLD family implements a familiar *sum-of-products* logic by using a programmable AND array whose output terms feed a fixed OR array. The sum of these can be expressed as a Boolean transfer function.

The programmable I/O offered in the Cypress PLD family allows product terms to directly control outputs of the device. One product term is used to directly control the three-state output buffer, which then gates the summation of the remaining terms to the output pin. The output of this summation can be fed back into the PLD as an input to the array. A number of Cypress PLDs offer registered outputs that allow any of these circuits to function as a state sequencer. The programmable macrocell for Cypress PLDs provides the capability of defining the architecture of each output individually—either as registered or combinatorial. The polarity of each output may also be individually selected. Additional flexibility is provided through "array" configurable "output enable" for each potential output. This feature allows the outputs to be reconfigured as inputs on an individual basis or, alternatively, used as a bidirectional I/O controlled by the programmable array.

Figure 2.15*a* shows the I/O macrocell for Cypress CY7C335 Universal Synchronous EPLD. The architecture of CY7C335 consists of user-configurable output macrocell, bidirectional I/O capability, input registers, and three separate clocks to enable the design of high-performance state machines that can communicate with each other, or with microprocessor-based systems, over bidirectional buses. The user configurable macrocells enable the designer to designate D, T, JK, or SR type flip-flops so that the number of product terms required for logic implementation is minimized. Figure 2.15*b* shows the architectural block diagram of CY7C335 EPLD.

Multiple array matrix high-density EPLDs. The Cypress multiple array matrix (MAX) family of EPLDs provides user configurable CMOS EPLD CY7C34x, where the suffix "x" denotes various device types, ranging in density from 32 to 192 macrocells. These devices are based on the Cypress 0.8 μm double-layer metal CMOS EEPROM process,

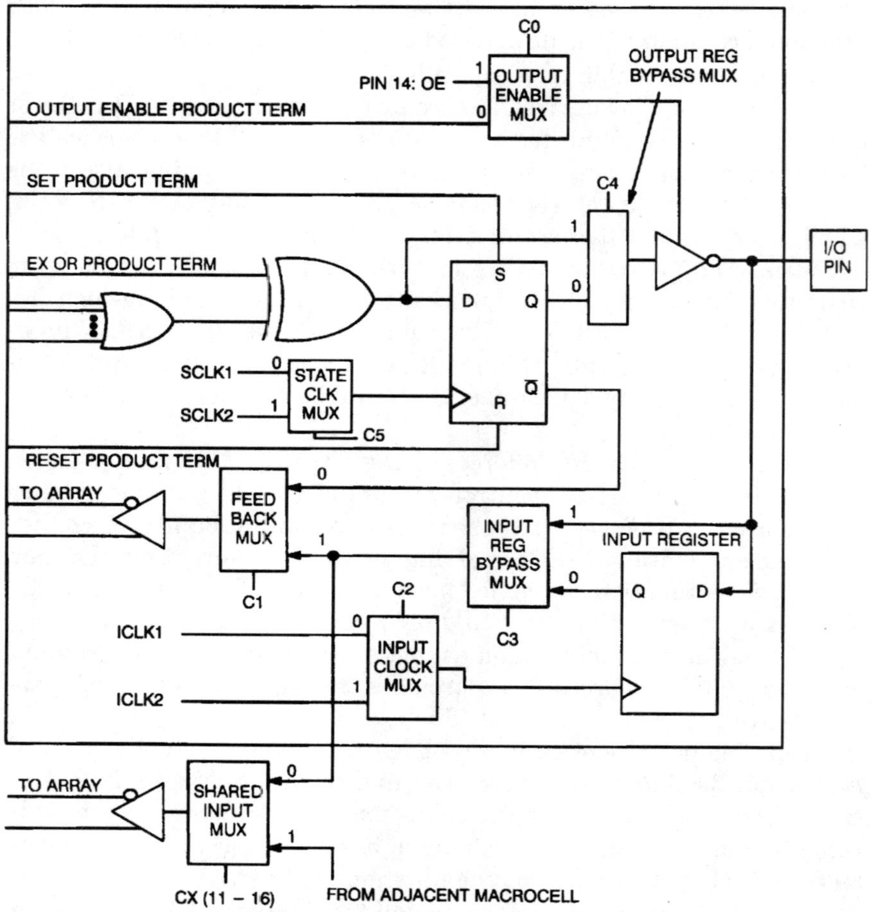

Figure 2.15a Universal synchronous EPLD CY7C335 I/O macrocell *(from Ref. 8 with permission of Cypress Semiconductor.).*

whereas an enhanced version of this family, CY7C34xB devices offering faster speed, are based on a 0.65 μm process. The MAX architecture is based on flexible macrocells grouped together into logic array blocks (LABs). Each of these LABs consists of a macrocell array, expander product-term array, and an I/O block. The number of macrocells, expanders, and I/Os vary, depending on the device type used. A global feedback of all signals is available within a LAB, providing each function block complete access to all the LAB resources. In the devices with multiple LABs, the LAB itself is fed by the programmable interconnect array (PIA) and dedicated input bus. The feedbacks of the macrocells and I/O pins feed the PIA, providing access to them through other LABs in the device.

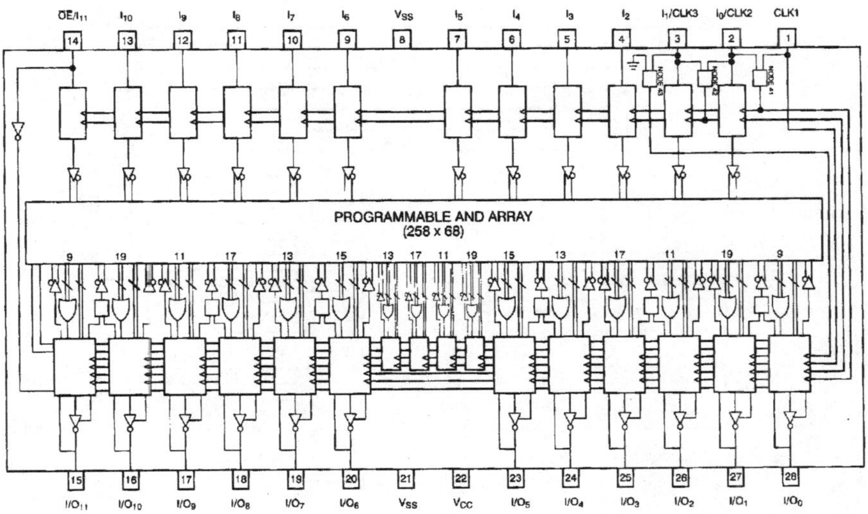

Figure 2.15b Universal synchronous EPLD CY7C335 architectural block diagram *(from Ref. 8 with permission of Cypress Semiconductor.)*

Figure 2.16 shows the logic block diagram of CY7C341B, a 192-macrocell MAX EPLD consisting of 12 LABs. Each of these LABs consists of a macrocell array containing 16 macrocells, an expander product-term array containing 32 expanders, and an I/O block. The LAB is fed by the PIA and the dedicated input bus. All macrocell feedbacks go to the macrocell array, the expander array, and the PIA. The expanders feed themselves and the macrocell array. All I/O feedbacks go to the PIA to make them accessible to macrocells in the other LABs, in addition to the macrocells in the LAB in which they are located.

Each macrocell consists of a product-term array and a configurable register. In the macrocell, combinatorial logic is implemented with three product terms ORed together, which then feeds an XOR gate. The second input to the XOR gate is also controlled by a product term that provides the ability to control active high or active low logic, and for implementing T and JK types of flip-flops. If more product terms are needed to implement a complex logic function, such as address decoders, adders and state machines, they may be added to the macrocell from expander product array. The registers within the macrocell may be programmed for either D, T, JK, or SR operation.

The expander product terms are fed by the dedicated input bus, the PIA, the macrocell feedback, the expander themselves, and the I/O pin feedbacks. The outputs of the expanders then go to each and every product term in the macrocell array, which allows expanders to be "shared" by the product terms in the LAB. One expander may feed all

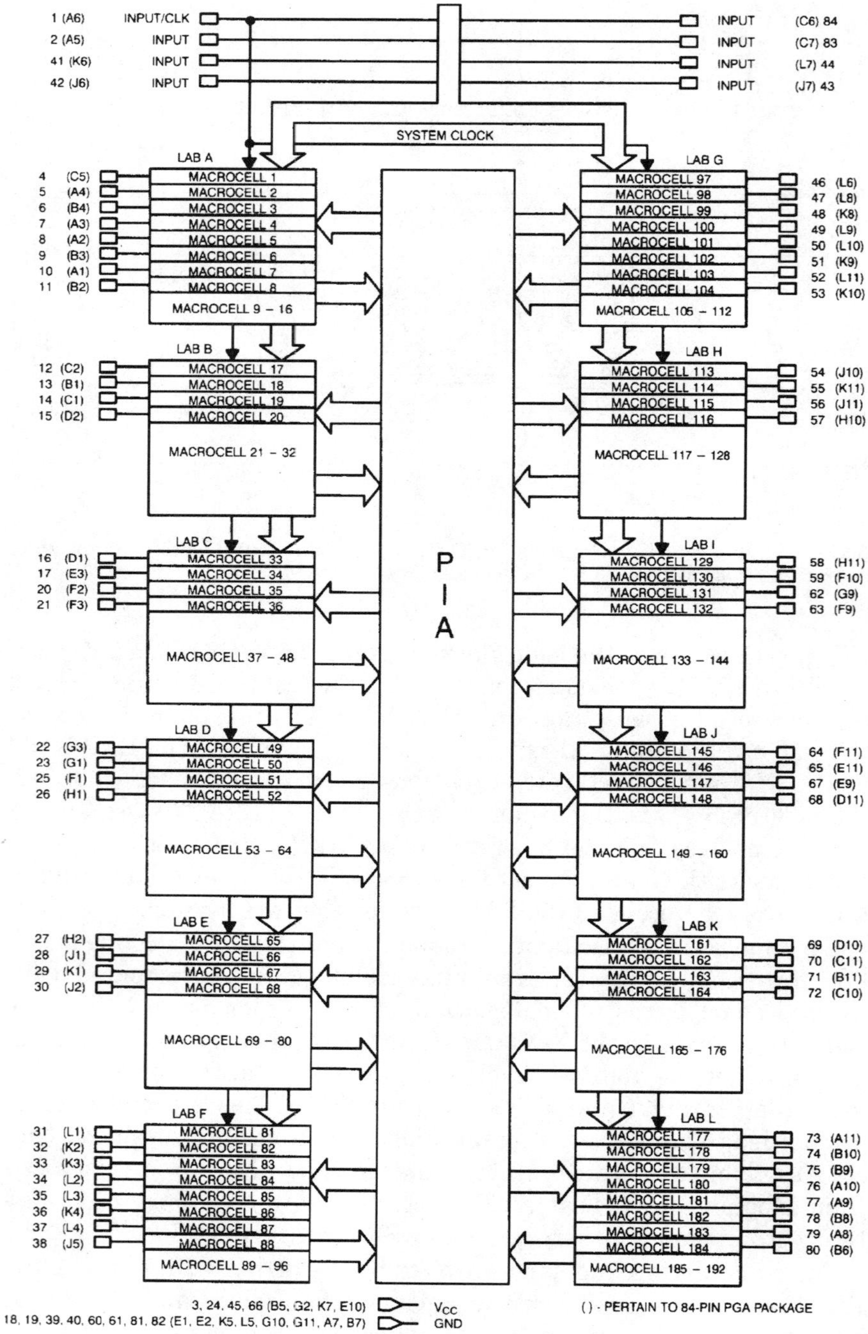

Figure 2.16 Logic block diagram of MAX CMOS EPLD CY7C341B *(from Ref. 8 with permission of Cypress Semiconductor).*

macrocells in the LAB, or even multiple product terms in the same macrocell. Since these expanders feed the secondary product terms (preset, clear, clock, and output enable) of each macrocell, complex logic functions may be implemented without utilizing another macrocell.

UltraLogic high-density flash CPLD family. The UltraLogic 37000 family of CPLDs, based on Cypress Flash technology, are in-system reprogrammable (ISR) devices compliant with the JTAG IEEE 1149.1 standard, operating at 3.3 V or 5 V, and are fully compatible with the PCI local bus specification. This family consists of devices ranging in density from 32 to 512 macrocells. The architecture is based on a number of logic blocks that are connected by a programmable interconnect matrix (PIM). Each logic block features its own product-term array, product term allocator and 16 macrocells. The PIM distributes signals from the logic block outputs and all input pins to the logic block inputs. All Ultra37000 family devices feature four dedicated inputs/clocks and provide synchronous/asynchronous clocking capabilities.

An example of this family is Ultra37256 device with 256 macrocells that is available in register-intensive and I/O-intensive versions. The Ultra37256P160, featuring 128 buried and 128 I/O macrocells, is suitable for register-intensive operations, whereas Ultra37256P256, with 192 I/O macrocells and 64 buried macrocells, is targeted for I/O-intensive applications. Figure 2.17 shows the logic block diagram of Ultra37256 device.

FLASH370i ISR CPLD family. The Cypress UltraLogic FLASH370i is an electrically erasable and in-system reprogrammable (ISR) family based on 0.65 μm CMOS flash technology that provides density ranging from devices with 32 to 128 macrocells. The ISR function is implemented through a four-pin serial interface. The data is shifted in and out through the SDI and SDO pins, respectively, using the programming voltage pin (V_{pp}). These pins are dual function, providing a pin-compatible upgrade to earlier version of EPROM-based FLASH370 (i.e., CY7C37x series) devices. The FLASH370i family offers two packaging options selection—one that is I/O intensive and another as register intensive.

The basic building block of FLASH370i architecture is the logic block, which consists of a product-term array, a product-term allocator, 16 macrocells, and a number of I/O cells. Two types of logic blocks are used: (a) logic blocks with an equal number [16] of I/O cells and macrocells, suitable for I/O intensive applications, and (b) logic blocks featuring a buried macrocell along with each I/O macrocell. Figure

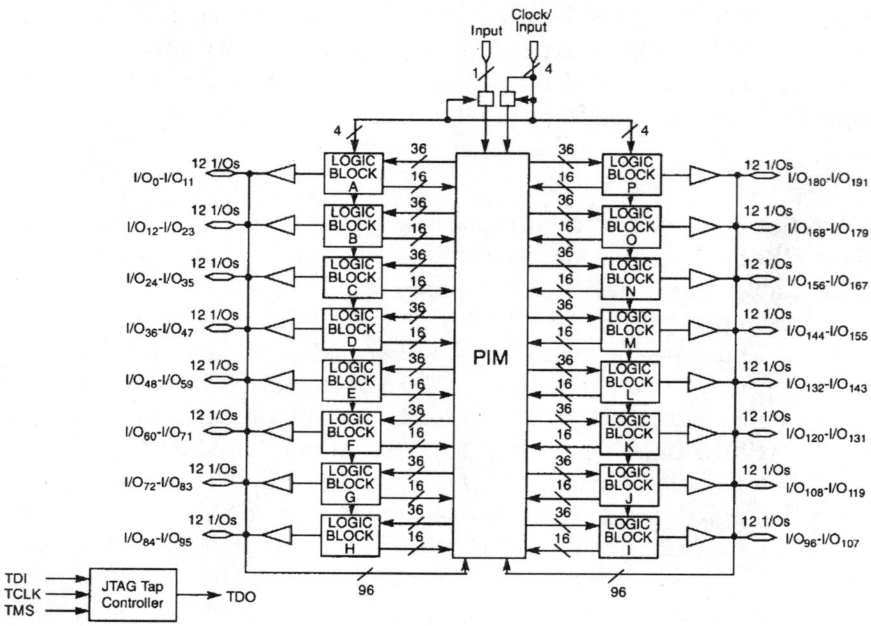

Figure 2.17 Logic block diagram for Cypress Ultra37256 CPLD *(from Ref. 8 with permission of Cypress Semiconductor).*

2.18*a* shows the logic block diagram for I/O-intensive CY7C series devices such as 371i, 373i, and 375i. Figure 2.18*b* shows the logic block diagram for register-intensive CY7C series devices such as 372i and 374i.

The programmable interconnect matrix (PIM) consists of a completely global routing matrix for signals from the I/O pins and feedback from the logic blocks. The inputs to PIM consists of all I/O and dedicated input pins and all macrocell feedbacks from within the logic blocks. The outputs from the PIM are signals routed to the appropriate logic blocks. Each logic block receives, from the PIM, 36 inputs and their complements, allowing for 32-bit operations to be implemented in a single pass through the device.

Each logic block features a 72 × 86 programmable product-term array, which is fed by 36 inputs from the PIM that originate from the macrocell feedbacks and device pins. Active low and active high versions of each of these inputs are generated to create the full 72-input field. The 86 product terms in the array can be created from any of the 72 inputs. The 80 product terms out of 86 are for general-purpose use for the 16 macrocells in the logic block. Four of the remaining six product terms in the logic block are output enable (OE) product terms. Each of these OE product terms controls up to 8 of the 16 macrocells

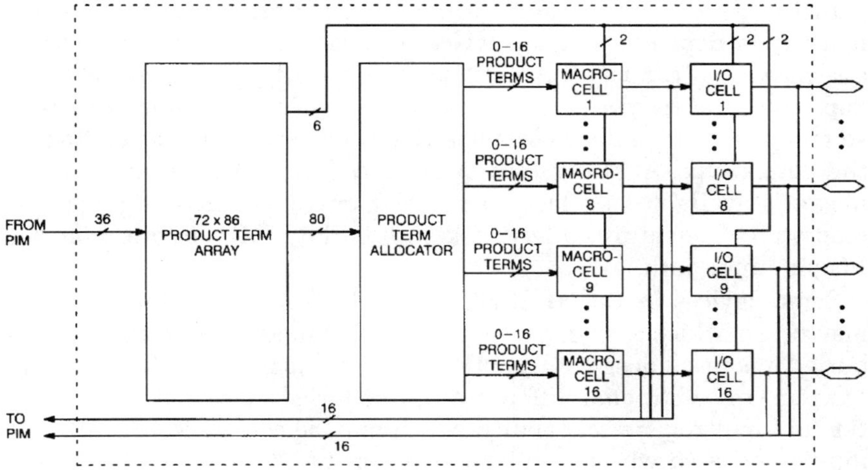

(a)

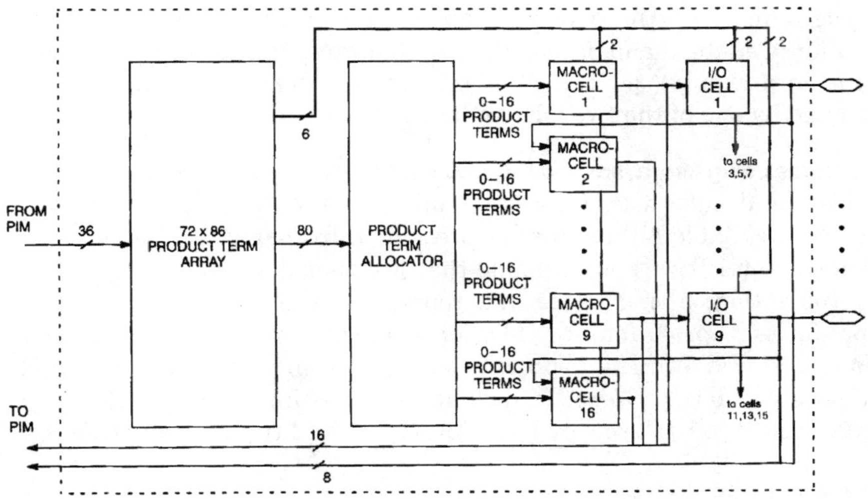

(b)

Figure 2.18 FLASH370i CY7C family logic block diagram: (a) I/O-intensive devices such as 371i, 373i, 375i and (b) register-intensive devices such as 372i and 374i *(from Ref. 8 with permission of Cypress Semiconductor).*

and is selected on an individual macrocell basis, i.e., each I/O cell can select between one of the two OE product terms to control the output buffer. The product-term allocator in the logic block distributes product terms among the 16 macrocells. For implementation of complex logic functions, the product-term allocator provides both the product-term steering and product-term sharing capabilities.

In the FLASH370i architecture, each logic block has 8 or 16 I/O macrocells, depending on the device type used. The macrocell features a register that can be configured as combinatorial, D flip-flop, a T flip-flop, or a level triggered latch. The clocking of the register is flexible, so two or four global synchronous clocks are available, depending on the device type. At the output of macrocell, a polarity control multiplexer is available to select active LOW or active HIGH signals. The macrocell also features a feedback path to the PIM, separate from the I/O pin input path.

Some devices in the FLASH370i family feature additional buried macrocells, which are very similar to the I/O macrocell except that the buried macrocell does not have the ability to output data directly to an I/O pin. One additional difference in the buried macrocell is the addition of input register capability, which may be configured as a D type flip-flop or latch whose input comes from the I/O pin associated with the neighboring macrocell. The output of the buried macrocell is sent directly to the PIM, regardless of its configuration. The I/O cell can be programmed by the user to change the way the three-state output buffer is enabled and/or disabled. Each output can be set permanently on (output only), permanently off (input only), or dynamically controlled by one of the two OE product terms.

Cypress design software and development tools. Many tools are available for design with Cypress Semiconductor's simple and complex PLDs, available either from Cypress or third-party tool vendors. Development software is available that provides design entry, synthesis, optimization, fitting, place and route, and simulation. The software produces a programming file for use with a device programmer. *Warp2*™ and *Warp2+*™ provide VHDL design description and functional simulation. *Warp3*™ includes *Warp2+* functionality plus schematic entry, VHDL source code simulation, and timing simulation.

2.3.4 ICT PEEL arrays

ICT products include standard 20- and 24-pin PLD architectures such as 20V8, 22V10, 22CV8, 22CV10A, and so on. In addition, ICT offers programmable electrically erasable logic (PEEL™) arrays as medium-density CPLDs based on CMOS EEPROM technology. The PEEL PA7000 family includes devices such as PA7024, PA7128, and PA7140 in 24-, 28-, and 40/44-pin packages, respectively. These devices offer 36 to 60 total registers (including buried registers) and 36 to 72 sum-of-products output functions. The PEEL arrays have a "zero power" feature in which the devices go to "sleep" when no inputs are active, thus reducing power consumption. An enhancement of the PA7000 family is the PA71000 series, under development, offering devices such as the

PA71500 (5 V) and PA71300 (3 V) with higher densities (96 registers) and clock rates of 120 MHz.

The PEEL arrays are based on a multilevel programmable logic array (PLA) architecture that includes input latches, buried registers, and sum-of-products logic functions. The PLA logic array structure provides programmable AND/programmable OR functions that allow product-term-sharing capabilities. Two major elements of PEEL array architecture include I/O cells (IOCs), input cells (INCs), logic control cells (LCCs), and a logic array. Figure 2.19*a* shows the block diagram of PEEL array architecture.

The input signals to the PEEL array are first fed through either I/O cells or input cells, each of them equipped with a user-configurable register/latch. The signals from the I/O and input cells, as well as feedbacks from the logic control cells, are fed to the logic array in both true and complement signal forms. The logic array that incorporates a PLA structure with true product-term sharing allows full sum-of-product logic functions to be fed (in groups of four) to each logic control cell.

The logic array implements all logic functions and provides interconnection and control of the cells. The number of product terms among PEEL arrays (PA7000 family) range from 67 to 125. All product terms (with the exception of some of those that are fed to the global cells) can be programmed for connection to any of sum-terms of the logic control cells. The product-terms and sum-terms are also routed to the global cells for control purposes.

Figure 2.19*b* shows the block diagram of an LCC used to allocate and control the logic functions created in the logic array. Each LCC has four primary inputs and three outputs. The inputs to each LCC are complete sum-of-product logic functions from the array, which can be used to implement combinatorial and sequential logic functions, and to control LCC registers and I/O cell output enables.

The LCC consists of three signal routing multiplexers and a register with synchronous or asynchronous D, T, or JK registers. EEPROM memory cells are used for programming the desired configuration. Four sum-of-product logic functions (sum-terms A, B, C, and D) are fed into each LCC from the logic cell arrays. The sum-of-products functions provided to the logic cells can be used for clocks, resets, presets, output enables, and so on. An important feature of the LCC is its ability to have multiple output functions per cell, each operating independently.

In addition to LCC, other major elements of PEEL array architecture are input cells, I/O cells, and global cells. Input cells (INCs) are included on dedicated input pins with PEEL array device types PA7128 and PA7140. Each INC consists of a multiplexer and a register/transparent latch, which can be clocked from various sources selected by the global cell. All PEEL arrays have the I/O cells (IOCs),

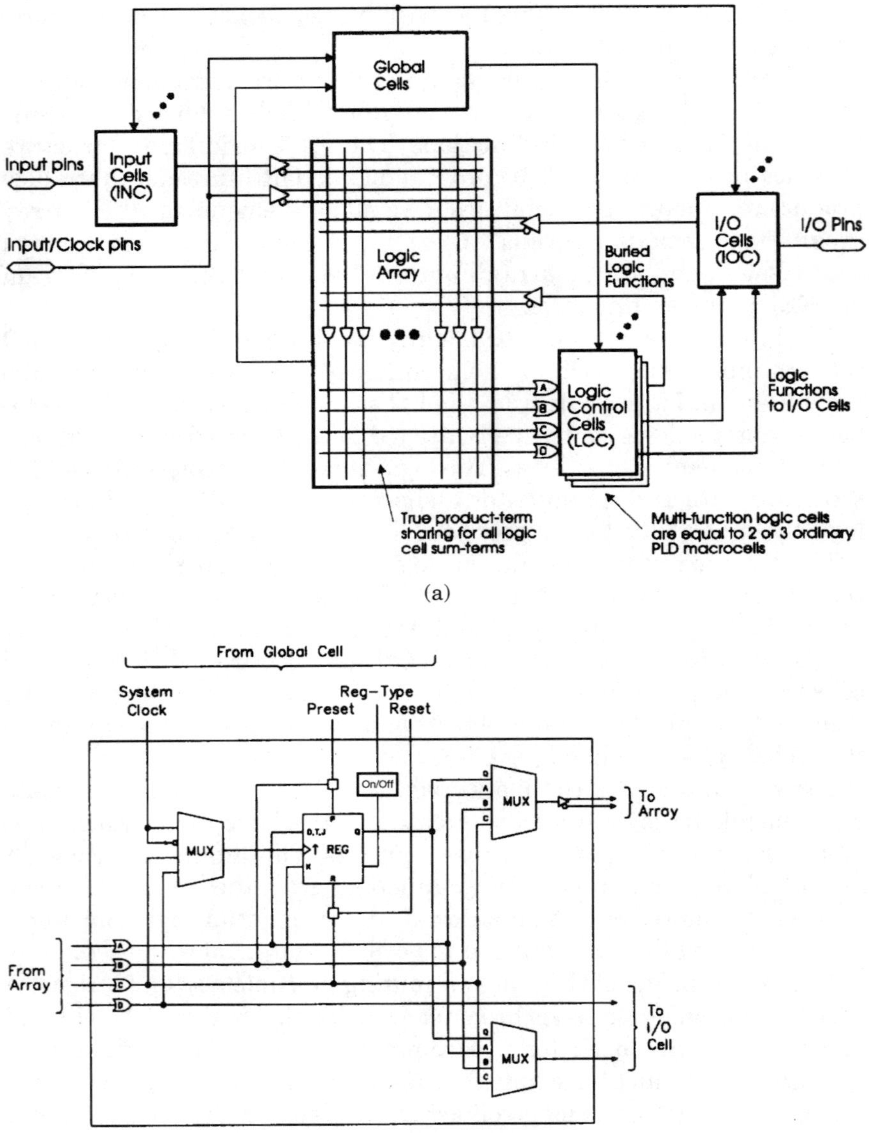

Figure 2.19 PEEL array block diagram: (a) architecture and (b) logic control cell (LCC) *(from Ref. 9 with permission of ICT).*

each consisting of routing and control multiplexers, an input register/transparent latch, a three-state buffer, and an output polarity control. The register/latch can be clocked from a variety of sources determined by the global cell.

The global cells are used to direct global clock signals and/or control terms to the LCCs, IOCs, and INCs. They also provide polarity control for INC and IOC clocks, enabling rising or falling clock edges for input register/latches. The global cell for LCCs include sum-of-products control terms for global set and preset, and a fast product-term control for LCC register-type, used to save product terms for loadable counters and state machines.

The design and development support for PEEL arrays is provided by ICT and popular third-party tool suppliers such as ABEL, CUPL, Synario, and OrCAD. ICT provides PEEL Array fitters for the third-party development tool environment. ICT also offers PLACE development software that includes an architectural editor, logic compiler, waveform simulator, documentation utility, and a programmer interface.

2.3.5 Lattice Semiconductor CPLD and GAL families

CPLD families: general description. Lattice Semiconductor's high-density in-system programmable large scale integration (ispLSI®) family CPLDs are based on nonvolatile EECMOS technology.[10] There are six ispLSI families, as follows: 1000/1000E, 2000/2000E, 3000, and 6000 series devices, ranging in density from 1000 to 25,000 PLD gates. The core of ispLSI architecture is the global routing pool (GRP), which connects all of the internal logic so as to provide complete interconnectivity with fixed and predictable delays.

The key element in this architecture is the generic logic block (GLB), which provides high input-to-output ratios for best logic efficiency. The GLB used in ispLSI 1000/2000 families feature 18 inputs that drive an array of 20 product terms (PTs). These PTs feed four outputs, which handle both wide and narrow gating functions. The ispLSI 3000 family utilizes a twin GLB that delivers wider logic functionality. The twin GLB accepts 24 inputs and feeds two arrays of 20 product terms and, eventually, drives two sets of four outputs. The ispLSI6000 family combines high-density general programmable logic with dedicated memory and register/counter modules.

An additional element of architectural flexibility is the product-term-sharing array (PTSA) scheme, which allows 20 PTs from the AND array to be shared with any or all of the four GLB outputs. Each of these four outputs from the PTSA feed into the flexible output logic macrocell (OLMC), consisting of a D-type flip-flop with an exclusive-OR gate on the input. The OLMC allows each GLB output to be configured as either combinatorial or registered. The combinatorial mode is available as AND-OR or the exclusive-OR functions. The registered mode is available as a D, T, or JK flip-flop. A flexible clock distribution network provides a choice of the clock signals to each GLB: global synchronous clock signals or internally generated asynchronous product-term clock signals. The output routing pool (ORP) is an ispLSI architectural feature, which provides flexible connections between the GLB outputs and the output pins.

ispLSI 1000/E families. The key element of ispLSI 1000/E families is the GLB, each of which has 18 inputs, a programmable AND/OR/XOR array, and four outputs that can be configured to be either combinatorial or registered. The ispLSI 1000E devices are functional supersets of the ispLSI 1000 devices and are architecturally similar except that the 1000E family features two new global output enable pins per device (only one for the 1016E) and programmable output slew rate control. Figure 2.20*a* shows the functional block diagram for ispLSI 1032E, which has 32 GLBs identified as A0, A1...D7. The inputs to the GLB come from the GRP and dedicated inputs. The ispLSI 1032E has 64 I/O cells, each of which is directly connected to an I/O pin. Each I/O cell can be individually programmed to be a combinatorial input, registered input, latched input, output, or bidirectional I/O pin with three-state control.

The GRP is an interconnect structure that allows the outputs from the GLB or the I/O cell inputs to be connected to the inputs of the GLBs. Any GLB output is available to the input of all other GLBs and, similarly, an input from an I/O pin is available as an input to all of the GLBs. The ORP routes signals from the GLB outputs to I/O cells configured as outputs or bidirectional pins. The purpose of the ORP is to allow greater flexibility when assigning the I/O pins.

A megablock consists of eight GLBs, an ORP, 16 I/O cells, two dedicated inputs and common product term OE. For the 1000/E families, the eight GLBs within the megablock share two dedicated input pins, which are not available to GLBs in any other megablock. The product term OE signal is generated within the megablock and is common to all 16 of the I/O cells in the megablock.

The ispLSI 1000/E families range in density from devices with two megablocks (i.e., 16 GLBs) to devices with six megablocks (i.e., 48 GLBs). Figure 2.20*b* shows the megablock layout.

The internal logic of the GLB is divided into four separate sections: the AND array, the product-term-sharing array (PTSA), the reconfigurable registers, and the control functions. Figure 2.21 shows the block diagram of PTSA. The AND array in PTSA consists of 20 product terms that can produce logical sum of any of the 18 GLB inputs. Sixteen of the inputs come from the GRP and are either feedback signals from any of the GLBs or inputs from the external I/O cells. The two re-

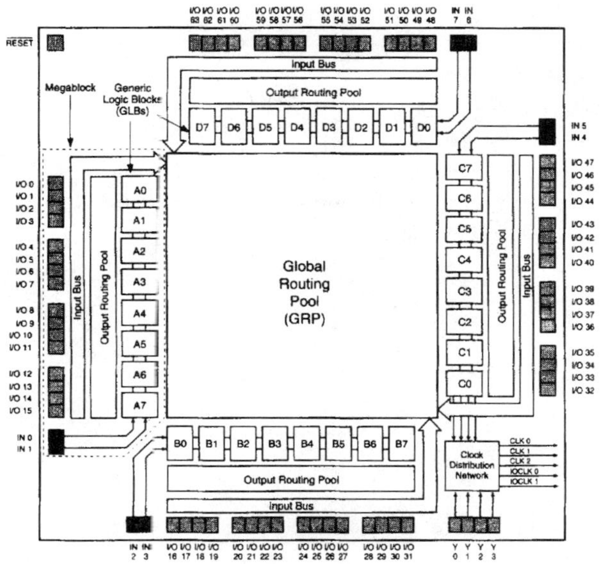

(a)

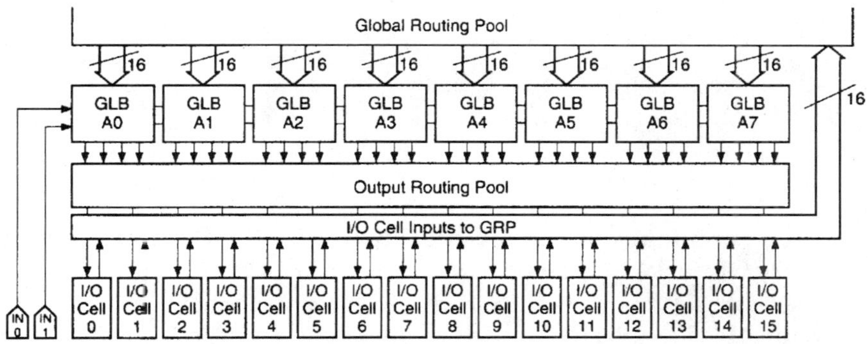

(b)

Figure 2.20 Lattice ispLSI 1000 family: (a) functional block diagram (ispLSI 1032 E) and (b) megablock layout *(from Ref. 10 with permission of Lattice Semiconductor).*

maining inputs come directly from the dedicated input pins. The PTSA takes 20 product terms and routes them to the 4 GLB outputs. There are four OR gates with four, four, five, and seven product terms each. The output of any of these OR gates can be routed to any of the four GLB outputs and, if more product terms are needed, the PTSA can combine as well as share them as necessary.

The reconfigurable registers consist of four D-type flip-flops with an XOR gate on the input. The XOR gate can be used either as a logic element or to reconfigure the D-type flip-flop to emulate a JK- or T-type

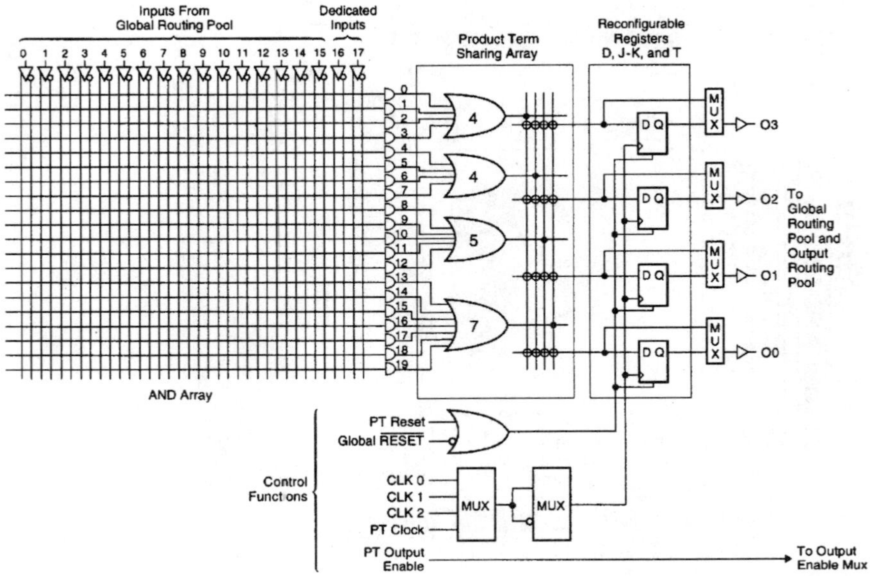

Figure 2.21 Lattice ispLSI 1000/E GLB product-term-sharing array (PTSA) *(from Ref. 10 with permission of Lattice Semiconductor).*

flip-flop. The registers can be bypassed if the user needs a combinatorial output. Each register output is brought back to the GRP and also to the I/O cells via ORP.

The input signals are handled in two ways within the device:

1. Each I/O cell has its input routed directly to the GRP, which gives every GLB within the device access to each I/O cell input.
2. Each megablock has two dedicated inputs, which are directly routed to the eight GLBs within the megablock. The I/O cell is used to route input, output, or bidirectional signals connected to the I/O pin. The two logic inputs come from the ORP. A multiplexer selects the input signal that will be used and its polarity. The output enable of the I/O cell is controlled by the OE signal generated within each megablock.

The clock distribution network generates five global clock signals as CLK0, CLK1, CLK2, IOCLK0, and IOCLK1. The first three clocks (CLK0, CLK1, and CLK2) are used for clocking all GLBs in the device. Similarly, IOCLK0 and IOCLK1 signals are used for clocking all I/O cells in the device.

ispLSI 2000/V families. The ispLSI 2000/V is the low-voltage version of the ispLSI 2000 family. The basic unit of logic for ispLSI 2000 family is the GLB, which is essentially same as that for the 1000/E families. However, there are some specific architectural differences, briefly described as follows:

- *GLB and clock multiplexer scheme.* The clock GLB distribution network of 1000/E families are replaced by three dedicated global GLB clock input signals CLK0, CLK1, and CLK2. These three clocks are used for clocking all the GLBs configured as registers in the device. They feed directly to the GLB clock input via a clock multiplexer. CLK0 corresponds to system clock pin Y0, CLK1 is associated with system clock pin Y1, and CLK2 with clock pin Y2. The GLB global clocks and the global product term clocks feed to their corresponding clock multiplexer, as shown in Figure 2.22*a*.
- *I/O cell and OE structure.* In 2000/V families, the reconfigurable input register or latch has been removed to simplify the I/O cell architecture. Each I/O cell can be individually programmed to be combinatorial input or output, or a bidirectional I/O pin with three-state control. This simplified I/O cell architecture has enabled I/O clocks to be removed. In addition to the standard output configuration, the outputs of the ispLSI 2000/V families are individually programmable either as standard totem pole outputs or as open-drain outputs (the open-drain option is not available on the 2032/V device).
- *Output routing pool (ORP).* For 2000/V devices with full I/O, each megablock contains two ORPs to increase output routability. A set of four GLBs is associated with one of the two ORPs within the megablock. The 16 outputs of the 4 GLBs within a megablock feed to any of the 16 associated I/O cells. Therefore, 32 GLB outputs of a megablock can feed 32 I/O cells, instead of 32 GLB outputs feeding only 16 associated I/O cells in the 1000/E family. This architectural enhancement doubles the routability of 2000/V devices as compared to the 1000/E family. For ispLSI devices with half I/O (i.e., one I/O per two GLB outputs), each megablock contains only one ORP, which feeds 16 I/O pins. Figure 2.22*b* shows the block diagram of output routing pool for ispLSI 2000/V families.

  The ispLSI 2000/V family ranges in density from the devices with 1000 PLD gates offering 32 macrocells to devices with 6000 PLD gates offering 128 macrocells.

ispLSi 3000 family. The Lattice ispLSI 3000 family ranges in density from devices with 7000 PLD gates, offering 192 macrocells, to devices with 14,000 PLD gates, offering 320 macrocells. The basic logic structure for ispLSI3000 family is similar to that of the ispLSI1000/E fam-

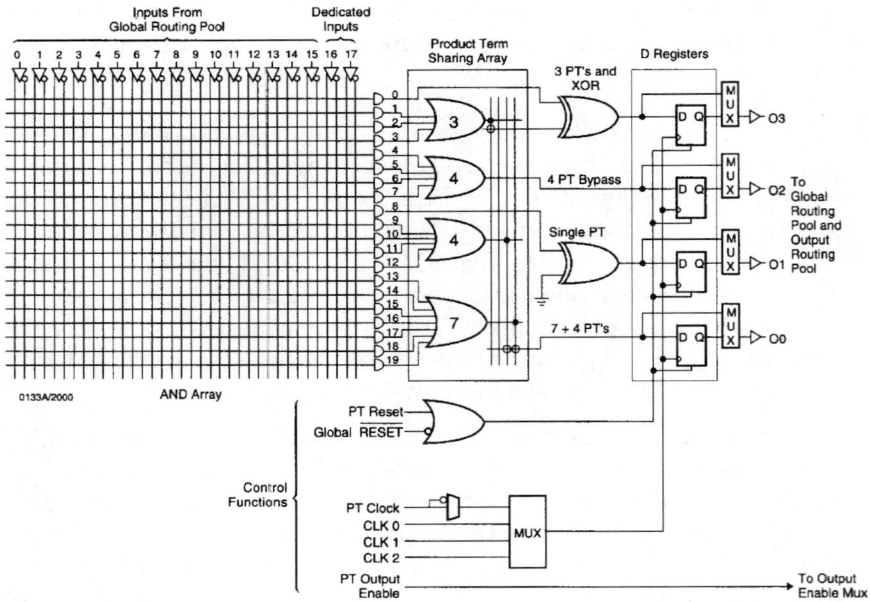

(a)

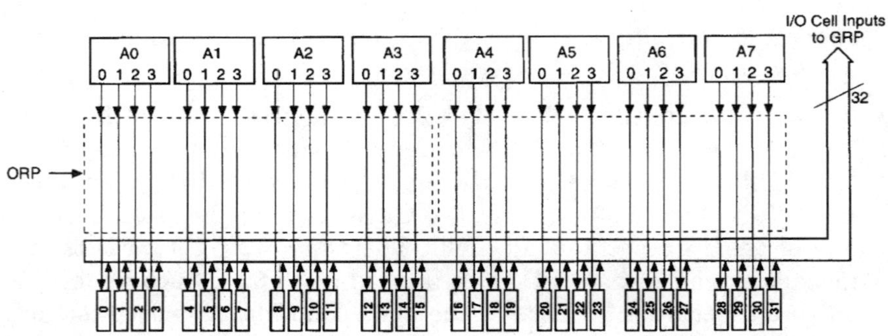

(b)

Figure 2.22 Lattice ispLSI 2000/V families, block diagram: (a) GLB with clock multiplexer and (b) output routing pool *(from Ref. 10 with permission of Lattice Semiconductor).*

ily except for some notable architectural differences briefly described below.

- *Generic logic block (GLB).* The standard logic block for the ispLSI 3000 family is the twin GLB with 24 inputs, 8 outputs, and the associated logic necessary to implement most standard logic functions. The internal logic of twin GLB is divided into four separate sections as follows: the AND array, the PTSA, the reconfigurable registers, and the control section. The AND array consists of 20 PTSAs, which can produce the logical sum of any of the 24 twin GLB inputs. These inputs all come from the GRP, and are either feedback signals from any of the 32 twin GLBs or inputs from the external I/O cells. The two PTSAs take 20 product terms each and allocate them to four twin GLB outputs. There are four OR gates, with four, four, five and seven inputs, respectively. The output of any of these gates can be routed to any of the four twin GLB outputs and, if more product terms are needed, the PTSA can combine them as necessary. To increase the cell performance, the PTSA can use the bypass circuit with four product terms.

- *Megablock structure.* In ispLSI 3000 family, a megablock consists of four twin GLBs. Each GLB has a maximum fan-in of 24 inputs, and no dedicated inputs associated with any megablock. A GLB has eight associated outputs. A total of 32 GLB outputs are fed to the ORP of *single-I/O* devices that have one I/O pin for every two GLB outputs (e.g., ispLSI 3256A). However, only 16 of the 32 outputs feed to 16 I/O cells. For the *double-I/O* ispLSI 3000 family devices that have 1 I/O pin for each GLB output (e.g., ispLSI 3192), 16 GLB outputs are fed to each ORP, which drives the 16 I/O cells. Therefore, *double-I/O* devices employ two ORPs per megablock. Figure 2.23 shows the ispLSI 3000 family megablock layout for single I/O devices and double I/O devices.

- *Global clock structure.* In the ispLSI 3000 family, the clock GLB generation network of the ispLSI 1000/E families have been removed so that all clock input signals can be fed directly to the GLB clock input via a clock multiplexer. The global clock structure consists of five global clock input pins as Y0, Y1, Y2, Y3, and Y4. Three of the clock pins are dedicated for GLB clocks, and the remaining two clock pins are dedicated for I/O register clocks. The GLB global clocks do not have inversion capability, but the product-term clock does have inversion capability before it reaches the clock multiplexer.

- *I/O cells.* The I/O cell architecture is essentially same as for the 1000/E families, except that it contains boundary scan registers to allow the implementation of JTAG IEEE 1149.1 on-chip test capability during the PCB testing. An input pin has only one scan register. A global test OE signal is hard-wired to all I/O cells and is useful for static testing of all the three-state output buffers within the device. In addition to this test OE signal, two global OEs are connected to all I/O pins. The product-term OE is shared between two megablocks. The megablock OE signal and global OE signal are fed to an OE multiplexer.

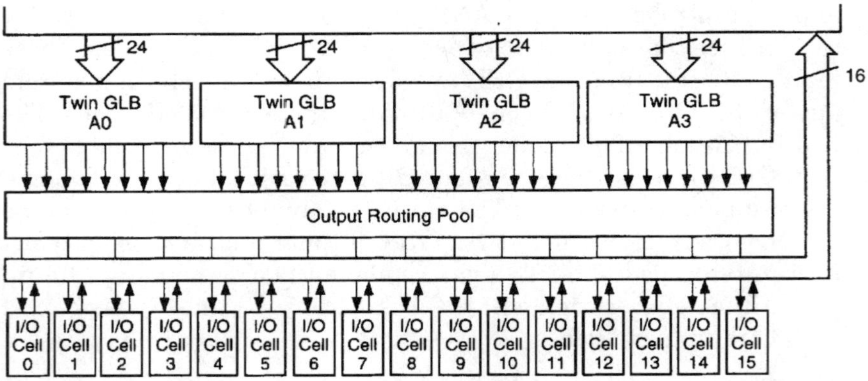

(a)

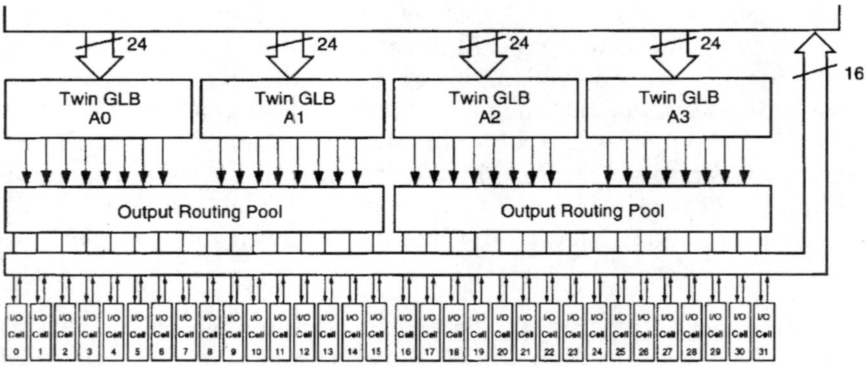

(b)

Figure 2.23 Lattice ispLSI 3000 family megablock layout: (a) single-I/O devices and (b) double-I/O devices *(from Ref. 10 with permission of Lattice Semiconductor).*

ispLSI 6000 family. The ispLSI 6000 family is based on high-density, cell-based programmable logic that contains dedicated memory module, a dedicated register/counter module, and an 8000-gate general-purpose programmable logic block. The family consists of following three devices: ispLSI 6192FF (FIFO), ispLSI 6192SM (single-port RAM) and the ispLSI 6192DM (dual-port RAM). The global routing pool (GRP) and output routing pool (ORP) provide flexible connectivity between all elements. All devices incorporate the same general-purpose programmable logic modules and register/counter modules, with only the memory module changing between the devices. Figure 2.24 shows the functional block diagram of the ispLSI 6192. The major architectural elements are briefly described below.

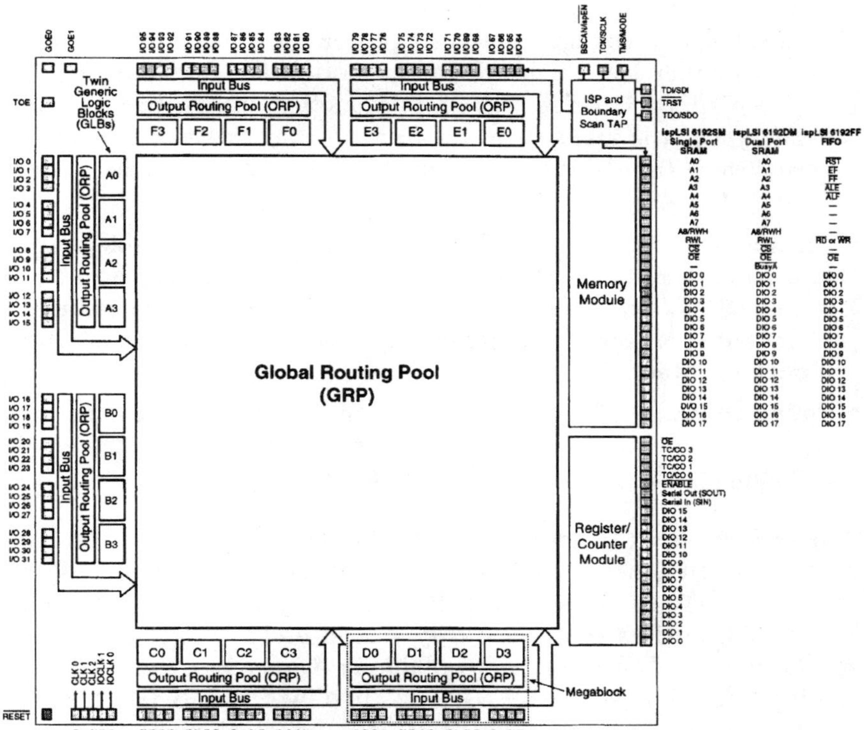

Note:
Since certain signal names are duplicated on Memory Module and Register/Counter Module pins (OE, DIO), the notation:
OE (RAM)
OE (RC)
DIO (RAM)
DIO (RC)
will be used periodically in this data sheet to differentiate signals.

Figure 2.24 The ispLSI 6000 family, ispLSI 6192 device functional block diagram *(from Ref. 10 with permission of Lattice Semiconductor).*

- *Programmable logic module.* The basic structure for the ispLSI 6000 family is the twin generic logic block (GLB). The ispLSI 6192 includes 24 twin GLBs, each containing 24 inputs, a programmable AND array, 2 OR/exclusive-OR arrays, and 8 output blocks that can be independently configured to be combinatorial or registered. A logic megablock consists of 4 twin GLBs, 16 I/O cells, and 1 ORP. The ispLSI 6192 contains 6 megablocks and 96 I/O cells.
- *Memory module.* The memory module for ispLSI 6000 family can take one of following three forms: dual-port FIFO, single-port memory, or dual-port memory. In the ispLSI 6192FF, the FIFO memory module is user configurable as 256 × 18 or 512 × 9 and is accessible through dedicated FIFO I/O pins. There are various FIFO control flags available, such as full (FF), almost full (ALF), almost empty

(AE), and empty (EF). The single-port memory module (ispLSI 6192SM) can be configured as a single 256 × 18 or 512 × 9 memory or as two smaller 128 × 8 or 256 × 8 memories. The dual-port memory module (ispLSI 6192DM) has similar functionality as the single-port memory, but the memory access from GRP or external pins is supported concurrently. It can be configured as either a single 256 × 18 or 512 × 9 memory.

- *Register/counter module.* Eight 16-bit blocks are available to function as registers or shift registers, while four of these can be programmed as loadable up/down counters. This block also includes a 16-bit data port connected to the GRP, along with a variety of control inputs and status flag outputs.
- *Boundary scan.* The ispLSI 6000 family includes boundary scan capability, since all I/O pins have associated boundary scan registers and support all IEEE 1149.1 mandatory instructions.

Generic array logic (GAL) family. In 1985, Lattice introduced a new type of PLD, called generic array logic (GAL) device, which had a significant advantage over its bipolar PAL counterparts. These EECMOS-based GAL devices could be easily erased and reprogrammed. Currently, these GAL devices are available in five basic architectures, including GAL16V8, GAL20V8, GAL22V10, GAL20RA10, GAL20XV10, and their several extension products, to provide enhanced functionality. Additionally, Lattice continues to develop its rapidly growing families of low-power, quarter-power and "zero-power" products for low-voltage (3.3 V) operation.

- *GAL16V8 and GAL20V10.* The GAL16V8 and GAL20V8 devices typically consume 75 mA I_{CC}, which is 50 percent compared to their bipolar counterparts, and quarter-power versions typically consume 45 mA I_{CC}. The GAL16V8 is a 20-pin device which contains eight dedicated input pins and eight I/O pins. The GAL20V8 is a 24-pin version of the 16V8 device with 12 dedicated input pins and 8 I/O pins. Their generic architecture provides design flexibility by allowing the output logic macrocell (OLMC) to be user configurable. There are three OLMC configuration modes possible: registered, complex, and simple. These modes cannot be mixed. In the registered mode, the output can be combinatorial or registered.

  The output of the AND array are fed into an OLMC, where each output can be individually set to active high or active low, with either combinatorial (asynchronous) or registered (synchronous) configurations. A common output enable is connected to all registered outputs, or a product term can be used to provide individual output enable control for the combinational outputs in the registered mode or combinatorial outputs in the complex mode. There is no output enable control in the simple mode. Figure 2.25 shows GAL16V8 and GAL20V8 block diagram.

- *GAL 22V10 family.* This family uses industry standard 22V10 universal architecture and includes the GAL22V10 (24-pin), GAL18V10 (20-pin), and GAL26CV12 (28-pin) devices. The GAL22V10 device typically consumes 90 mA I_{CC}, whereas the quarter-power versions typically consume 45 mA. The three device types differ in the number of I/Os, pins, and product terms offered, and are briefly described below.

  1. The 24-pin GAL22V10 contains12 dedicated input pins and 10 macrocells and I/O pins. It has variable number of product terms per OLMC, ranging from 8 to 16 per output.
  2. The GAL18V10 is a 20-pin version of the popular 22V10 device, but it provides a smaller footprint. It has 8 dedicated input pins and 10 macrocells and I/O pins.

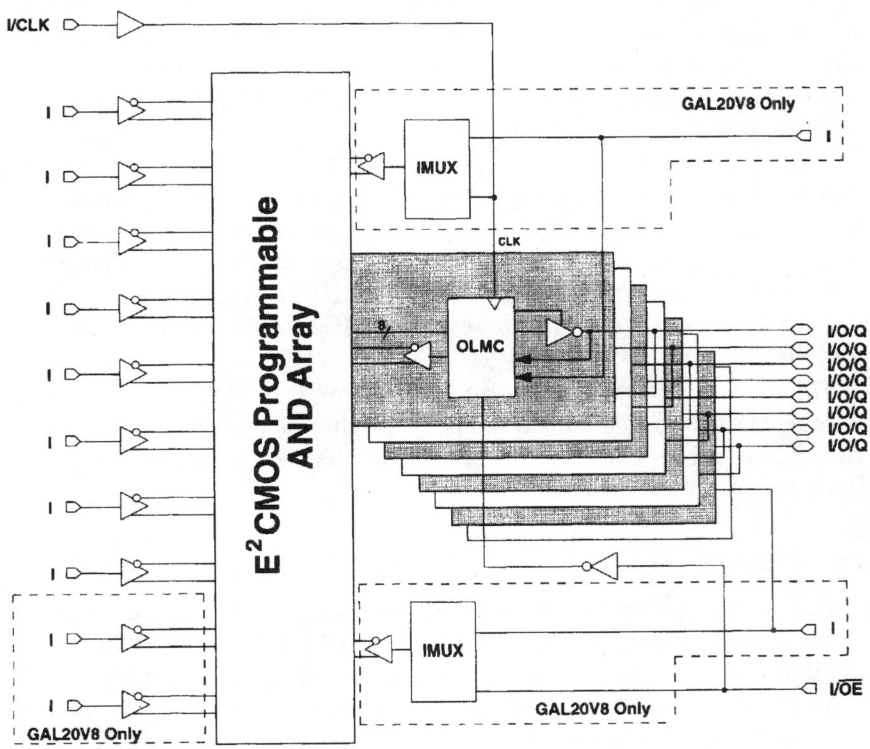

Figure 2.25 Lattice generic array logic GAL16V8 and GAL20V8 block diagram *(from Ref. 10 with permission of Lattice Semiconductor).*

3. The GAL26CV12 is a 28-pin version of the device featuring more inputs and outputs to provide greater functionality and increased I/Os. It has 14 dedicated input pins and 12 macrocells and I/O pins.
4. The ispGAL22V10 features a popular 22V10 architecture combined with in-system programmability via a four-wire serial programming interface.

The GAL22V10, 18V10, and 26CV12 each have a variable number of product terms per OLMC. The output polarity of each OLMC can be individually programmed to be true or inverting, in either combinational or registered mode. GAL22V10 family devices have a product term for asynchronous reset (AR) and a product term for synchronous preset (SP), both of which are common to all registered OLMCs.

Lattice ispLSI and GAL families: design and development tools. Lattice pdS® software is a self-contained PC-based package that uses familiar Boolean equations and macro design entry. The pDS device fitter provides interactive partitioning, automatic place and route, and simulation timing tables for design verification. Lattice provides Viewlogic's PROsim package, which is compatible with pdS for functional and timing simulation. Additionally, the designs implemented in pDS can be simulated using Viewlogic's ViewSim and OrCAD's simulation packages.

Lattice ispDS+™ HDL synthesis-optimized logic fitter supports a wide range of CAE vendor design environments such as Cadence, Mentor Graphics, Synario, OrCAD, Synopsys, Viewlogic, Exemplar, Synplicity, and others. The design entry is typically performed through schematic capture or by using HDLs such as ABEL, VHDL, and Verilog-HDL.

Figure 2.26 shows the general design flow, including schematic capture or synthesis, through the ispDS+ architectural synthesis and optimization process to design verification, using any of the supported CAE vendor simulation tools.

2.3.6 Philips Semiconductors' CPLD family

Philips Semiconductors' CoolRunner™ CPLD family ranges from devices with 32 macrocells (1000 gates) to 128 macrocells (4000 gates) available for low-voltage (3.0 to 3.3 V) and standard 5 V applications.[11] This family is based on fast zero power (FZP™) design techniques to reduce static power consumption (e.g., less than 100 μA for 128 macrocell device PZ128). The power reduction is achieved by utilizing Total-CMOS™ CPLD, i.e., a CPLD that is CMOS both in process and design

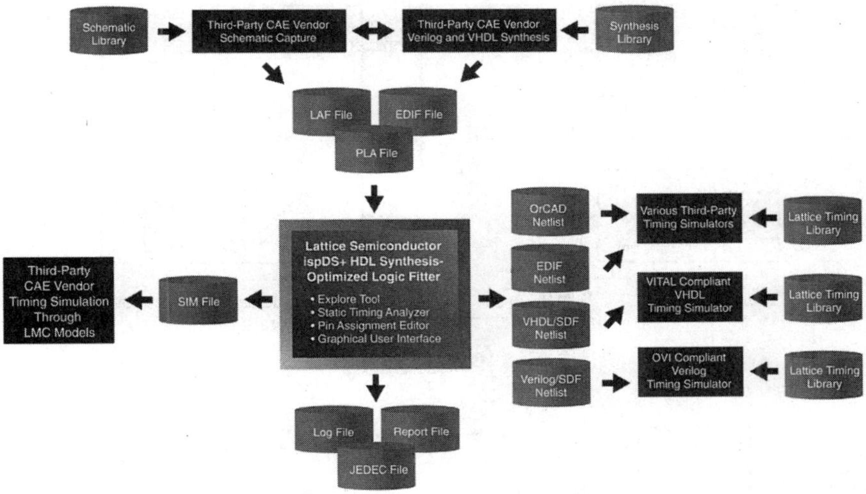

Figure 2.26 ispDS+ general design flows for Lattice ispLSI families *(from [Ref. 10 with permission of Lattice Semiconductor).*

techniques. The conventional bipolar design techniques for PLDs use sense amplifiers to get a full voltage swing in a very short time and require high current for full product-term array (e.g., 160 mA for a 128 macrocell CPLD). In the FZP design technique, the product-term array is implemented by cascading the logic in a tree of full-CMOS gates. These full-CMOS gates switching in 200 ps are cascaded in a manner to achieve speed performance comparable to the sense amplifier approach.

This family is based on the extended programmable logic array (XPLA™) architecture, consisting of logic blocks that are connected by a zero-power interconnect array (ZIA). Each block is essentially a 36V16 device with 36 inputs from the ZIA and 16 macrocells. Each logic block also provides 32 ZIA feedback paths from the macrocell and I/O pins. The number of logic blocks contained within a device determines the macrocell count of a device. For example, devices containing 2, 4, and 8 logic blocks are 32, 64, and 128 macrocell devices, respectively. Figure 2.27 shows an XPLA block diagram and logic block architecture.

Each logic block contains control terms, a PAL array, a PLA array, and 16 macrocells. The 6 control terms can individually be configured as either AND or SUM product terms and are used to control the preset/reset and output enables of the 16 macrocell's flip-flops. The PAL array consists of a programmable AND array with a fixed OR array, while the PLA array consists of a programmable AND array with a programmable OR array. The PAL array provides a high-speed path

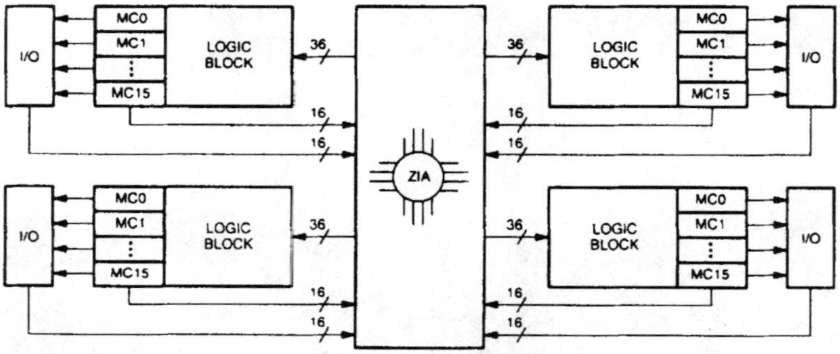

(a)

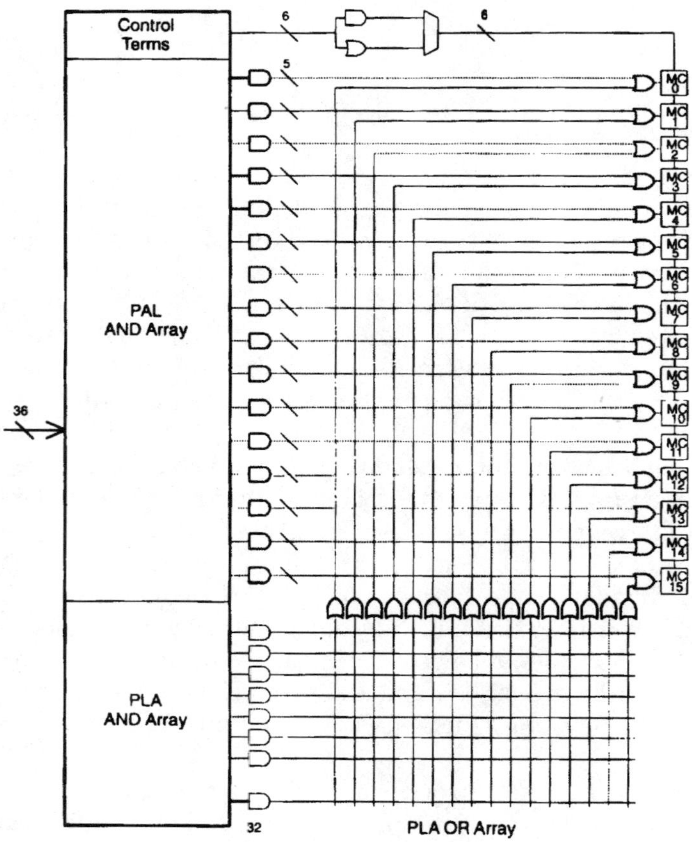

(b)

Figure 2.27 Block diagram of CoolRunner CPLD family: (a) XPLA architecture and (b) logic block architecture *(from Ref. 11 with permission of Philips Semiconductors).*

through the array, whereas the PLA provides increased product-term density.

Each macrocell has five dedicated product terms from the PAL array, and if a macrocell needs more than five product terms, the additional product terms are obtained from the PLA array. The PLA array consists of 32 product terms, which are available for use by all 16 macrocells.

Figure 2.28 shows the block diagram of macrocell architecture used in PZ5128C/PZ5128N, a 128-macrocell CPLD with enhanced clocking. The macrocell can be configured as either a D- or T-type flip-flop, or as combinatorial logic. Each of the flip-flops can be clocked from any of six sources. Clock 1 (CLK1), clock 2 (CLK2), and clock 3 (CLK3) can be used as a "synchronous" clock that is driven by an external source, or as an "asynchronous" clock driven by a macrocell equation. The other clock sources are two of the six control terms (CT2 and CT3) provided in each logic block. These clocks can be individually configured as a PRODUCT term or SUM term equation created from the 36 signals available inside the logic block.

Control terms CT0 and CT1 are used to control the asynchronous preset/reset of the macrocell's flip-flop. Control terms CT2 and CT3 can be used as clock signals to the macrocell's flip-flops and as the output enable of the macrocell's output buffers. Control terms CT4 and CT5 can be used to control the output enable of the macrocell's output buffers. All CoolRunner devices include a global tristate ($\overline{\text{GTS}}$) pin to support the in-circuit (or bed-of-nails) testing approach.

There are two feedback paths to the ZIA: one from the macrocell and one from the I/O pin. The ZIA feedback path before the output buffer is the macrocell feedback path, while the ZIA feedback path after the

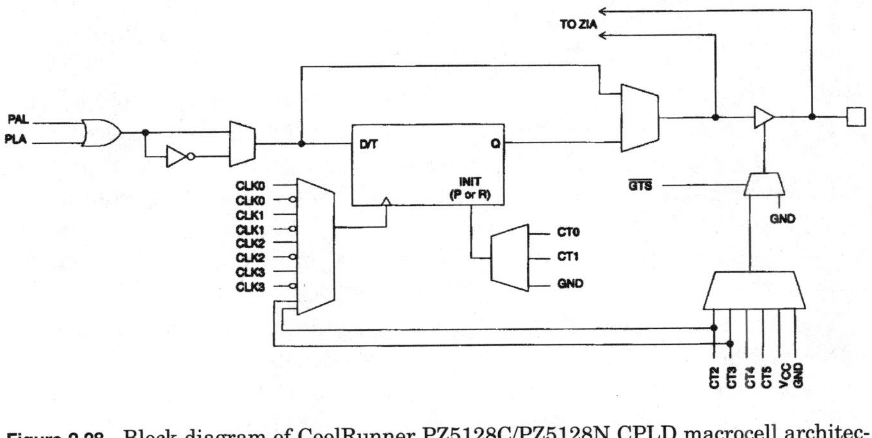

Figure 2.28 Block diagram of CoolRunner PZ5128C/PZ5128N CPLD macrocell architecture *(from Ref. 11 with permission of Philips Semiconductors).*

output buffer is the I/O pin feedback path. When the macrocell is used as an output, the output buffer is enabled, and the macrocell feedback path can be used to feed back the logic implemented in the macrocell. When the I/O pin is used as an input, the output buffer will be three-stated, and the input signal will be fed into the ZIA through the I/O feedback, and the logic implemented in the buried macrocell can be fed back to the ZIA via the macrocell feedback path.

Philips Semiconductors' family of FZP CPLDs is supported by industry standard CAE tools (Cadence, Mentor Graphics, Synopsys, Viewlogic, Synario, MINC, and others) using schematic entry or HDLs. Philips offers XPLA Designer tool for design definition, functional simulation, device fitting, post layout (timing) simulation, and programming. The XPLA Designer uses Philips' hardware description language (PHDL) to support the following design entry formats: Boolean equations, state machines, and truth tables. The XPLA Designer also produces VHDL and Verilog timing models, which can be used in board-level simulations. The CPLD family is electrically reprogrammable using industry standard device programmers and includes the IEEE 1149.1 JTAG interface for in-system programming (ISP).

2.3.7 Xilinx fastFlash CPLD family

The XC9500 FastFlash family, introduced by Xilinx in October 1995, is a series of 5 V, in-system programmable (ISP) devices based on 0.6 μm, double-polysilicon, two-layer metal CMOS FastFlash™ technology.[12] The XC9500 CPLD family ranges in density from devices with 36 macrocells (800 usable gates) to devices with 288 macrocells (6,400 usable gates). This family is supported by IEEE 1149.1 boundary scan with extended JTAG instruction set that allows control of programming patterns and in-system debugging.

Figure 2.29*a* shows XC9500 architectural block diagram. Each XC9500 is a subsystem consisting of multiple function blocks (FBs) and I/O blocks (IOBs), which are fully interconnected by the fastCONNECT™ switch matrix. The IOB provides buffering for the device inputs and outputs. Each FB provides programmable logic capability with 36 inputs and 18 outputs. The fastCONNECT switch matrix connects all FB outputs and input signals to the FB inputs. For each FB, 12 to 18 outputs (depending on the package pin count) and associated output enable signals drive the IOBs directly.

The functional block consists of 18 independent macrocells, each capable of implementing a combinatorial or registered function. The FB also receives global clock, output enable, and set/reset signals. It generates 18 outputs that drive the fastCONNECT switch matrix. The logic within FB is implemented using a sum-of-products approach

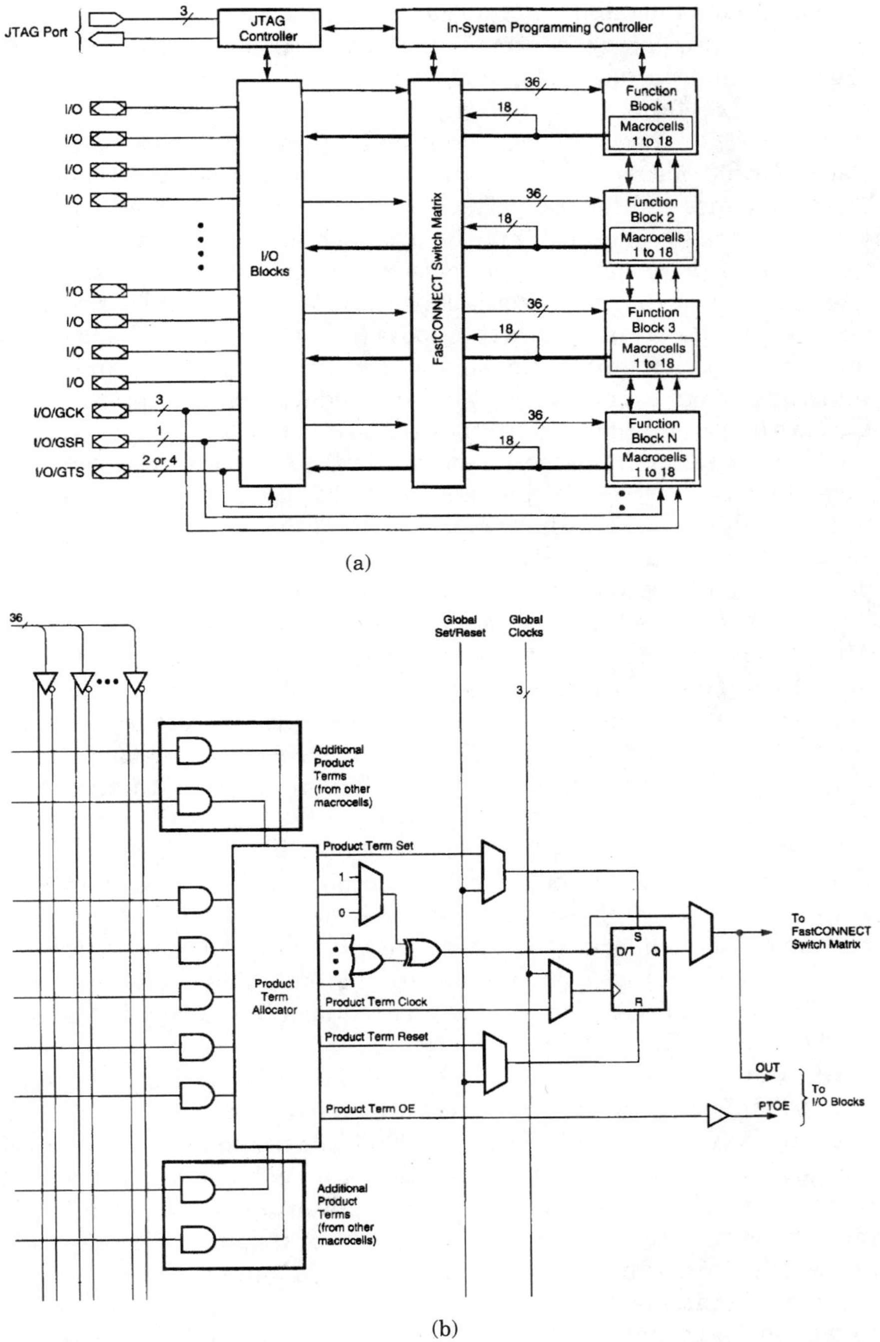

Figure 2.29 XC9500 CPLD family block diagram: (a) architecture and (b) macrocell with function block *(from Ref. 12 with permission of ©Xilinx Inc.).*

with 36 inputs providing 72 true and complement signals into the programmable AND-array to form 90 product terms. Any number of these product terms, up to 90 available, can be allocated to each macrocell by the product term allocator.

Figure 2.29*b* shows the macrocell and associated functional block. Each XC9500 macrocell may be individually configured for a combinatorial or registered function. The register can be configured as a D-type or T-type flip-flop, or it may be bypassed for combinatorial operation. During power-up, all user registers are initialized to a user defined preload state. Five direct product terms from the AND-array are available for use as primary data inputs (to the OR and XOR gates) to implement combinatorial functions or as control inputs including clock set/reset and output enable signals. The product-term allocator associated with each macrocell controls the selection of five product terms and can reassign the product terms within the FB to increase the logic capacity of a macrocell. It can make available up to 15 product terms to a single macrocell (with some incremental delay). It can also reassign product terms from any macrocell within the FB by combining partial sum of products over several macrocells.

The fastCONNECT switch matrix connects signals to the FB inputs. All IOB outputs (corresponding to the user pin inputs) and all FB outputs drive the fastCONNECT switch matrix. It is capable of combining multiple internal connections into a single wired-AND output before driving the destination FB. This provides additional logic capability and increases the effective logic fan-in of the destination FB without any additional timing delays. The I/O block (IOB) interfaces between the internal logic and the device user I/O pins. Each IOB includes an input buffer, output driver, output enable selection multiplexer, and user-programmable ground control. A useful feature provided is the pin locking capability, which allows the user to lock-in the predefined pin assignments during the preliminary design cycle. The XC9500 family devices can be programmed in-system via a standard four-pin JTAG protocol.

Xilinx is introducing a new 3.3 V, 0.35 μm, triple-layer metal based XC9500XL family, which is architecturally advanced compared to the original 5 V XC9500 family. Each XC9500XL device contains 18 macrocell function blocks (FBs) interconnected by an enhanced FastConnect II switch matrix. Each FB consists of an AND array, which provides 90 product term arrays arranged as groups of 5 assigned to each of 18 macrocells (this is change from the 5 V XC9500 family). Flexible product term allocators allow product terms to be allocated to specific macrocells on an individual product term basis, allowing any number of product terms (up to 90 maximum) to drive a macrocell. Each macrocell can also select between global or product term or signal clocks for clocks, three-state enables, set, and reset. This architecture allows for product terms based clock-enable signal into each register. The XC9500XL family has I/Os with mixed signal (5 V, 3.3 V, or 2.5 V) capabilities and will be footprint compatible with the XC9500 (for the same packages).

All Xilinx CPLDs are supported by the Xilinx Foundation Series tools (DS-FND) for PCs running Win95/NT4.0. The designs can be created and simulated using schematics, state diagrams, equations or a HDL (either VHDL or Verilog). For the third party environment on NT4/Sun/HP workstations (Cadence, Synopsys, Mentor, Viewlogic, Synplicity, and others), support is provided with the Xilinx Alliance package (DS-ALI).

References

1. Om Aggarwal, "PLD architectures, applications and benefits," Paper supplied by AMD.
2. Special report on Complex Programmable Logic Devices (CPLDs), *EDN,* July 1995.
3. Programmable Logic Technologies, *Computer Design,* October 1993.
4. Bryon Moyer, "Inside AMD CMOS PLD technology," Application Note, *PAL® Devices Data Book and Design Guide,* 1996.
5. AMD *PAL Devices Data Book and Design Guide,* 1996
6. Vantis *MACH® Data Book,* 1997.
7. Altera *Data Book 1995* and product data sheets supplied by Altera Corp.
8. Cypress *Programmable Logic Data Book 1995* and data sheets supplied by Cypress Semiconductor.
9. ICT Inc., *PEEL™ Products Data Book 1995/1996*.
10. *Lattice Handbook* and revised 1997 data sheets supplied by Lattice Semiconductors.
11. Philips Semiconductor *Complex Programmable Logic Devices Data Book,* 1997
12. Xilinx, *XC9500 CPLD Application Guide with Data Sheets, 1996,* and the latest figures with related text updates supplied, courtesy of ®Xilinx, Inc.

Chapter

3 Field Programmable Gate Arrays

3.1 General Architecture

The FPGA architecture can be broadly classified in two different ways, based on (a) granularity (size and flexibility) of the logic cell and (b) the routing architecture. An FPGA logic block can be as simple as a transistor or have the complexity of a microprocessor. Typically, it should be capable of implementing many different combinational and sequential logic functions. The current generation of FPGA architectures are based on one or more of the following logic elements as building blocks: transistor pairs; small gates, such as two-input NANDs or exclusive-ORs; multiplexers; look-up tables (LUTs); and wide fan-in AND-OR structures.

The FPGA logic blocks differ greatly in their size and implementation capabilities.[1] For example, a sea-of-tiles architecture based on small configurable units (called *tiles* in the Gatefield ProASIC family) is similar to gate array approach, as compared to the look-up table (LUT) based logic block used in Xilinx 3000 series FPGA, which can implement any five-input logic function but is relatively larger. These differences in basic cell structure for various logic block architectures are often classified by their granularity. The granularity of a logic block can be defined in various ways, such as the number of Boolean functions that the logic block can implement, the number of equivalent two-input NAND gates, the total number of transistors, or the number of I/Os. In the PLD industry, the commercial logic block architectures are often broadly categorized as fine-grain and coarse-grain.

The fine-grain logic blocks bear a close resemblance to MPGA basic cell structures and consist of few transistors with programmable interconnect resources. The major advantage of using fine-grain logic blocks is the high logic block utilization achievable. It is more efficient and easier to map complex functions using transistor-level logic blocks, since the logic synthesis techniques for such blocks are very similar to those for the conventional mask-programmed gate arrays and standard cells. The main disadvantage of fine-grain blocks is that they require many wire segments and programmable switches. These routing resources require additional chip area and timing delays. Therefore, the FPGAs using fine-grain block architecture are generally slower and achieve lower densities relative to the coarse-grain FPGA architectures.

An example of coarse-grain architecture is Actel FPGA logic block, which is based on the ability of a multiplexer to implement different logic functions by connecting each of its inputs to a constant or to a signal. The Actel family ACT 1 logic block consists of three multiplexers and one logic gate, which adds up to a total of eight inputs and one output. Multiplexer based logic blocks have the advantage of providing a high degree of functionality for a relatively small number of transistors. However, this is achieved at the cost of larger number of inputs (e.g., eight in case of the ACT 1 logic block) requirements which, for complex logic implementation, can place a high demand on routing resources. Therefore, coarse-grain logic blocks are more suitable for FPGA architectures that use programmable switches of small size, such as antifuses.

The routing architecture of an FPGA is the manner in which the programmable switches and wiring segments are utilized to allow the programmable interconnection of the logic blocks. Therefore, the interconnection resources include the connection block, which appears in all architectures. A connection block existing in horizontal and/or vertical direction provides connectivity from the inputs and outputs of a logic block to the wire segments in the channel. The second structure is the switch block, which provides connectivity between the horizontal as well as vertical wire segments. In some FPGA architectures, the connection block and switching block are intermingled where, as in the others, they are combined into a single structure.

An FPGA routing architecture typically employs wire segments of varying lengths, which can be interconnected via electrically programmable switches. The choice of number of wire segments incorporated affects the density achieved by an FPGA. If an inadequate number of segments is used, only a small fraction of logic blocks can be utilized, resulting in poor FPGA density. However, the use of an excess number of wiring segments can increase the die size and result in poor silicon utilization efficiency. The distribution of the lengths of the wire segments in an FPGA architecture also affects its density and performance. A routing architecture for an FPGA must meet two criteria: routability and speed.

The routability of an FPGA is its capability to accommodate all nets for a typical application, even though the wiring segments (number, lengths, and locations) must be predefined for blank (unprogrammed) FPGA configuration.[2] Another major factor in FPGA performance is the propagation delay through routing, since some nets require longer routing paths than others. The FPGA interconnect delay is significant portion (roughly 40 to 60 percent) of the total delay. Any programmable switch (antifuse, SRAM-based devices, or EEPROM transistor) has a significant resistance and capacitance. This results in addition of an RC delay stage each time a signal passes through a programmable switch, which tends to increase the average net delay.

FPGAs are generally classified into three basic architectural groups based on logic block size, functionality, and the structure of the routing domain: (1) row-based FPGAs, (2) symmetrical FPGAs, and (3) fine-grain, cellular architecture. The architecture of row-based FPGAs basically consists of coarse-grain, programmable logic blocks or simple gates, organized in rows, that are divided horizontally by programmable routing channels. The programmable routing is accomplished by connecting together wire segments of variable length, spanning the routing channels. An example of this architecture is Actel FPGAs, in which antifuses are used as programmable elements for connecting the wire segments. Figure 3.1*a* shows the schematic diagram of row-based FPGA architecture.[3]

The second important architectural classification is the symmetrical FPGAs, which are built of large grain blocks called configurable logic blocks (CLBs). Figure 3.1*b* shows the generic block diagram of a symmetrical FPGA, for which the routing domain consists of vertical routing channels between the columns of logic blocks and horizontal routing channels between the rows. This architecture can be visualized as a net of programmable wires for direct (neighborhood) connections, along with general-purpose and long lines. The general-purpose and long interconnect lines consist of wire segments of variable length, usually controlled by SRAM cells or antifuses.

The third FPGA general classification is fine-grain or cellular architecture, which resembles the "sea-of-gates" approach used in FPGAs. An example of this approach is the Pilkington architecture used in Motorola FPGAs, in which all functional elements (cells) directly abut each other, as compared to more common channeled architecture (e.g., Actel and Xilinx devices). These are two-dimensional symmetrical arrays of reconfigurable logic blocks with a hierarchical interconnect structure. The FPGA is divided into sections and provided with wire segments that span each section. The logical blocks within the section can connect using a single wire segment, whereas the longer connections use a higher level of interconnect to reach one section from an-

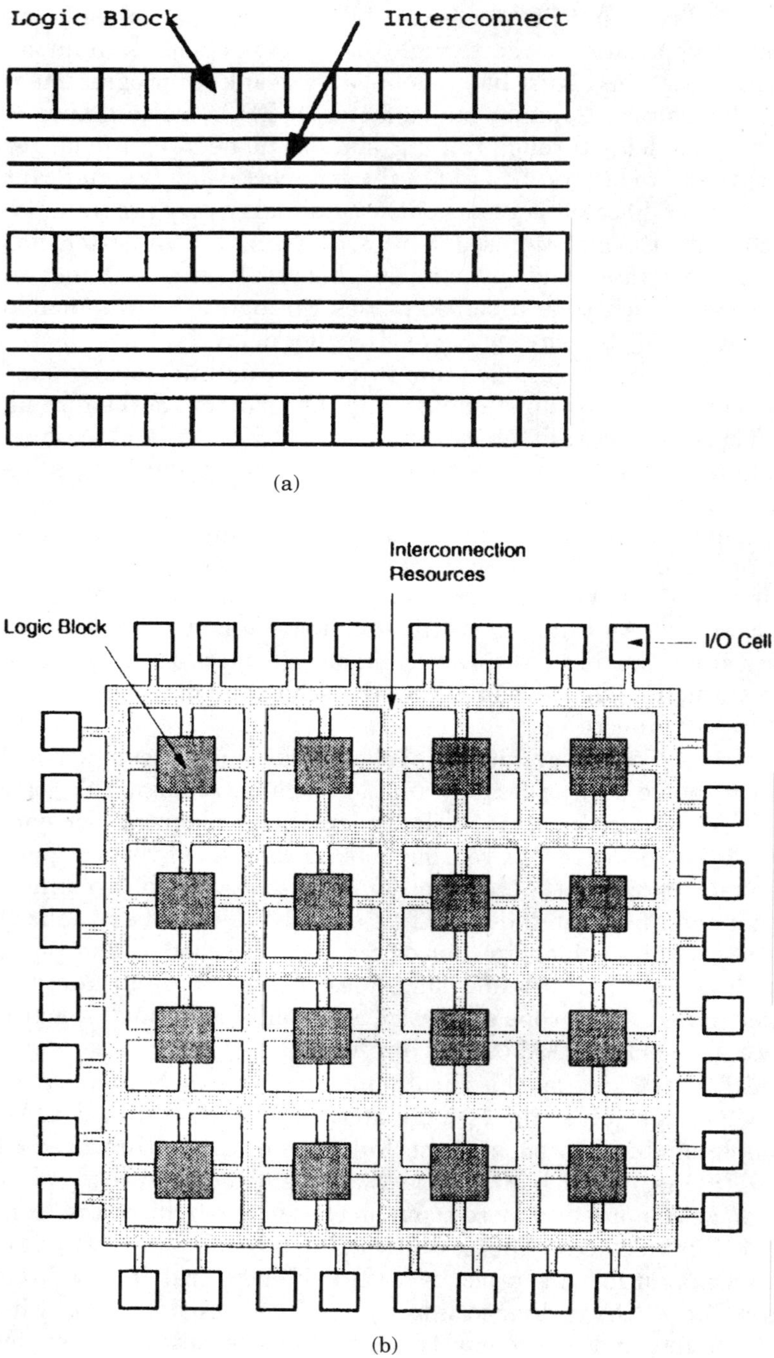

Figure 3.1 Generic block diagram of (a) row-based and (b) symmetrical FPGA.

other. The usual size of logic blocks used in this architecture is small (i.e., fine-grained).

Three general FPGA architectural categories are discussed in the following sections.

3.1.1 Row-based FPGAs

The Actel Family of FPGAs exemplifies the row-based FPGA model. This is also called a *segmented-channel array-based FPGA,* in which the rows of logic cells are located adjacent to the routing channels and combine the advantages of mask programmable gate arrays with the flexibility of users programmability. However, unlike the conventional standard cell design, the FPGA routing channels contain predefined wiring segments of various lengths that are connected using antifuses. For example, the antifuse used in Actel FPGAs represents a one-time programmable element, made up of a diffusion layer placed over polysilicon layer and separated by a dielectric between the two layers (see Sec. 3.2).

The row-based FPGA architecture has to deal with the problem of estimating the number of required tracks, where the tracks are straight sections of wire spanning the entire width or length of the routing channel and usually consist of number of wiring segments of varying length. The routing interconnections involve determining which segments are to be used and for what nets. The limited chip size constrains the number of available tracks, and this can affect the reconfigurability of the logic blocks. Therefore, it is important to have an accurate estimate of the number of tracks required for a given device size, since the interconnection topology (length and pattern of switching elements) strongly influences the circuit performance. A major objective is to minimize the number of fuses traversed by each net in order to optimize the timing characteristics.

Figure 3.2 illustrates some trade-offs involved in length of wiring segments in a channel, the area required, and the resistance-capacitance of the switch.[4] Figure 3.2*a* shows a set of nets routed in an unconstrained channel where there is a complete freedom for wiring configuration, such as for a masked programmed gate array. In an FPGA, achieving this freedom would require switches at every crosspoint. More switches would be required between each two crosspoints along a track to allow the track to be subdivided into segments of arbitrary lengths, as shown in Figure 3.2*b*.

Another alternative would be to provide sufficient number of continuous tracks to accommodate all the nets, as shown in Fig. 3.2*c*. This approach is used in many types of programmable logic arrays and interconnect array portion of certain PLDs. An advantage of this ap-

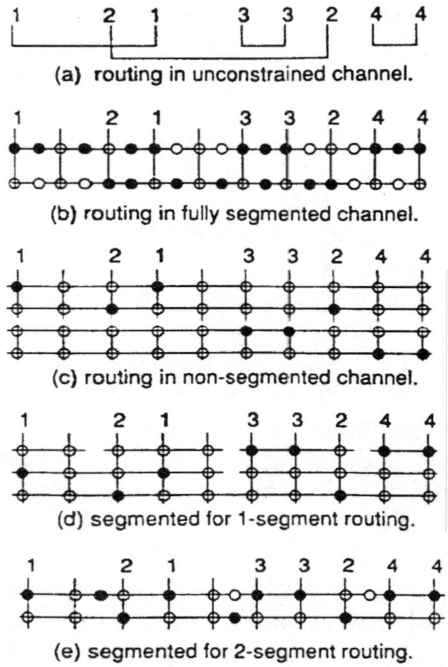

Figure 3.2 Segmented routing channel architecture and trade-offs *(from Ref. 4 with permission of IEEE).* Open circles represent unprogrammed antifuses and dark circles denote programmed antifuses.

proach is that only two RC stages are added, and the delay of each net is identical and predictable. However, it takes excessive chip area, and even short nets incur the capacitance of a full length track.

An intermediate approach is offered by segmented routing channel in which tracks are divided into segments of varying lengths, as shown in Fig. 3.2*d*. This allows each net to be routed using single segments of appropriate size. Greater routing flexibility is achieved by allowing multiple adjacent segments in the same track to be joined end to end by switches, as shown in Fig. 3.2*e*. The problem of routing a segmented channel, i.e., determining which segment(s) to assign each net, can be solved in linear time for single segmented channel, shown in Fig. 3.2*d*.

A study was performed to develop analytical models for the design of FPGA channel architecture and routability analysis based on generic characterization of routing algorithms for row-based FPGAs.[5] This study investigated the design of channel architecture, i.e., channel lengths, various segmentation patterns, and interconnections between the segments. The conclusions of this study, based on experimental results, are summarized as follows:

- The length of wire segments in the interconnect dramatically affects speed-performance characteristics.
- It is crucial to limit the number of programmable switches that signals pass through in series to minimize the RC delay factor.
- The impact of CAD routing tools is very significant, and the CAD tools should take into consideration both speed-performance and area utilization. Also, the CAD tools should focus on minimizing delays for time-critical nets to reduce number of tracks needed, and routability should be prioritized for noncritical nets.

3.1.2 Symmetrical (SRAM-based) FPGAs

An SRAM programmable FPGA contains the memory cells that control the logic required to perform the application function of the FPGA design. There is no separate RAM area on the chip. The memory cells are distributed among the logic elements that they control to implement logic and control interconnection paths for signal routing. The memory is built from static RAM cells, so the chips are often referred to as *SRAM programmable.* Since the FPGA memories do not change during normal operation, they are built for stability and density rather than speed. The major disadvantage for SRAM-based FPGAs is their volatility, which means that a device loses its programming configuration when turned off. Therefore, the chip must be reprogrammed every time it is powered up.

An example of this architecture is Xilinx family of FPGAs, *called logic cell arrays (LCAs),* which are programmed by loading set of configuration bits into on-chip SRAM that functions as a look-up table (LUT). The truth table for a K-input logic function is stored in a $2^K \times 1$ SRAM. The address lines of the SRAM function as inputs, and output of the SRAM provides logic function value. For example, consider the truth table of the logic function $f = ab + \bar{c}$ and its schematic representation shown in Fig. 3.3*a*. If this function is implemented using a three-input LUT, then the SRAM would have a 1 stored at address 000, a 0 stored at address 001, and so on, as represented by the truth table.

A big advantage of the LUT is its high functionality, which means a K-input LUT can implement any function of K-inputs. The disadvantage is that, since the number of memory cells needed for a K-input LUT is 2^K, the memory array size for more than an approximate five-input LUT gets unacceptably large. Although the number of functions

a	b	c	f
0	0	0	1
0	0	1	0
0	1	0	1
0	1	1	0
1	0	0	1
1	0	1	0
1	1	0	1
1	1	1	1

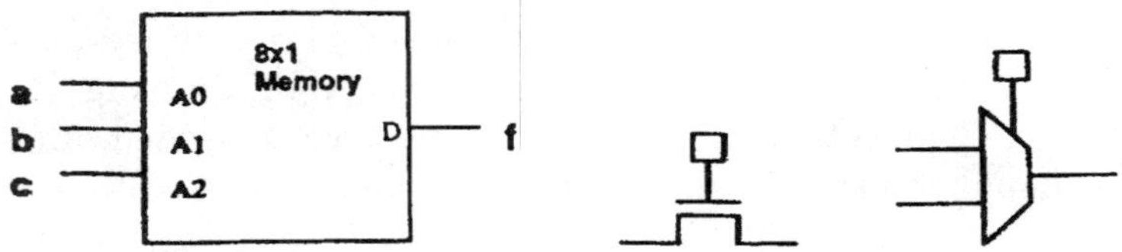

(a) (b) (c)

Figure 3.3 (a) Truth table for a logic function $f = ab + \bar{c}$ and its schematic representation, (b) pass transistor controlled by a memory cell, and (c) multiplexer controlled by a memory cell.

that can be implemented also increases, these additional functions are not commonly used and are difficult to handle using logic synthesis tools.

The second building block for SRAM-based FPGAs is called a *programmable interconnect point* (PIP), as shown in Fig. 3.3*b*. The PIP is a pass transistor controlled by a memory cell. The wire segments on each side of the pass transistor are connected (or not connected) depending on the logic value in the memory cell. This pass transistor introduces a resistive-capacitive (RC) element, and hence delay, in the routing path. The third building block is a multiplexer controlled by a memory cell, as shown in Fig. 3.3*c*. This multiplexer is a special-case, one-directional routing structure. However, the multiplexers may be of any width, with more configuration bits needed for wider multiplexers.

In an SRAM-based FPGA, the RC delay introduced by each PIP in the signal routing path adds up. Therefore, the total delay incurred due to routing in an FPGA is significantly greater than that in a traditional mask-programmed gate array where the wiring interconnect is a metallization track with low resistance and capacitance. In SRAM-based FPGAs, fewer PIPs in the interconnect path not only reduce the loading in the interconnect segments but also reduce the routing utilization. In comparison, more PIPs improve routability but add to the loading and increase in the die area, since each PIP includes a memory cell. The signal delay can be reduced by implementing longer interconnect resources without substantially increasing the load on those interconnects. Also, buffering in the interconnect can help reduce the effect of side-branch loading and improving delay.

A study was performed to evaluate optimization of SRAM-based FPGA routing architectures, such as those used by Xilinx, by examining two important metrics: (1) speed performance of implemented circuits and (b) effective use of available interconnect resources.[6] In addition to the routing architecture, the study also addressed CAD tools that select these routing resources for circuit implementation. The CAD tools suite was custom developed and parameterized to work with an FPGA model with many variable parameters. The key CAD tool used in this study was the Sega (segment allocator) detailed router, which has parameters that allow virtually any mixture of wire segment lengths in the routing channels.[7]

The benchmark circuits were used from the Microelectronics Center of North Carolina (MCNC) logic synthesis benchmark suite. The experimental procedure involved technology mapping the circuits, placing logic blocks with a min-cut algorithm implementation, and routing using the traditional two-stage method of global routing followed by the detailed routing. The test results of this study are summarized below:

- Wire segment lengths dramatically affected speed performance and in SRAM based FPGAs; this effect could be as much as 50 percent.
- For speed performance, an important issue is the number of switches that the signal must pass through in series. The capacitive loads associated with wire segments longer than the connections that use them are insignificant in comparison to the number of switches traversed.
- CAD routing tools should focus on timing delays only for time-critical nets. Such tools should route nets that are not time critical, with more consideration to routability.

In an FPGA architecture, design trade-offs have to be made between logic block structure versus size and speed. Simple logic blocks have some advantages but may lead to slow logic design if many of these blocks with their associated delay intensive wiring interconnect segments are required to implement a given function. A more complex logic block containing a larger LUT can implement more complex functions by using less interconnect but may not be fully utilized, which means poor logic efficiency. The capacity and speed of FPGAs is related to both the architecture and the software used to map a given design into that architecture, including partitioning, placement, and routing. These issues will be discussed in more detail in Chap. 4.

A study was performed to analyze the effect of logic block architecture on the speed of FPGAs using four logic block architecture classes:

NAND gates, multiplexer configurations, look-up tables (LUTs), and wide-input AND-OR gates.[8] An experimental approach was used in which a set of benchmark logic circuits was synthesized into FPGAs that use different logic blocks. Then, the speed of resulting FPGA implementation was measured. Although the results depend on the delay of the programmable routing, experiments indicated that five- and six-input LUTs and certain multiplexer configurations produced the lowest total delay over the realistic values of routing delay. A major reason is that these blocks can implement typical logic circuits using the fewest levels of logic block, and thus they use small number of programmable logic elements. In comparison, the fine-grain blocks, such as two-input NAND gates, exhibit relatively poor performance, because these gates require many levels of logic block for circuit implementation, and hence require a larger routing delay.

A recent development in FPGA architectural enhancement has been the incorporation of small-to-medium sized logic blocks of memory such as single-port or multiport SRAMs, FIFOs, and buffers.[9] These integrated memory blocks simplify system designs and allow systems to operate at faster speeds. Some examples of these FPGA families are the XC4000E from Xilinx, the ORCA families from Lucent Technologies (formerly AT&T Microelectronics), as well as families of arrays with dedicated blocks of RAM from Actel and Altera that allow the users to implement blocks of SRAM. These memory blocks approach gate array efficiencies and have access times ranging from 5 to 20 ns. Although gate arrays can offer better efficiencies for RAM than FPGAs that convert the distributed LUT memory into SRAMs, the FPGAs still provide the designers with the quick turnaround time for prototypes and low-volume production, which is usually difficult with the mask-programmed gate arrays.

The two basic approaches to embedded memory blocks are (a) Xilinx/ORCA-type arrays that convert the Boolean LUT memories into small blocks of RAM that are distributed across the chip, and (b) the Actel/Altera devices, which embed the dedicated blocks of RAM within the FPGA. A study performed by Altera designers for area versus memory block size trade-off has shown that, for memory blocks larger than about 32 words × 8 bits, such blocks would be more efficiently implemented as part of a dedicated RAM block than from the LUT blocks of RAM that are distributed across the FPGA. However, some applications need buffers, registers, and other memory blocks requiring less than 256 bits, such as small FIFOs and pipeline registers.

The latest families of FPGAs from Xilinx include the XC4000E series with LUT-based SRAM that has synchronous write and dual-port options, improved macrocell structure, and routing resources. The XC4000EX series, using a 0.35 μm advanced CMOS design process, employs additional routing resources, a more flexible I/O interconnection scheme called VersaRing, three or four metal layers for signal routing, and built-in JTAG test logic. In the XC4000EX, each of the configurable logic blocks on the array contains a 32-bit block of SRAM that serves as a LUT for the Boolean functions. The two 16-word × 1-bit sections of this LUT can be combined to form a 32-word × 1-bit RAM, a 16-word × 1-bit dual-port memory, two independent 16-word × 1-bit memories, or as a 16-word × 2-bit memory. Thus, multiple blocks of RAMs may be combined to form larger blocks of RAM.

Lucent Technologies ORCA FPGA family has basically the same capability of using LUT blocks in the configurable logic cells as the RAM building blocks. To implement dual-port blocks, more internal array logic must be used, which affects their performance. However, ORCA series has migrated to 0.35 μm design rules for higher performance and offers devices that have capacities of over 40,000 usable gates (see Sec. 3.3.4).

In the Altera FLEX 10K family, the SRAM is divided into sub-blocks of 2048 bits, and each sub-block is part of larger element called an *embedded array block* (EAB). Each of these EABs can be used to implement 2048 bits of SRAM and may be configured as 2048 × 1, 1024 × 2, 512 × 4, or 256 × 8 bits. Also, if an EAB is not used as RAM, it can be used to implement other logic functions such as multipliers, digital filters, state machines, decoders, and ROMs. One benefit of separating the EAB memory from the main logic array is that the EAB memory can be reconfigured at any time without reconfiguring the entire chip. The logic function or stored data can then be changed without interrupting the other functions of the chip.

The Actel 3200 DX family is targeted for applications that need a modest amount of memory on a chip—a maximum of 4 kb of dual-port RAM and from 6500 to 40,000 gates for logic. The Actel 3200DX memory blocks are not as flexible as Altera's FLEX 10K array blocks, so that when they are not being used as RAM or ROM (simulated), they are disabled and provide no contribution to the array. However, they do provide higher access times. The 3200DX series includes three different types of logic elements that are distributed across the chip, as compared to Xilinx/ORCA arrays that use identical logic blocks. The basic SRAM block is a true dual-port structure with separate 6-bit read and write address ports that have their own clock inputs for independent operation. The memory can be organized as either 32-words × 8-bits, or 64-words × 4-bits, for synchronous or asynchronous applications. Thus, multiple SRAM blocks may be concatenated to form larger memory blocks without additional delay penalty due to local interconnection resources included in each block.

A study was performed to examine the effects of the memory/logic connection block flexibility on the overall routability and speed of a circuit implemented on an FPGA with embedded array.[10] It was found that the routability of circuits with memory is not strongly affected by a low memory/logic interconnect flexibility, especially for architectures with fewer than eight embedded arrays. For such architectures, the logic routing sources are flexible enough to compensate for the low flexibility in memory/logic interconnect. However, this becomes less true as the number of arrays (and hence the number of connections to memory) is increased.

3.1.3 Fine-grain (or cellular) FPGAs

The fine-grain architecture allows direct connection between the neighboring cells, enabling the users to combine cells to form compact local functions. An example of fine-grain FPGAs is the Motorola programmable array (MPA) family based on Pilkington architecture (see Sec. 3.3.5). A key feature of this approach is the Pilkington array, which is partitioned into 100 cell zones, each of which can be considered as a separate array with the port cells forming an interface between the zone interconnect and the global interconnect. The global interconnect joins individual zones together to form the larger array.

The interconnect structure is hierarchical so that, at the lowest level, the fast connections are used to join local cells to form macros. These can efficiently implement small functions such as complex combinatorial logic, counters, and comparators. These functions are then combined using the medium interconnect within the zone and the highest-level global interconnect throughout the array for signal routing and complete layout. The short range of local connections and their limited loading means they can be quite fast.

The Motorola MPA architecture also differs from other coarse-grained, channeled-array devices in functionality of the cell, which provides only a small set of two input functions. All cells have a basic AND function along with a subsidiary function such as an XOR or D-type flip-flop. In the channeled array architecture, such a cell would be very inefficient, with a high routing overhead between each level of logic. However, using a hierarchical routing structure, these small cells can be very efficiently combined into macros with negligible routing overhead. There are two basic advantages, as follows:

1. There is no redundancy, i.e., if a cell is used to perform simple function, there are no additional logic resources in the cell to be wasted.

2. The cell can be highly optimized, since the critical paths through the cell are limited in number.

The functions of the cells in the array are no longer homogeneous, but vary over a group of cells called a *tile*. These are repeated uniformly over the array. Thus, the fine-grained structure of this architecture is well suited to logic synthesis tools, and automatic design and layout.

3.2 Programming Technologies

A high-performance FPGA requires a programmable interconnect switch that occupies a small area and, at the same time, has a low parasitic resistance and capacitance. The other major attributes for programming technology consideration are series on-resistance of the programmed switch, its volatility and reprogrammability, and process complexity. Several different programming technologies are used to implement the programmable switches in FPGAs. The three most commonly used programming technologies are as follows:

- Antifuse switch of dielectric or amorphous silicon composition, which on electrical programming forms a low-resistance interconnect path
- SRAM-based technology in which the switch is a pass transistor controlled by the state of a RAM bit in a look-up table (LUT)
- EPROM-based technologies (either UVEPROM or EEPROM) in which the switch is a floating-gate transistor that can be turned off by injecting charge on to the floating gate

In all these cases, the programming switch element occupies a larger area and exhibits higher parasitic resistance-capacitance compared to a typical contact or via used in MPGAs. Therefore, the performance achievable by current generation of FPGAs is usually about an order of magnitude lower than that for the MPGAs manufactured using the same process technology.

The floating-gate EPROM technology (UVEPROM, EEPROM, and flash EEPROM) has been used in complex programmable logic devices (CPLDs) such as Altera, Lattice, and Xilinx families, discussed in Chap. 2. The floating-gate technology has also been used for Gatefield ProASIC family fine-grained FPGAs, which are based on a proprietary flash-EPROM controlled switch that is both nonvolatile and reprogrammable (see Sec. 3.3.3). However, for majority of FPGAs, the antifuse and SRAM are the dominant programming technologies. The following sections discuss only these two technologies and their relative trade-offs.[11]

3.2.1 Antifuse programming technology

An antifuse is basically a two-terminal device with an unprogrammed state representing a very high resistance (several hundred megohms) between its terminals. On the application of a high voltage (from 11 V to 21 V), depending on the device and antifuse type used, the antifuse is "blown" to create a low-resistive, permanent link (or connection). The two most commonly used antifuse technologies are: (a) oxide-nitride-oxide (ONO) dielectric based and (b) amorphous silicon, or metal-to-metal antifuse structures. The programming of antifuses requires extra on-chip circuitry to deliver the high programming voltage (and relatively high current) through the pass transistors.

The dielectric antifuses consist of a layer of dielectric material placed between N^+ diffusion and polysilicon. This dielectric breaks down upon an application of sufficient voltage. Early dielectric antifuses used a single-layer oxide dielectric. A *programmable low impedance circuit element* (PLICE), which is a multilayer oxide-nitride-oxide (ONO) dielectric fuse, was developed for use in Actel FPGAs. An application of programming pulse (16 to 21 V) across the PLICE melts the dielectric, creating a conductive link of polycrystalline silicon between the electrodes. Figure 3.4*a* shows the cross section of PLICE antifuse structure. A thin layer of oxide is thermally grown on top of the N^+ surface, followed by a layer of low-pressure chemical vapor deposition (LPCVD) nitride and the reoxidized top oxide. The PLICE adds three masks to a conventional double-metal CMOS process.

The manufacturers of antifuse-based FPGAs provide their own programming algorithms as part of the device specification and application information. In all types of antifuses, the smaller the thickness of a given antifuse material, the lower the link resistance for given current, and higher the leakage current for unprogrammed state.[12] Also, the more current applied during programming, the lower the resistance of the link, and link resistances can be improved by appropriate programming waveforms and soaking times. Since the programming current amount has a large effect on link resistance, the programming circuits for antifuses need to supply high currents (e.g., 15 mA for Actel devices) for reliable configuration of high-performance interconnects.

In its simplest form, an antifuse array is a collection of vertical and horizontal wires with an antifuse at every crossing (or intersection), as shown in Fig. 3.4*b*.[13] Let us define V_{pp} as the programming voltage needed to program any fuse. Then, typically, the programming of an antifuse in the array requires three steps, as follows:

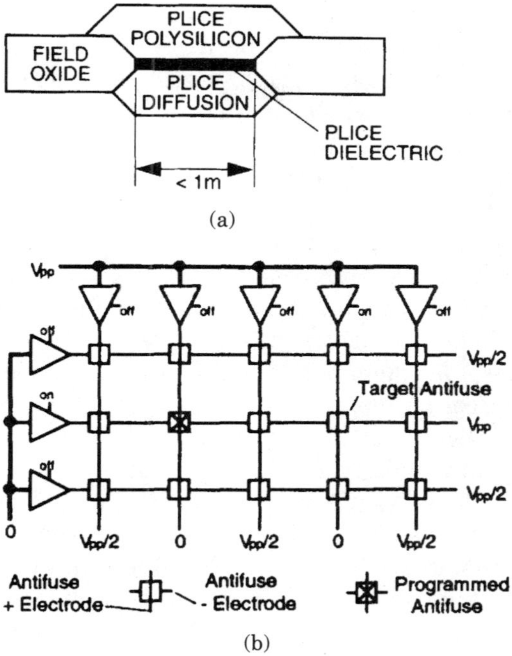

Figure 3.4 (a) Cross section of a PLICE antifuse structure *(from Ref. 15)* and (b) an array of antifuses *(from Ref. 13)*.

1. Precharge all vertical/horizontal wires to $V_{pp}/2$.
2. Select the antifuse to be programmed (address) and drive the vertical (horizontal) wire connected to the target antifuse to 0 V, and select (address) and drive the horizontal (vertical) wire connected to the target antifuse to V_{pp}. This ensures that only the target antifuse is stressed to V_{pp}, and the remaining antifuses remain at 0 V or $V_{pp}/2$ (see Fig. 3.4*b*).
3. On antifuse breakdown, soak the antifuse with high current to stabilize the newly formed conductive link and to lower the antifuse resistance.

The programming voltage and current are generally controlled by analog circuits (programmers) external to the FPGA chip. The addressing requires independent horizontal and vertical decoding controls, and these schemes vary from one FPGA vendor to another. Actel FPGAs use an indirect addressing scheme to apply V_{pp} and 0 V to the target antifuse electrodes after precharge to $V_{pp}/2$.[14] For Actel antifuses, the programming algorithm specifies the peak programming voltage (V_{pp})

as 21 V, I_{max} (clamp current) = 15 mA, and number of soak pulses from 30 to 800. A confirmation that an antifuse has been programmed is made through monitoring the current on V_{pp} pin, which is typically <10 μA for an unprogrammed antifuse. Once an antifuse is programmed and an electrical connection is made between V_{pp} and ground, 3 to 15 mA current may be observed on V_{pp}. Once the antifuse is considered programmed and enters "soak" cycle, extra pulses are applied to the antifuse to achieve minimum resistance.

In contrast to Actel's indirect addressing scheme, QuickLogic uses direct addressing to program the antifuses. Since no pass transistors are used, QuickLogic uses a dedicated driver plus shift register bit(s) per wire segment. The number of segments addressed is equal to the number of shift register/driver cells.

An alternative to dielectric based antifuse has been the development of amorphous (noncrystalline) silicon antifuse technology. This is based on the principle that a layer of amorphous silicon placed between two metal layers undergoes a phase change when a current is passed through it, and it becomes conductive. In its unprogrammed state, the amorphous silicon is an insulator with resistance greater than 1 GΩ. The user can program an antifuse link by applying relatively high current (roughly 20 mA) signals that effectively convert the insulating amorphous silicon into conductive polysilicon link.

An example of amorphous silicon antifuse application is QuickLogic pASIC FPGA families, based on ViaLink, which provides a low-resistance, low-capacitance programmable connection directly from one metal layer to another. The ViaLink element is formed by depositing a very high resistance (>1 GΩ) of amorphous silicon above a tungsten via plug that would otherwise bridge the insulation between the two metal layers. On the application of a programming voltage to a selected via, a direct metal-to-metal link is formed with typical low-resistance values of less than 50 Ω. In a 0.65 μm CMOS process, the size of a ViaLink is roughly 1 μm^2, which is orders of magnitude smaller than the active elements, resulting in low capacitive loading [<1 fF (femtofarad, $10E^{-15}$ farads)] and improved speed performance. Figure 3.5 shows (a) a three-layer metal ViaLink structure, (b) an unprogrammed ViaLink element, and (c) a programmed ViaLink element.[16]

3.2.2 SRAM-based programming technology

SRAM-based FPGA architecture was discussed in Sec. 3.1.2. The SRAM programming technology uses static RAM cells to control pass gates or multiplexers. The speed of an FPGA is a measure of the delay required to implement a function and to propagate signals to the neighboring functions. The speed of the logic part of the block is the

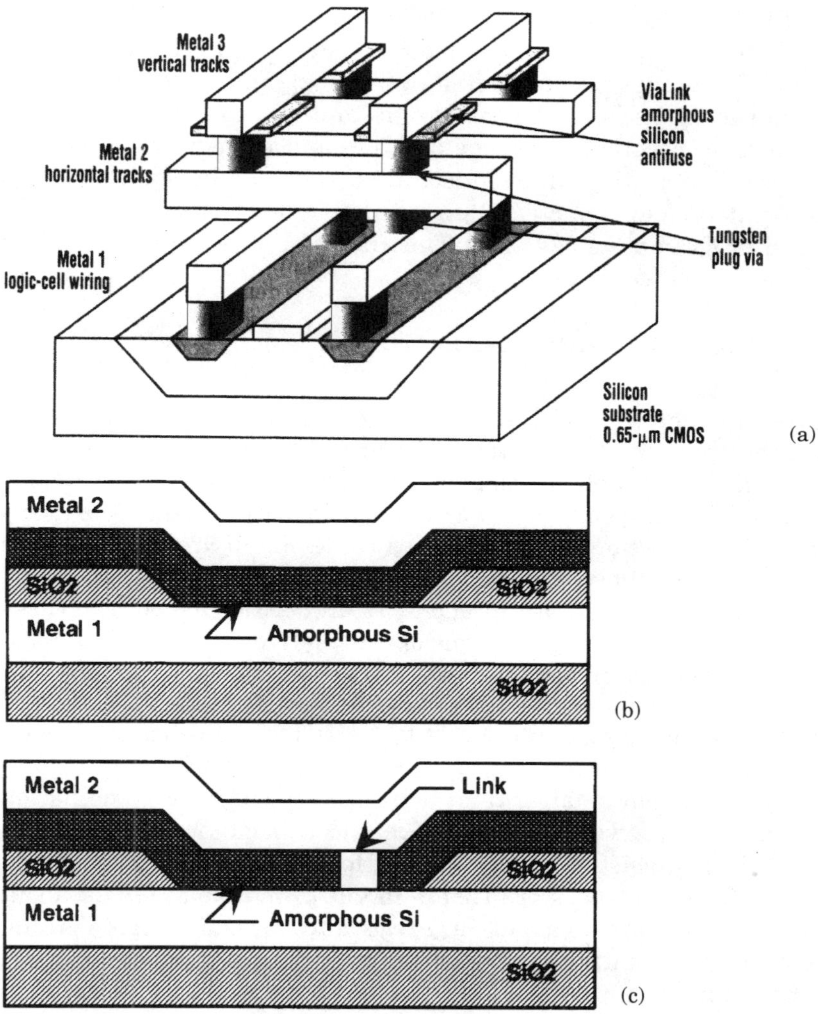

Figure 3.5 (a) Three-layer metal ViaLink structure, (b) unprogrammed ViaLink element, and (c) programmed ViaLink element *(from Ref. 16)*.

sum of delays from the input selection multiplexers, the look-up table, and the output drivers. The speed and density trade-offs are based on the user's design requirements. An architecture with more programmable interconnect points (PIPs) provides more routing flexibility, but each PIP adds to the size of FPGA and the routing delay.

Therefore, the FPGA size and performance trade-offs need to take into consideration not only the logic elements but also the interconnect segments, PIPs, and the multiplexers required to connect the elements. The LUT width is also an important factor. For example, a four-input LUT has sixteen memory cells, and a three-input LUT has only eight. If the logic being implemented in a four-input LUT breaks naturally into three input gates, half of every four-input LUT area will be wasted, leading to an inefficient design implementation. A study was performed to investigate a range of LUTs and their effect on overall logic block and routing interconnect area.[1] The results showed that a three-input or four-input LUT provided the best density for a wide range of programming cell sizes. The larger LUTs were preferred for high-speed architecture.

3.2.3 Antifuse versus SRAM-based technology trade-offs

A review of general FPGA architectures and programming technologies shows that both the antifuse- and SRAM-based devices have relative advantages as well as disadvantages. The selection of a particular programming technology for an FPGA is primarily based on the user's application requirements and involves trade-offs in areas such as design logic utilization capacity, performance, and in-system programmability features. There are proponents and critics for both of these approaches. The following are some of the relative advantages and disadvantages of each technology:

- Antifuse technology is inherently faster than SRAM-based technology because of lower interconnect resistance and capacitance values. Antifuse is much smaller than a transistor-RAM cell combination and has much lower "on" resistance and capacitance than a minimum-sized transistor. As an example, for 0.8 μm CMOS technology, the impedance of pass transistor in SRAM-based FPGA is about 250 Ω, as compared to about 50 Ω resistance for QuickLogic ViaLink metal-to-metal antifuses. As for the capacitance values, it is only 1 fF for unprogrammed ViaLink element compared to 50 fF for pass gate transistor capacitance. Since the net delays are determined by RC time constraints, the lower resistance and capacitance of the antifuse networks provides significant smaller delays than the SRAM-based pass gate transistor networks.
- For equivalent silicon area per gate, antifuse FPGAs provide more flexibility and are easier to route compared to the SRAM-based FPGAs. SRAM pass-gate interconnects require at least four transistors for the flip-flop and one transistor for the pass gate. This structure implies limited use of interconnect resources and forces macrocell architecture grain size to be coarse.
- A disadvantage of antifuse-based technology is that it requires more process layers and mask steps than the SRAM-based FPGAs and

uses on-chip, high-voltage programming transistors. SRAM-based FPGAs are easier to migrate to the next process generation and, unlike antifuse-based devices, the performance of both logic elements and programmable interconnect benefit from scaling to smaller geometries. However, the proponents of antifuse-based FPGAs claim that the smaller physical size of antifuse results in less expensive die, and the area required to hold a pass gate transistor and associated memory cell is certainly larger than the area required for an antifuse.

- A survey of FPGA market shows that the SRAM-based FPGAs offer higher capacity than the antifuse-based devices. Some SRAM-based FPGAs feature a power-down mode in which the power is supplied only to the memory cells that hold the configuration program.
- One of the major advantages of SRAM-based FPGAs is the flexibility offered by their in-system reconfigurability, which makes them a good candidate technology for prototype development. This feature can significantly ease the debugging process and reduce overall system design and development costs. In contrast, the antifuse technologies are one-time programmable (OTP) configurations.
- Another advantage often cited for the SRAM-based FPGAs reprogrammability feature is that the parts can be fully tested at the factory, as compared to the antifuse based devices, which are tested as "blanks," since the antifuses are user programmable per design requirements.
- A disadvantage of the SRAM-based FPGAs is that the programming is volatile, i.e., when the power is turned off, the FPGA loses its configuration program information. Therefore, the SRAM-based FPGAs must be reprogrammed each time power is applied, as part of turn-on initialization sequence. This initialization sequence requires an external memory for permanent (nonvolatile) storage of the program.

3.3 FPGA Vendor Families and Development Tools

3.3.1 Actel FPGAs

Actel FPGAs are based on the channeled array segmented architecture and PLICE antifuse technology discussed in Secs. 3.1 and 3.2, respectively. Actel FPGA offerings consist of following device families:

- ACT 1 family, with up to 2000 gate array gates (6000 PLD equivalent gates)

- ACT 2 family, with 2500 to 8000 gates capacity
- Recently introduced ACT 3 family, with capacities ranging from 1500-10,000 gate array equivalent gates
- 1200XL and 3200DX Integrator series FPGAs targeted toward low-cost and mid- to high-range performance applications

These families are briefly described in the following sections.[15]

ACT 1 family. ACT 1 family devices are implemented in silicon gate, double-level metal CMOS process and use PLICE antifuse technology. There have been several die revisions in this family, with L_{eff} ranging from 2.0 to 1.0 μm. The user-definable I/Os are capable of driving at both the CMOS and TTL levels. A security fuse may be programmed to disable all further programming and to protect the device from being copied or reverse engineered. The basic general-purpose logic module for an ACT 1 device is an eight-input, one-output logic circuit that can implement logic functions (such as NAND, AND, OR, NOR, and so on) in gates of two, three, or four inputs. Basically, the logic module consists of three two-to-one multiplexers with select lines and a two-input OR gate.

Figure 3.6*a* shows the schematic diagram of an ACT 1 logic module. The select line of the two multiplexers, M1 and M2, are individual inputs, whereas the select line of multiplexer M3 is an OR function of the other two inputs. One of these logic modules can implement all functions of two-input variables, most functions of three-input variables, and some functions of four-input variables. In all, 702 distinct combinational macros are possible. The ACT 1 family ranges from device A1010 with 295 logic modules to A1020 with 547 logic modules. ACT 1 devices are specified for a maximum flip-flop toggle rate of 100 MHz and system speed (maximum) of 40 MHz.

Figure 3.6*b* shows the basic layout of ACT 1 FPGA logic modules and the interconnect structure. The rows of logic modules are located adjacent to the horizontal routing channels containing predefined wiring segments of variable lengths placed at various locations. There are also wiring segments that run vertically through the modules and across the channels (only few out of many possible vertical segments shown). Each logic module computes a single output function of several inputs. Also, each logic module input is connected to a dedicated vertical wiring segment spanning either the channel above or below the module. Each output signal appears on a dedicated vertical wiring segment of somewhat longer length. An antifuse is provided at each intersection of a horizontal and vertical segment, permitting an interconnection. The output of the driver, module 3, is

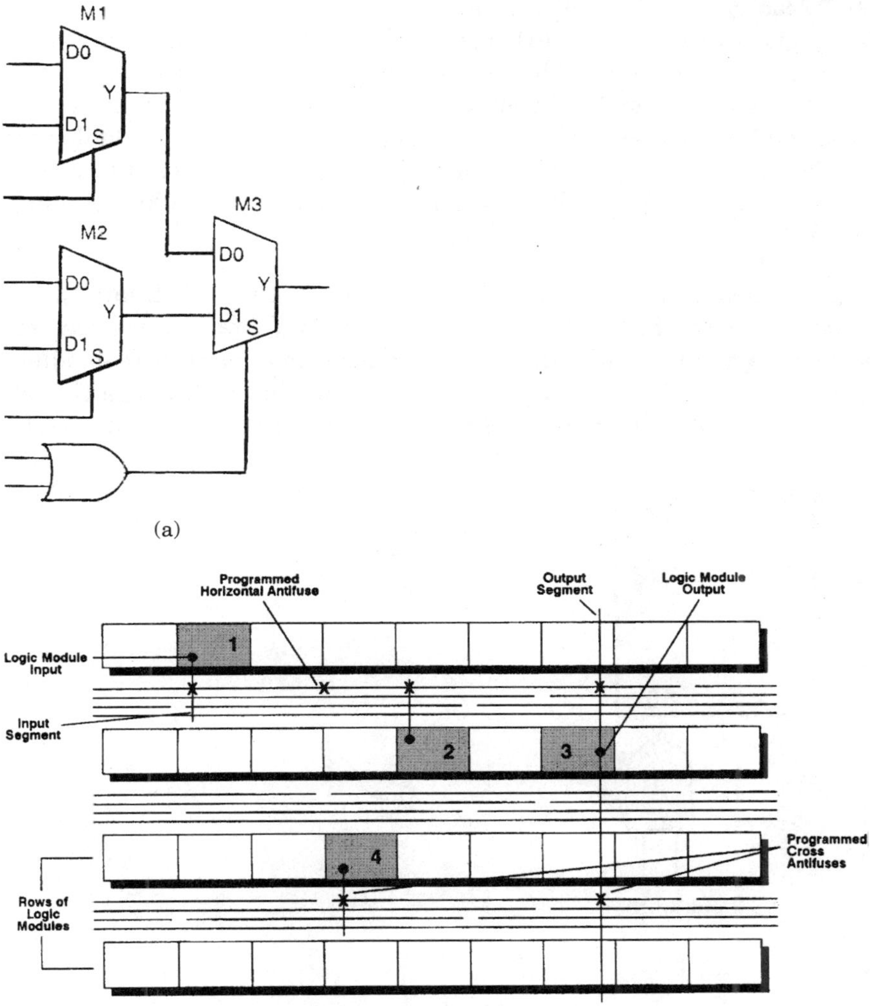

(b)

Figure 3.6 ACT 1 family: (a) logic module and (b) logic module layout and basic interconnect structure *(from Ref. 15 with permission of Actel Corp.).*

connected by the programmed antifuses to horizontal segments, which in turn are connected to input segments. In the top channel, an antifuse is used to link two adjacent horizontal segments end to end, which makes it possible to reach input of the module 1 as well as module 2.

This central array of modules and channels is surrounded by the I/O pads and buffers. Each I/O buffer can be connected to internal logic through a special module near the edge of the array.

ACT 2 family. ACT 2 family devices range in capacity from A1225 with 2500 gates capacity and 451 logic modules to A1280 with 8000 gates and 1232 logic modules. The ACT 2 family architecture consists of five key building blocks: logic modules, I/O modules, routing tracks, global clock networks, and probe circuits. The basic structure is similar for all device types in the family, differing only in the number of rows, columns, or I/Os, depending on the device capacity. These key building blocks are briefly described below:

- *Logic modules.* In the second generation of the ACT 2 family, the single logic module of the first-generation ACT 1 family has been replaced with two different logic modules: combinatorial (C-module) and sequential (S-module). These are shown in Fig. 3.7. The C-module is an enhanced version of the ACT 1 family logic module, which

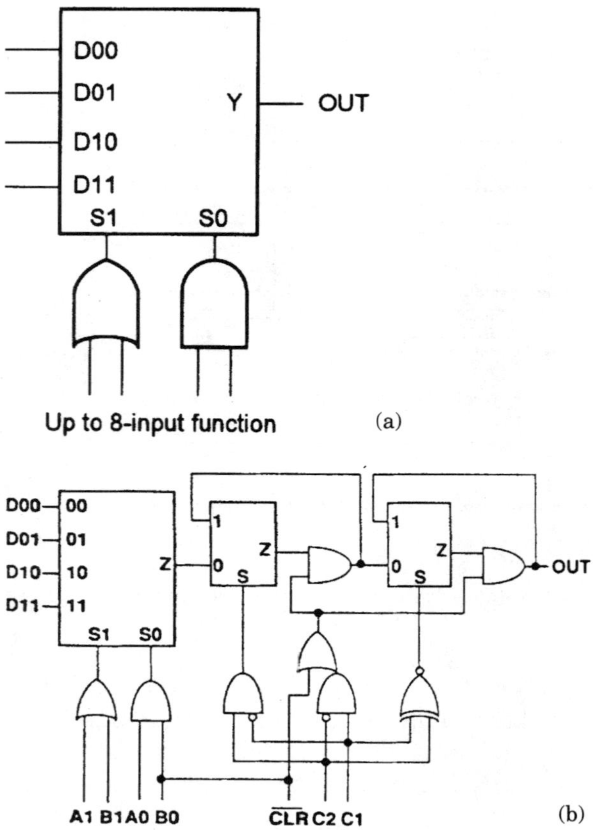

Figure 3.7 Actel ACT 2 family (a) combinatorial logic C-module and (b) sequential logic S-module (from Ref. 15 with permission of Actel Corp.)

has been optimized for implementing high fan-in combinatorial macros, such as five-input AND and five-input OR gates. The S-module also contains combinatorial logic, which allows an extra level of logic implementation without incurring additional propagation delay. The S-module shown consists of a front-end equivalent to C-module followed by a sequential block built around two latches. The sequential block can be used as a rising- or falling-edge D flip-flop, or a transparent-high or transparent-low latch, by tying the C1 and C2 inputs to a clock signal, logical zero, or logical one in various combinations.

The C-module and S-module are arranged in pairs called *module-pairs,* which are arranged in alternating pairs and make up bulk of the programmable array. This tiling arrangement allows placement software to support two module-macros of four types (CC, CS, SC, and SS). The I/O modules are arranged around the periphery of the programmable array. The array of logic and I/O modules is surrounded by the test and programming circuits controlled by following external pins: MODE, SDI, and DCLK. While the MODE is in low state (GND), the device is in normal or user mode. When the MODE pin is high (V_{cc}), the device is placed into one of several programming or test states. The SDI pin (when MODE is high) is used to input serial data to the MODE register and various registers surrounding the array. The data is clocked into these registers using the DCLK pin. The MODE register determines the test or programming state of the device. Many of the test modes are used during wafer sort and final test at the factory, but others are used during programming with the Actel Activator 2®, and some modes are available only after programming.

- *Input/output (I/O) modules.* The I/O architecture consists of pad drivers located near the bonding pads and I/O modules located in the array. There are two types of I/O modules: side and top/bottom. The side I/O modules occupy the left-most and right-most two columns of the array, whereas the top/bottom I/O modules are located in the top and bottom rows, respectively. The I/Os implement a variety of functions determined by the library macro selection. There are some special-purpose I/O pads, temporarily used for programming and testing the devices, such as SDI (serial data in), DCLK (serial data clock in), PRA (probe A output), and PRB (probe B output).

- *Clock networks.* The ACT 2 architecture uses two low-skew, high-fanout clock distribution networks, as shown in Fig. 3.8*a*. These networks are referred to as CLK0 and CLK1. Each network has a clock module (CLKMOD), which selects the source of the clock signal and may be driven as follows: (a) externally from CLKA pad, (b) externally from CLKB pad and internally from the CLK1NA input, and

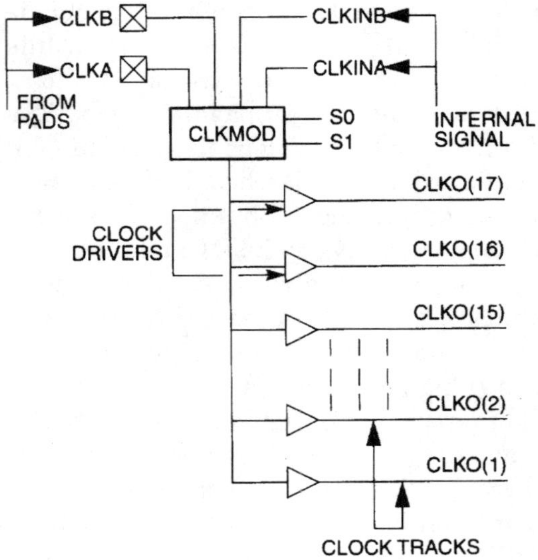

(a)

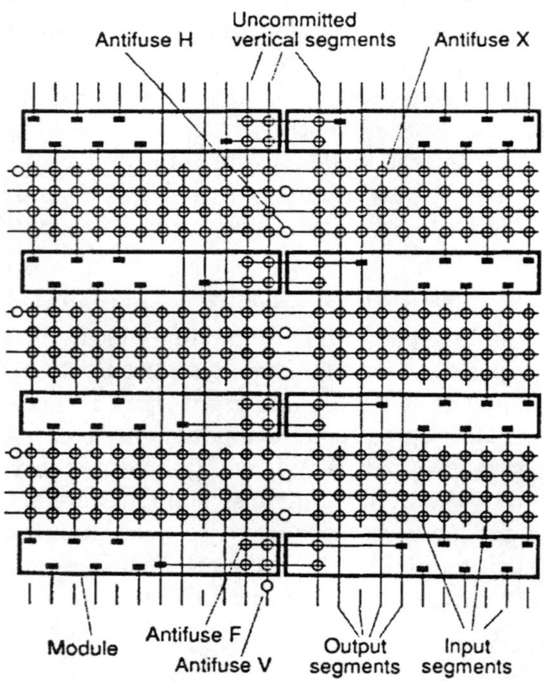

(b)

Figure 3.8 Actel ACT 2 family (a) clock networks and (b) routing architecture global view *(from Ref. 15 with permission of Actel Corp.).*

(d) internally from the CLK1NB input. The clock modules are located in the top row of I/O modules. Clock drivers and a dedicated horizontal clock track are located in each horizontal routing channel. The user controls the clock module by selecting one of two clock macros from the macro library. The clock input pads may also be used as normal I/Os, bypassing the clock network. To minimize loading effects on the clock networks, only a few of the C-modules and S-module inputs can be connected to the clock networks.

- *Routing architecture.* The ACT 2 routing architecture is an enhancement of the ACT 1 architecture and provides more flexibility of interconnection. The ACT 2 family also uses horizontal and vertical routing tracks to interconnect the various logic and I/O modules. Figure 3.8*b* shows a global view of the routing architecture and logic modules (only a representative sample of input and uncommitted segments shown). These routing tracks are metal interconnects, which may either be of continuous length or broken into pieces called *segments.* The segments can be joined together at the ends, using antifuses to increase their length up to the full track length.

  The horizontal channels are located between the rows of modules and consist of several routing tracks. These horizontal routing tracks within the channel are divided into one or more segments. The minimum horizontal segment length is the width of a module-pair, and the maximum horizontal segment length is the full length of the channel. Any segment that spans more than one-third of the row length is considered to be a long horizontal segment. The non-dedicated horizontal routing tracks are used for routing signal nets, whereas the dedicated routing tracks are used for the global clock networks and for power/ground tie-off tracks.

  The vertical routing tracks are of three types: input, output, and long. The vertical tracks are also divided into one or more segments. Each segment in an input track is dedicated to the input of a particular module, and each segment in an output track is dedicated to the output of a particular module. The long segments are uncommitted and can be assigned during routing. Each output segment spans four channels (two above and two below), except near the top and bottom of the array. The assignment of uncommitted segments in each horizontal and vertical channel to those nets that need them constitutes a segmented routing channel problem for which several algorithms have been developed.

  The ACT 2 family uses the PLICE antifuse element (see Sec. 3.2.1). Four types of antifuse connections are used in the routing structures of the ACT 2 array. The physical structure of these antifuses is identical in each case—only the usage differs. The four antifuses, their functional designation, and application are listed below:

XF	Cross-connected antifuse	Connects the perpendicular tracks of most horizontal and vertical tracks' intersection
HF	Horizontally connected antifuse	Connects adjacent segments in the same horizontal track, end to end
VF	Vertically connected antifuse	Connects adjacent segments of long vertical tracks (divided into two segments), end to end
FF	"Fast fuse" antifuse	Connects a module output directly to a long vertical track

- *Hard macros.* The ACT 2 design library provides over 350 logic function macros. The hard macro logic functions range from simple SSI gates such as AND, NOR, and exclusive-OR to more complex functions such as flip-flops with 4:1 multiplexed data inputs. These hard macros can be implemented in the ACT 2 architecture by using one or more C-modules or S-modules. Two-module hard macros always utilize a module-pair, either SS, CC, CS, or SC. Since one- and two-module macros have small propagation delay variations, their performance can be predicted quite accurately. The propagation delays for the hard macros are specified in the data sheets. The term *soft macros* refers to multiple hard macros that are connected to form complex MSI functions, such as 16-bit counters and accumulators. Soft macro delays are not specified in the data sheets.

ACT 3 family. The ACT 3 family, representing third generation of Actel FPGAs, is based on the PLICE antifuse technology and 0.8 μm double-metal, double-poly CMOS process with capabilities ranging from 1500 to 10,000 gate array equivalent gates. The ACT 3 family is basically an improvement over the ACT 2 family architecture, consisting of logic modules, I/O modules, I/O pad drivers, routing tracks, clock networks, and programming and test circuits. The ACT 3 logic modules are enhanced versions of the ACT 2 family logic C-module and S-module. The C-module in ACT 3 is identical to the C-module in the ACT 2 family and is designed to accommodate high-fan-in combinatorial macros.

The ACT 3 S-module includes a combinatorial block, which is identical to the C-module, and a sequential block. One of the enhancements in ACT 3 logic module design is the addition of a dedicated CLEAR input to the sequential module, which significantly increases the number of "combinable" functions, especially of the high-fan-in type. Another enhancement to the sequential circuit design is the use of a novel nonrouted clocking scheme, in addition to the two routable clock networks available in the ACT 2 architecture. The nonrouted clocking scheme consists of two lightly loaded fast clocks, one of which clocks the input and output registers of the I/O module, with the other clocking the flip-flop in each sequential module. This fast clocking scheme allows for high speed on-chip and off-chip delays. Figure 3.9*a* shows the ACT 3 family S-module block diagram.

In the ACT 3 family, improvement has also been made in the architecture of the I/O module, as shown in Fig. 3.9*b*. The module includes an output register and an input register. Each I/O module contains two D-type flip-flops, and each flip-flop is connected to the dedicated I/O clock (IOCLK). Each flip-flop can be bypassed by nonsequential I/Os. In addition, each flip-flop contains a data enable input that can be accessed from the routing channels (ODE and IDE). The asynchronous preset/clear input is driven by the dedicated preset/clear network (IOPCL). Either preset or clear can be selected individually or on an I/O module by module basis. The I/O module output Y is used to bring pad signals into the array or to feed the output register back into the array. This allows the output register to be used in high-speed state machine design applications.

The ACT 3 architecture contains four clock networks: two high-performance dedicated clock networks for functions up to 200 MHz, and two general-purpose routed networks for functions up to 150 MHz. The dedicated clock networks contain no programming elements in the path from I/O pad driver to the input of S-modules or I/O modules, to support high performance. There are two dedicated clock networks: one for the array register (HCLK), and one for the I/O registers (IOCLK). The clock networks are accessed by special I/Os.

Integrator series 1200XL and 3200DX families. The Actel 1200XL and 3200DX Integrator series are targeted toward low-cost and mid- to high-range performance applications. They are based on Actel's PLICE antifuse technology, channeled array architecture implemented in 0.6 μm double-metal CMOS process. The basic architecture uses two module types: combinatorial (C-modules) and combinatorial-sequential (S-modules). The 1200XL family is pin and functionally compatible with the ACT 2 family and is design compatible with the ACT 1 and ACT 3 families. The 1200XL family provides significant performance enhancements as compared to the ACT 2 family. The I/O modules have been redesigned to improve pin-to-pin and clock-out delays. Also, the clock distribution networks have been improved for both propagation delays and timing skews for highly loaded designs.

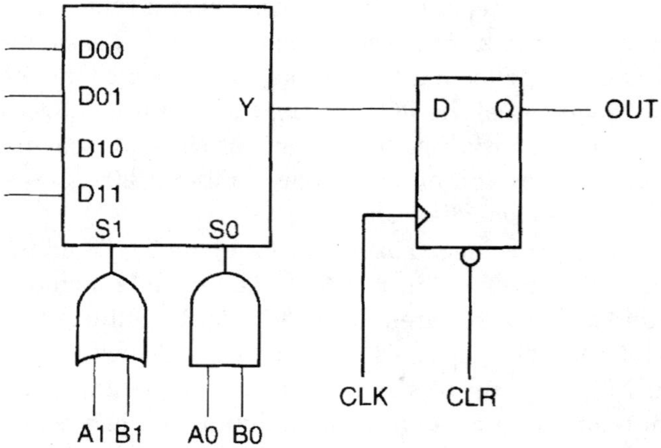

(a)

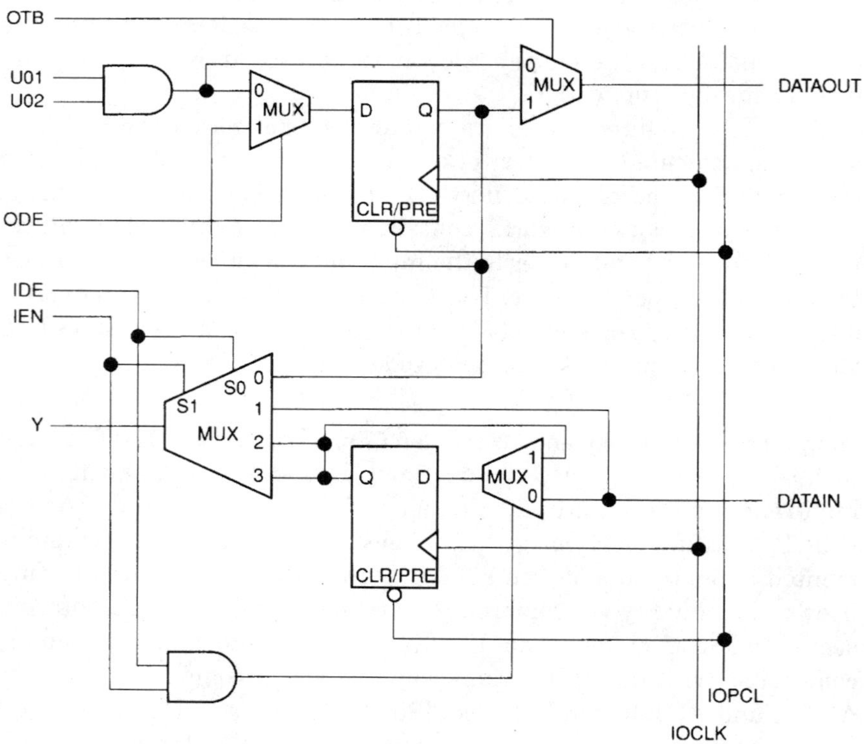

(b)

Figure 3.9 Actel ACT 3 family (a) S-module and (b) functional diagram for I/O module *(from Ref. 15 with permission of Actel Corp.)*.

The 1200XL family density ranges from A1225XL with 2500 gate array equivalent gates with up to 83 users I/O (maximum) to A1280XL with 8000 gates and 140 users I/O (maximum). The 3200 DX family offers dual-port SRAM for implementing fast FIFOs, LIFOs, and temporary data storage. The 3200 DX product family ranges from devices with 10,000 gates and 2048 bits to devices with 40,000 gates and 4096 bits. The large number of storage elements help implement efficiently the applications requiring wide data path manipulation and transformation functions such as telecommunications, networking, and DSP. The 3200DX family FPGAs provide basic building blocks, such as C-modules, S-modules, dual-port SRAM modules, and wide decode modules.

Figure 3.10 shows 3200DX dual-port SRAM block. The SRAM modules are arranged in 256-bit blocks that can be configured as 32×8 or 64×4, with independent clocking for the read and write ports. Each SRAM module contains six bits of read and write addressing (RDAD[5:0] and WRAD[5:0], respectively) for 64×4 bit blocks. When configured in byte mode, the highest order address bits (RDAD5 and WRAD5) are not used. The read and write ports of the SRAM block contain independent clocks (RCLK and WCLK) with programmable independent polarities, offering active HIGH or LOW implementation. The SRAM block contains eight data inputs (WD[7:0]) and eight outputs (RD[7:0]), which are connected to segmented vertical routing tracks.

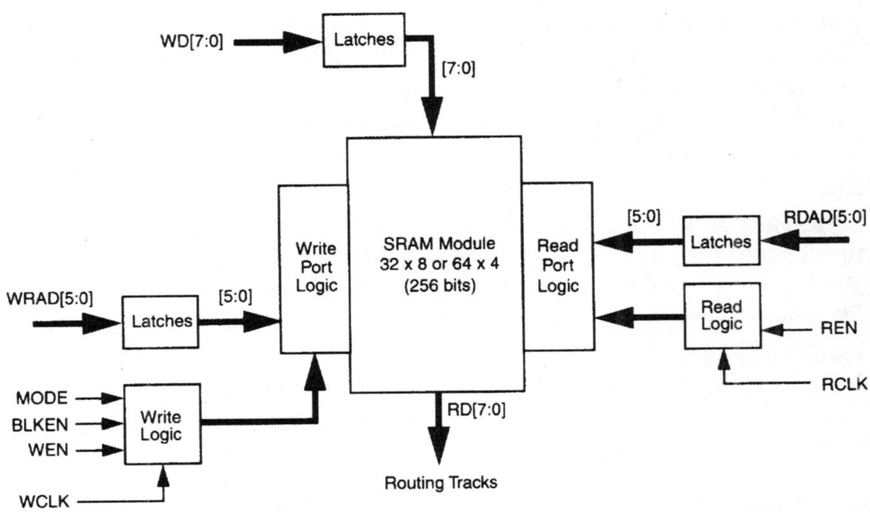

Figure 3.10 Actel 3200DX dual-port SRAM block *(from Ref. 15 with permission of Actel Corp.).*

The SRAM module's block architecture allows multiple modules to be cascaded together to form wider (e.g., ×16, ×32) and deeper (×128-bit, ×256-bit) memory arrays. The basic decoding element is a 7-input AND gate with programmable inversion on the output. This configuration allows a 35-input AND, OR, AND-OR (PLD type) logic block implementation in two layers of logic. Both 1200XL and 3200DX families provide means for post programming device testing and debugging through Actel's Actionprobe® facility. In addition, the 3200DX family also contains JTAG 1149.1 compliant boundary scan test (BST) circuitry as a design debugging tool and board level test support.

Actel FPGA design and development tools. Actel Designer Series development system is a complete package for design of Actel FPGAs. This consists of Designer Series tools represented by shaded blocks in the Actel FPGA design flow chart shown in Fig. 3.11. The Actel FPGA design process consists of following four steps:

Step 1: Design specification The first design phase is to define the functionality of the design. Actel supports four basic design specification methods: schematic capture, high-level design (HLD), ACTmap VHDL synthesizer/optimizer, and ACTgen Macro Builder. There are extensive Actel libraries available for design specification with all these methods.

ACTmap VHDL synthesis provides a user interface for synthesis, logic optimization, retargeting among Actel FPGA families, compiling of VHDL source files, and importing of existing PAL designs. ACTmap can optimize a design either for maximum operating speed or minimum area. It represents a design by using binary decision diagrams, which can map efficiently on to the Actel's multiplexer-based logic modules. ACTmap VHDL synthesis can map a design on to a fixed library of logic functions, or it can map Boolean equations directly into the logic modules for more efficient resource allocation.

ACTgen Macro Builder allows creation of macros that may be customized to the user design needs. It allows selection of the desired FPGA family, and the range of macro categories include various types of general logic including counters, registers, comparators, and RAM blocks. ACTgen Macro Builder can generate its output in EDIF, Verilog, and VHDL for interface with popularly used design tools. The Verilog and VHDL formats allow ACTgen created macros instantiation into high-level descriptions for logic synthesis.

Step 2: Design implementation Actel Designer 3.1 tools provide an FPGA placement and routing environment, which consists of following components:

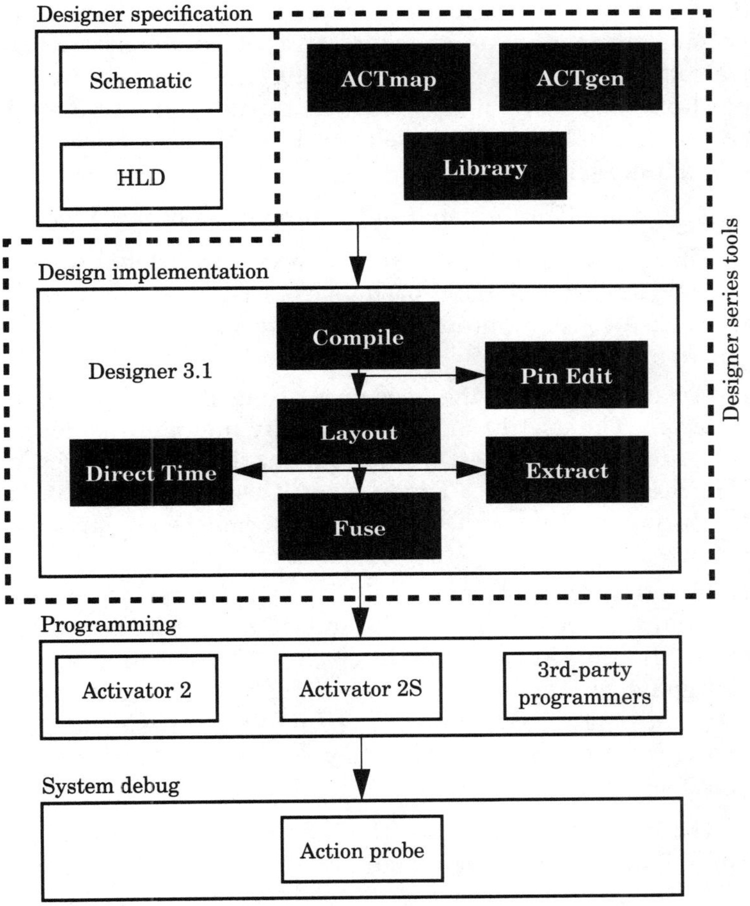

Figure 3.11 Actel FPGA design flowchart *(from Ref. 15 with permission of Actel Corp.).*

- *Compile.* The Compile program initiates all necessary programs for importation of a design, verification of the macros used in the design process, and preparation for layout. The Compile program will list the macros, I/O, and clock usage and report any errors or warnings found, such as fanout violation.
- *Layout.* The Designer series supports two ways to place and route a design. The standard layout uses Actel's layout algorithms to automatically place and route designs utilizing 90 to 100 percent of FPGA logic modules, even with all pins assigned. Standard layout automatically invokes the fastest placement and routing solution. Direct Time layout uses a set of timing specifications determined by the designer during the layout process.

- *Pin Edit.* Pin Edit is an optional program that allows the user to preassign pins or view I/O assignments after the automatic layout tools have run. Any pins that are explicitly fixed by this method will be assigned automatically by the software during the layout step. After the layout file has been completed, Pin Edit can be used to view the final pin assignments.
- *Direct Time.* Direct Time optional software consists of three tools:
  1. *Direct Time Edit* enables the user to specify the global system clock period and/or individual path delays with a simple user interface. These constraints are used in determining the module placement and routing.
  2. *Direct Time Layout* gives the user direct control over the timing of the design. The user has to simply specify the required design performance or specify delay requirements for each of the design's critical paths. Direct Time Layout then makes the tradeoffs in routing to achieve the specified performance or timing requirements.
  3. *Direct Time Analyzer* is timing analysis tool that provides critical feedback on the timing results of standard or Direct Time place and route. It provides a spreadsheet view of the timing for selected signals, automatically sorting them and listing the ones with longest delay first. Direct Time Analyzer can also display the timing results of a Direct Time Layout, i.e., tabulate both, the requested, and achieved timing.
- *Extract.* The extract function uses the information created during the layout and modifies it to a form suitable for back-annotated simulation.
- *Fuse.* The final stage in the design implementation step is to generate a fuse file that contains all the necessary information required by the programmer to program the device.

Step 3: Programming Device programming is accomplished by using Actel's Activator 2, Activator 2S, or third-party programmers (e.g., Data I/O).

Step 4: System debug The Actionprobe function in Actel devices provides the ability to probe any internal node during the normal device operation. The internal node is selected by scanning a serial address from the computer/workstation into the device on the pins SDI and DCLK. Once the addressing is complete, a path from the internal node to either the PRA or PRB I/O pin is created. A logic analyzer connected to those pins then can be used to observe the internal operation of the device. The hardware required to support the Actionprobe diagnostic tool includes an Activator 2 or Activator 2S programmer and the Action-probe pod, which connects the device under test to the Activator.

3.3.2 Atmel FPGA family

AT6000 series FPGA architecture. The Atmel AT6000 series SRAM-based FPGAs are designed to implement Cache Logic™, which offers the ability to implement adaptive hardware for computation intensive designs and perform hardware acceleration.[17] The AT6000 series utilizes a 0.8 μm single-poly, double-metal CMOS process and ranges in density from devices with 2000 to 20,000 usable gates, and 1024 to 6400 registers. The patented 6000 series architecture is based on a symmetrical array of many small, identical cells combined with a high-speed busing network to provide efficient communication over medium and long distances. The array is continuous and completely uninterrupted from one edge to the other, except for bus repeaters spaced every eight cells. In addition to implementing the logic and storage functions, the cells can also be used as wires for connecting functions together over short distances and for routing in tight spaces.

Figure 3.12 shows (3.12*a*) the busing network and (3.12*b*) cell-to-cell and bus-to-bus connections. There are two types of buses: local and express. The local buses provide the link between the array of cells and the busing network. There are two local buses: North-South 1 and 2 (NS1 and NS2) for every column of cells, and two local buses, East-West 1 and 2 (EW1 and EW2), for every row of cells. In a sector, each local bus is connected to every cell in its column or row, which provides every cell in the array with read/write access to two North-South and two East-West buses. In addition, each cell provides the ability to route a signal on a 90° turn between the NS1 bus and EW1 bus, and between the NS2 bus and EW2 bus.

The express buses are not connected directly to the cells and are the fastest way to cover long, straight line distances within the array. Each express bus is paired with a local bus so that there are two express buses for every column and two express buses for every row of cells. The connector units, *called repeaters,* are spaced every eight cells and divide each bus, both local and express, into segments spanning eight cells. The repeaters are aligned in rows and columns, partitioning the array into an 8 × 8 sector of cells. Each repeater is associated with a local/express pair, and on each side of the repeater are connections to the local-bus segment and an express-bus segment. The repeater can be programmed to provide any one of 21 connecting functions, which are symmetric to both the two repeater sides and the two types of buses. These functions have the ability to (a) isolate bus

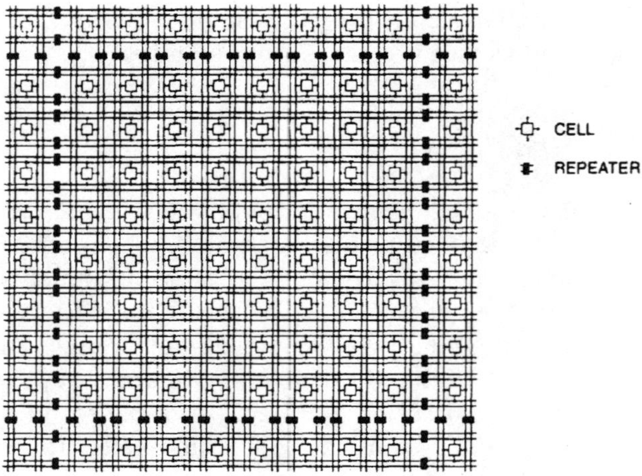

(a)

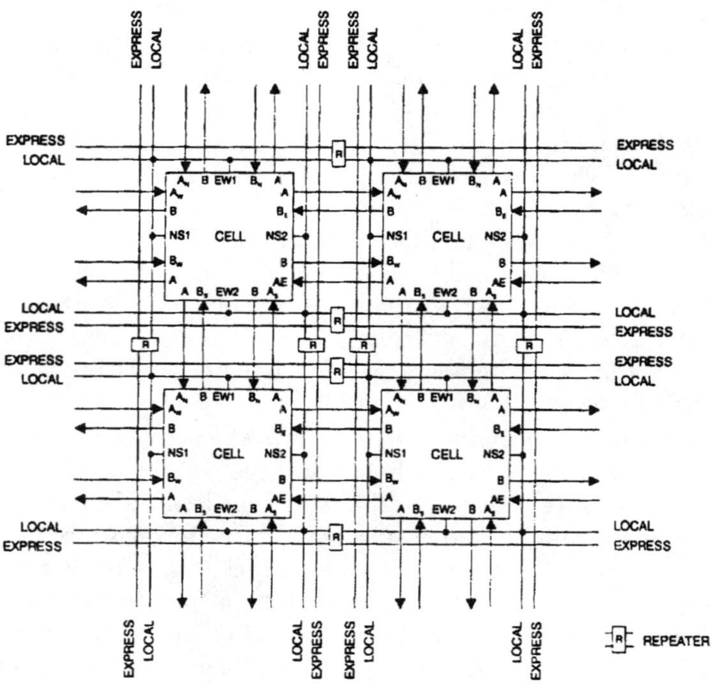

(b)

Figure 3.12 Atmel AT6000 family: (a) busing network and (b) cell-to-cell and bus-to-bus connections *(from Ref. 17 with permission of Atmel Corp.) (continues).*

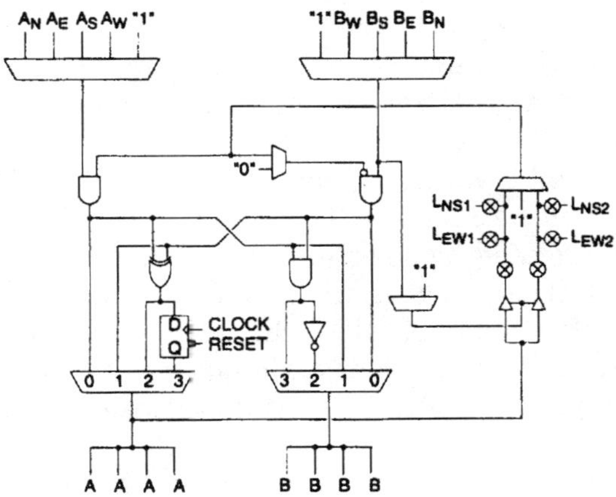

(c)

Figure 3.12 (continued) Atmel AT6000 family: (c) cell structure *(from Ref. 17 with permission of Atmel Corp.)*

segments from one another, (b) connect two local-bus segments, (c) connect two express-bus segments, and (d) implement a local/express transfer.

Figure 3.12*c* shows the cell structure, whose four sides are functionally identical so that each cell is completely symmetrical. The read/write access to four local buses (NS1, EW1, NS2, and EW2) is controlled, in part, by four bidirectional pass gates connected directly to the buses. To read a local bus, the pass gate for that bus is turned on, and the three-input multiplexer is set up accordingly. To write to a local bus, the pass gates for that bus and the associated tristate driver are both turned on. The two-input multiplexer supplying the control signal to the drivers allows either (1) active drive or (2) dynamic tristating controlled by the B input. The turns between L~NS1 and L~EW1 or between L~NS2 and L~EW2 are accomplished by turning on the two associated pass gates.

In addition to the four local bus connections, a cell receives two inputs and provides two outputs to each of its North (N), South (S), East (E), and West (W) neighbors. These I/Os are divided into two classes, "A" and "B." There is an A input and a B input from each neighboring cell, and an A output and a B output driving all four neighbors. Also, between the cells, an A output is always connected to an A input, and a B output to a B input. Inside the cell, the four A inputs and the four B inputs enter two separate, independently configurable multiplexers. The two multiplexer outputs enter the two upstream AND gates. Downstream from these two AND gates are an exclusive-OR (XOR) gate, a register, another AND gate, an inverter, and two four-input multiplexers, producing A and B outputs. These multiplexers are controlled in tandem (unlike the A and B input multiplexers) and determine the function of the cell.

The Atmel cell, with its 44 cell states, can implement a fairly wide variety of logic functions. Some states use both A and B inputs, whereas the other states are created by selecting the "1" input on either or both of the input multiplexers. There are 20 purely combinatorial functions such as NOR, AND, NAND, OR, and two-input multiplexers; 11 register states ranging from a simple register to a register preceded by a two-input multiplexer; 5 constant states capable of producing all combinations of constant values at the two cell outputs; and 5 tristate states. More complex functions can be created by using various combinations of cells. The array architecture provides a flexible interface to the configuration control logic and I/O pins. Two adjacent cells—an "exit" and an "entrance" cell—on the perimeter of the logic array are associated with each I/O pin.

The top edge of the array contains the logic for distributing the clock signals to the D flip-flop in each logic cell. The distribution network is organized by column so that the columns of cells can be independently clocked. A user-configurable multiplexer at the head of each column provides the clock signal for that column. It has four inputs: (1) global clock supplied through the CLOCK pin, (2) express bus adjacent to the distribution logic, (3) "A" output of the cell at the head of the column, and (4) logical constant "1" to conserve power (no clock).

The bottom of the array contains the logic for asynchronously resetting the D flip-flops in the logic cells. The asynchronous reset network has similar organization to clock distribution network and permits columns to be independently reset. A user-configurable multiplexer at the bottom of each column provides the reset signal for that column. It has four inputs: (1) global asynchronous reset supplied through the $\overline{\text{RESET}}$ pin, (2) express bus adjacent to the distribution logic, (3) "A" output of the cell at the foot of the column, and (4) logical constant "1" to conserve power.

Atmel design and development system. The Atmel Integrated Development System, available for use on the PC and Sun/Sparc workstations, uses industry standard software for design entry and simulation in conjunction with Atmel's proprietary software for component generation, automatic and interactive placement and routing, timing analysis, and bit stream generation. The chip configuration is the process of loading a design into AT6000 series SRAM-based FPGAs. Atmel's Integrated Development System generates the SRAM bit pattern required to configure an AT6000 series device. A PC parallel port, microprocessor, EPROM, or serial configuration memory can be used to download configuration patterns. The user can select from several configuration modes, depending on the board area requirements, configuration speed, the number of designs implemented in parallel, and so forth. The configuration is controlled by the dedicated configuration pins and dual-function pins that double as I/O pins while the device is in operation. The number of dual function pins required for each mode varies.

3.3.3 Gatefield FPGAs

Gatefield offers ProASIC™ product families GF250F and GF260F embedded versions that are based on advanced non-volatile, flash-based 0.5 μm CMOS process technology.[18] These two product families and their design and development system are briefly described below.

GF250F ProASIC family. The GF250F family utilizes a unique sea-of-tiles architecture similar to the gate-array approach, which is based on an array of small configurable units called *tiles*. Each tile contains resources for implementing logic and routing functions. Figure 3.13*a* shows a three-input/one-output core logic cell that can implement almost all one, two, and three input logic functions.

Therefore, each tile may be configured as a specific logic function (e.g., inverter, NAND gate, NOR gate) by programming interconnect switches. Simple functions such as flip-flops and multiplexers fit in one tile; however, more complex functions can take two or more tiles. Any tile can serve as combinatorial logic, flip-flop, and/or routing. The local interconnect has no boundaries, so the individual macrocells can be of any size, using any number of tiles. The macrocells can also be implemented in any shape from a single row to a rectangle, and to a single column; they can also be placed in any location.

Figure 3.13*b* shows interconnect routing resources for GF250F proASIC architecture. It uses four levels of routing hierarchy for higher performance and flexibility: (1) local lines, (2) long lines, (3) very long lines, and (4) global lines. The GF250F proASIC family ranges in density from devices with 12,000 usable gates and 176 I/O pins to the devices with 70,000 gates and 424 I/O pins. Each I/O pad can be configured as an input, an output, a three-state driver, or a bidirectional buffer. Also, each output can be configured for PCI compliance.

In the GF250F family, each device contains four primary clock inputs. Each clock input has programmable access to every tile throughout the chip. These four inputs can be individually programmed as separate clocks operating within particular regions of the core or programmed for operation throughout the entire core. One group of I/O

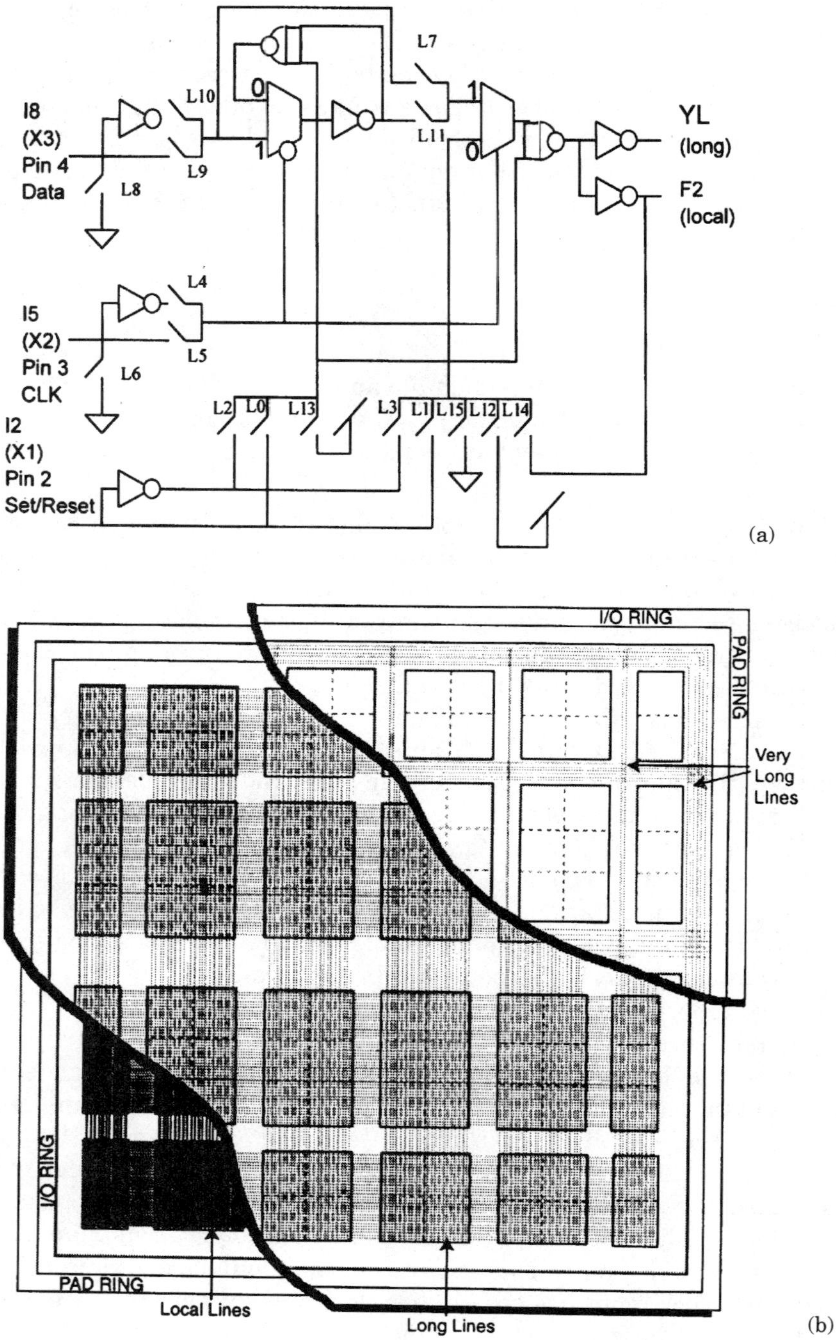

Figure 3.13 GF250F proASIC family architecture: (a) core logic cell and (b) interconnect routing resources *(from Ref. 18 with permission of Gatefield Corp.)*.

pins, called bus I/Os (BI/Os), are provided with extra power and ground pin connections. These pins are intended for high-speed buses that have multiple, simultaneously switching outputs. Local resources allow the output of each tile to directly connect to every output of its eight closest neighbors with high-speed, dedicated lines. The local routing mesh is homogeneous across the entire programmable ASIC and typically provides more than 60 percent of the user's circuit interconnections.

Long lines provide routing for longer distance and higher fanout connections. These long line resources are of variable length (typically 1, 2, and 4 tiles) and run both vertically as well as horizontally across the entire programmable ASIC. Each tile can drive signals onto the long resources, and long resources can access every input of a tile. In comparison, very long line resources travel greater distances across the chip and are used to route very long or very high-fanout nets. These run vertically and horizontally across the chip and provide multiple access to each group of 16 tiles throughout the chip.

Global resources are used to distribute the clocks, resets, and other nets requiring high fanout with guaranteed minimum skew. Global resources are implemented as four H trees, and signals can be introduced at each junction. These can be used hierarchically, with signals accessing every input on all tiles. Any portion of the global resources not required for the four primary global nets is made available to any other net requiring the distribution of high-fanout signals.

The GF250F proASIC macrocell library includes simple gates, complex gates, and storage elements such as flip-flops and latches. This library allows small register files and FIFOs to be compiled, utilizing functionality of the core tiles for distribution of these memory elements, where needed in relation to the surrounding logic. The ASICmaster™ software suite includes MEMORYmaster™, which accepts inputs from the designer regarding the number of memory words and bits, preferred origin and orientation, and the targeted device.

GF260F embedded ProASIC family. The GF260F proASIC product family is an enhancement of its logic gate "only" GF250F family and includes blocks of embedded SRAM memory, which are available as programmable logic functions. These memory blocks include hardwired decoder, I/O circuits, parity generation and detection circuits, FIFO flow generation logic, and timing and control circuits. The GF260F family shares the distributed memory architecture for GF250F plus the embedded memory elements for implementation of fast, dedicated memories. The GF260F family ranges in density from devices with logic gate count (utilized in a typical design, exclusive of memory) of 20k and 14k embedded RAM bits to devices with 57k logic gate count and 46k embedded RAM bits.

The embedded memory section is located across the top of the programmable gates. There are 6 to 20 blocks of memory (depending on the device type) available to support variety of possible memory configurations. Figure 3.14*a* shows the memory block architecture for a GF260F180 device in which 10 individual and unique memories are available. Each of these memories may be configured as synchronous

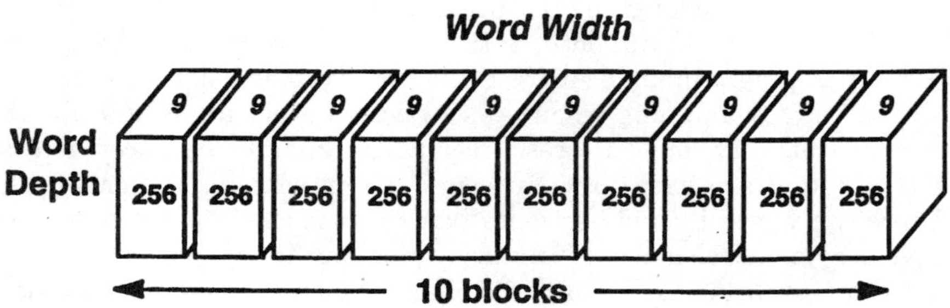

(a)

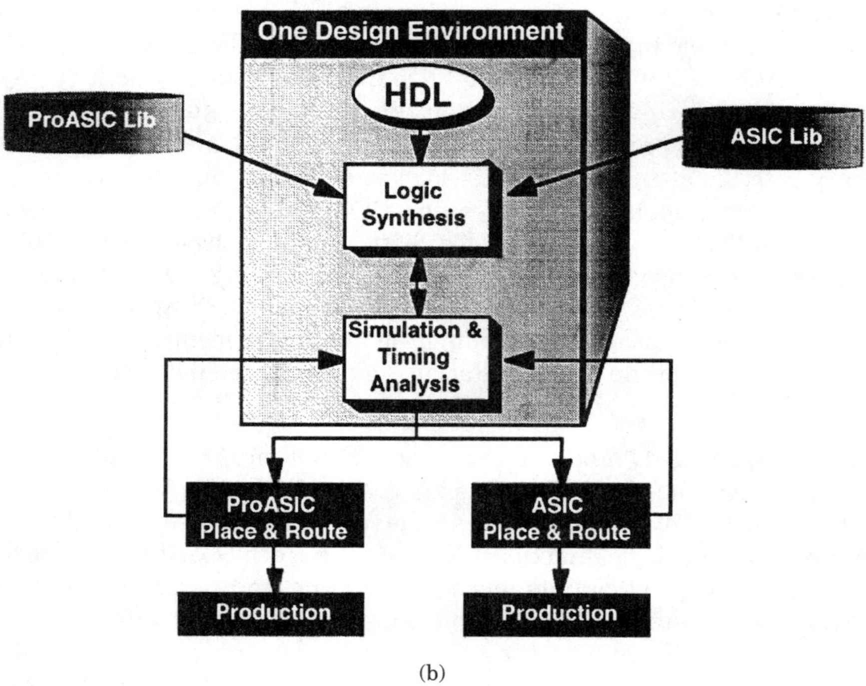

(b)

Figure 3.14 (a) Memory block architecture for GF260F180 device and (b) ProASIC families for ASIC migration (from Ref. 18 with permission of Gatefield Corp.).

or asynchronous, FIFO or SRAM, and single-port or multiport. There are other programmable characteristics including FIFO flags, parity control, depth, and width. Each of these 10 blocks contains a 9-bit wide × 256-word deep memory. The memory blocks may be paralleled to form wider memories or stacked to form deeper memories. Each GF260F block is designed and optimized as a two-port memory. This provides 23,040 total memory bits for optimal dual-port and single-port memory usage.

Gatefield design and development system. The GF250F and GF260F proASIC product families are supported by standard ASIC tools such Design Compiler, Verilog XL, QuickSim II, Motive, Leonardo, and so on. It utilizes the same VHDL and Verilog HDL descriptions that are targeted for the gate arrays and standard cells. Once the design has been synthesized and simulated, and timing analysis performed, it is ready for place and route. Gatefield provides ASICmaster™ as part of proASIC design flow, which accepts standard ASIC formatted netlists to perform design place-and-route operation into the selected device and provide back annotated delay information for simulation. Figure 3.14*b* shows the EDA environment that facilitates design conversion to ASIC for high-volume production.

3.3.4 Lucent Technologies ORCA FPGAs

Lucent Technologies (formerly AT&T Microelectronics) FPGAs, known as *optimized reconfigurable cell arrays* (ORCAs), are SRAM-based user-programmable gate arrays with basic architecture consisting of an array of programmable logic cells (PLCs), surrounded by programmable I/O cells (PICs).[19] Each PLC consists of the programmable function unit (PFU) and the programmable routing. The PLCs and PICs also contain routing resources and configuration RAM. The routing resources carry logic signals between PFU and I/O pads. All logic functions are performed in the PFU. Each PFU contains four 16-bit look-up tables (LUTs) and four latches/flip-flops (FFs). The ORCA FPGA functionality is determined by the internal configuration RAM. FPGAs internal initialization/configuration circuitry loads up the configuration data at power up. The configuration data is usually stored in an external serial ROM or EPROM.

The first-generation OR1C family ranges in density from devices such as OR1C03, with 3,500 to 4,300 usable gates and 160 user I/Os, to OR1C09, with 9,000 to 11,400 usable gates and 256 I/Os. The ORCA 1C family is based on 0.6 μm CMOS process. The second-generation ORCA 2C series represents an enhancement in usable gate capacity, speed, and power dissipation over the OR1C family. The ORCA 2C family is based on a 0.5 μm CMOS process and includes devices such as OR2C04, ranging in capacity from 4800 to 11,000 usable gates and 160 user I/Os, to OR2C40, with 43,200 to 99,000 usable gates and 480 user I/Os. It should be noted that the specified lower number in usable gate range is based on logic-only designs and assumes 48 gates per PFU, whereas the higher gate utilization number assumes 30 percent of the design as RAM-based and PFUs used as RAM are counted at four gates per bit. ORCA 2CA/OR2TA series are architectural migrations of older OR2C/OR2T families to 0.35 μm process, individually optimized for 5 V (2CA family) and low-power 3.3 V (2TA family).

ORCA OR2C/OR2T families. The basic architecture of ORCA OR2C/OR2T families consists of an array of 20 rows × 20 columns of programmable logic cells (PLCs) surrounded by programmable input/output cells (PICs) located on all four sides of the array. Figure 3.15 shows a block diagram of an ORCA OR2C15 array. The location of each PLC is identified by its row and column, e.g., the PLC in second row, and third column is BC. Similarly, each PIC is identified by its respective row/column, with PT (top) and PB (bottom) designating rows, and PL (left) and PR (right) designating columns, followed by a letter. Each PIC contains I/O buffers for interfacing with bonding pads and also contains the routing resources needed to connect signals from the bonding pads to/from PLCs. The combinatorial logic functions are implemented in look-up tables (LUTs). The LUTs medium- to large-grain architecture allows implementation of one to four logic functions. The major building blocks of OR2C are described below.

- *Programmable logic cells (PLCs).* The PLC consists of a programmable function unit (PFU) and routing resources. All PLCs in the array are identical. Figure 3.16 shows a block diagram of a PFU, which consists of four LUTs and four latches/FFs, and its I/O ports. The PFU has 19 external inputs and 6 outputs and can operate in several modes. The PFU uses three input data buses (a[4:0], b[4:0], wd[3:0]), four control inputs (c0, ck, ce, lsr), and a carry input (cin). There is a five-bit output bus (o[4:0]) and a carry-out (cout). The ports are referenced with a two- to four-character suffix to a PFU's location. The LUT can be configured to operate in one of the three general modes: (a) combinatorial logic mode, (b) ripple mode, and (c) memory mode.

  The functionality of the LUT is determined by its operating mode, as shown in Table 3-1 for combinatorial logic, ripple, and memory functions. The LUT can be divided into sub-LUTs, depending on the operating mode. The LUT consists of 32-bit half look-up tables called

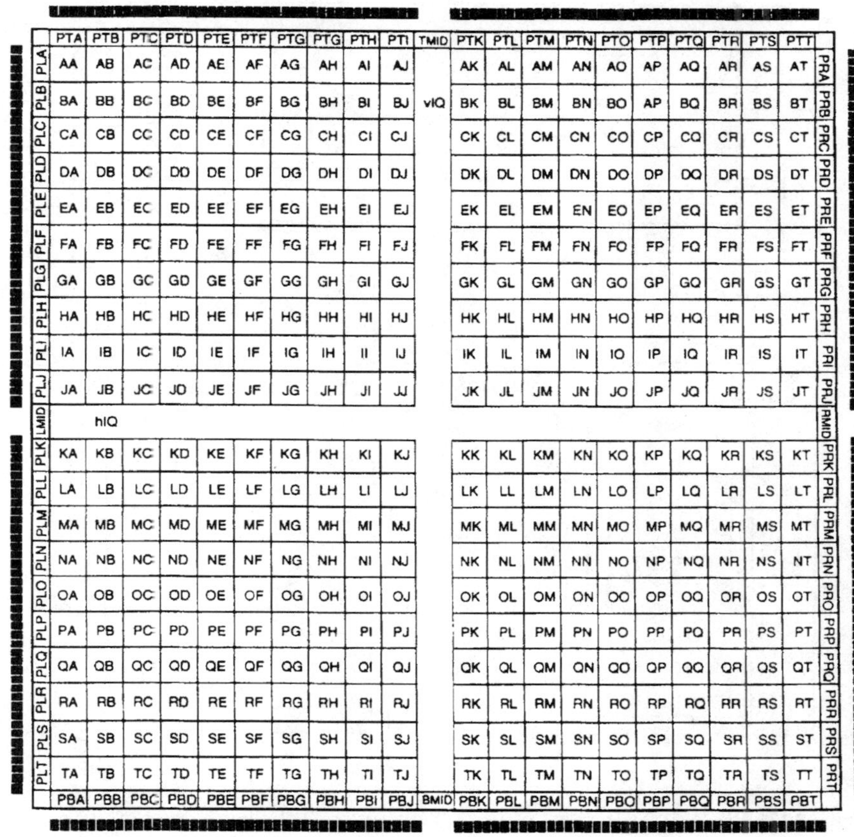

Figure 3.15 Block diagram of ORCA OR2C15 array *(from Ref. 19 with permission of Lucent Technologies).*

TABLE 3.1 ORCA2C FPU Look-Up Table Operating Modes *(from Ref. 21 with permission of Lucent Technologies)*

Mode	Function
F4A	Two functions of four inputs, some inputs shared (QLUT2/QLUT3)
F4B	Two functions of four inputs, some inputs shared (QLUT0/QLUT1)
F5A	One function of five inputs (HLUTA)
F5B	One function of five inputs (HLUTB)
MA	16×2 memory (HLUTA)
MB	16×2 memory (HLUTB)
R	Ripple—LUT

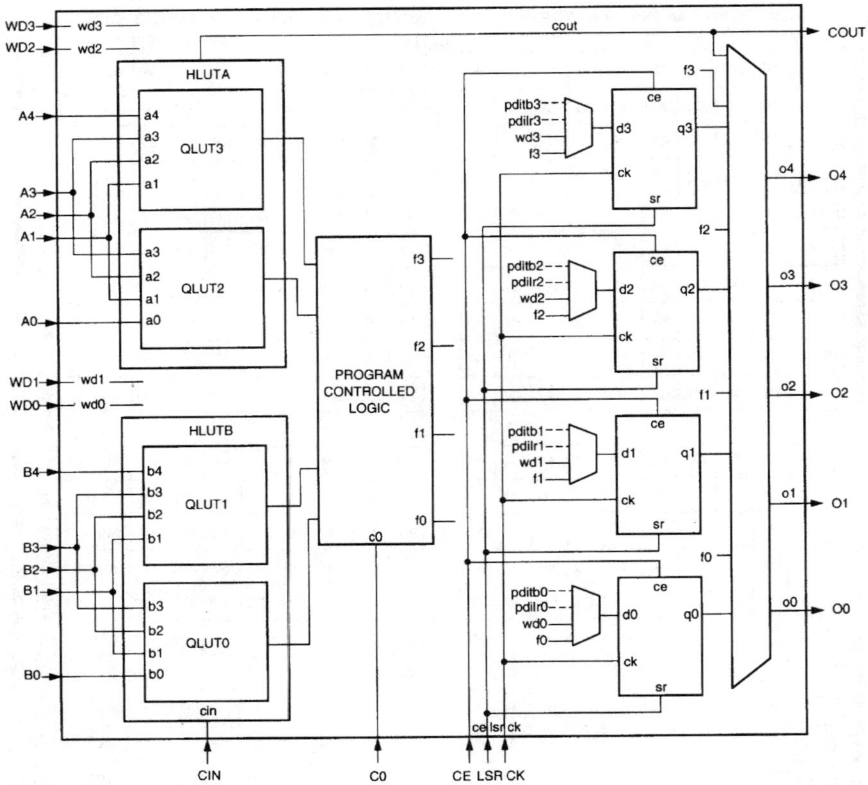

Figure 3.16 ORCA OR2C programmable function unit (PFU) block diagram *(from Ref. 19 with permission of Lucent Technologies).*

HLUTA and HLUTB. Each of these half look-up table is composed of two quarter look-up tables (QLUTs). HLUTA consists of QLUT2 and QLUT3, whereas HLUTB consists of QLUT0 and QLUT1. The outputs of QLUT0, QLUT1, QLUT2, and QLUT3 are f0, f1, f2 and f3, respectively.

The combinatorial logic mode uses a 64-bit LUT to implement any of the following Boolean functions: any single function of six inputs; any two functions of five inputs; four functions of four inputs (with some inputs shared); and three special functions based on two-input functions and C0.

In the ripple mode, the LUT can perform nibble-wide ripple functions with high speed carry logic. Each of the QLUTs have a dedicated carry out net for routing the carry to/from the adjacent QLUT. The fast arithmetic and counter functions can be implemented in one PFU, using the internal carry circuits. Similarly, each PFU has carry-in and carry-out ports for fast carry routing between the adjacent PFUs. The ripple mode is generally used in operations on two four-bit buses.

In the memory mode, the LUT in the PFU may be configured as either read/write or read-only memory. In the memory mode, each LUT can be used as a 16×2 memory. Each LUT is configured independently, allowing functions such as a 16×4 memory or a 16×2 memory in one HLUT and logic function of five input variables or less in the other HLUT. When the LUTs are used in memory mode, there are independent address, input data, and output data buses. To increase memory address locations (e.g., 32×4), two or more PLCs can be used.

The four latches/FFs in the PFU can be used in various configuration options, which are applicable to set of all four latches/FFs, or each of them as being independently programmable. The latches/FFs can be configured as either positive- or negative-level sensitive latches, or positive- or negative-edge triggered flip-flops. All latches/FFs in a given PFU share the same clock, and this clock can be inverted. The input into each latch/FF is either from the corresponding QLUT output (f[3:0]) or the direct data input (wd[3:0]).

- *PLC routing resources.* The routing resources architecture consists of intra-PLC routing and inter-PLC routing. The function of the intra-PLC routing resources is to connect the PFU's input and output ports to the routing resources used for entry to and exit from the PLC. These are the nets for providing PFU feedback, turning around corners, or switching from one type of routing resource to another. The inter-PLC routing is used to route signals between the PLCs.

  The routing resources consist of switching circuitry and metal interconnect segments. The metal lines that carry the signals are called *routing nodes* (R-nodes), which are connected to provide one or more of following three basic functions: signal switching, amplification, and isolation. A net running from a PFU or PIC output (source) to a PLC or PIC input (destination) consists of one or more R-nodes, which are connected by switching circuitry designated as configurable interconnect points (CIPs). The basic element in a CIP is one or more pass transistors, each of which is controlled by a configurable RAM bit. The process of connecting R-nodes requires three basic types of switching circuits: two types of CIPs, and bidirectional buffers (BIDIs). The two types of CIPs are the independent CIP or the multiplexed (or mutually exclusive) CIP, as shown in Fig. 3.17. A multiplexed set of CIPs contains two or more CIPs, only one of which can be on at a given time. An independent CIP has no such restrictions and can be on regardless of the state of other CIPs.

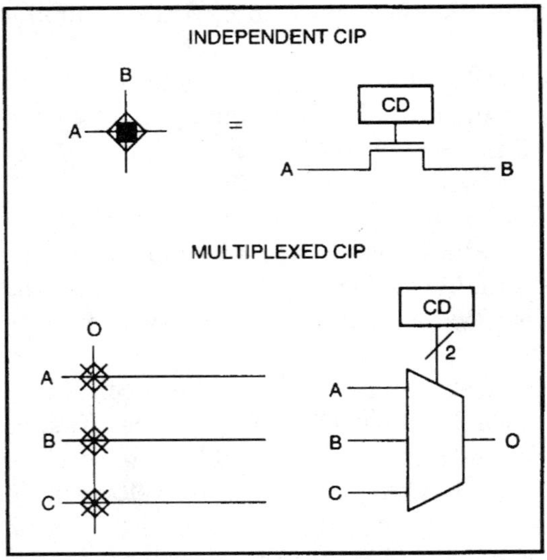

Figure 3.17 Configurable interconnect point (CIP) for connecting R-nodes *(from Ref. 19 with permission of Lucent Technologies).*

The bidirectional buffers provide isolation as well amplification for signals routed over long distance. These are also used to drive signals directly on to either the vertical or horizontal xL and xH-R nodes. The BIDIs can also be used to indirectly route signals through the switching R-nodes. In a given PLC, any number from zero to eight PLCs can be used. There are two sets of bidirectional R-nodes in the PLC, each set consisting of four bidirectional buffers, which are designated as BIDI and BIDIH, and have similar functionality. These BIDI R-nodes are used in conjunction with the xL R-nodes, and the BIDIH R-nodes are used in conjunction with the xH R-nodes.

Figure 3.18*a* shows the schematic of inter-PLC routing within one PLC, and Fig. 3.18*b* the global view of inter-PLC routing resources across the multiple PLCs. There are four sets of switching R-nodes in each PLC, one in each corner. Each set consists of five switching elements labeled as sul[4:0]. sur[4:0], sll[4:0], and slr[4:0], corresponding to the upper-left, upper-right, lower-left, and lower-right sections of the PFU, respectively. The switching R-nodes connect to the PFU inputs and outputs, as well as to the BIDI and BIDIH R-nodes. They also connect to both the horizontal and vertical x1 and x4 R-nodes (inter-PLC routing resources) in their specific corner.

One of the four sets of switching R-nodes can be connected to a set of switching R-nodes in each of the four adjacent PLCs or PICs. This

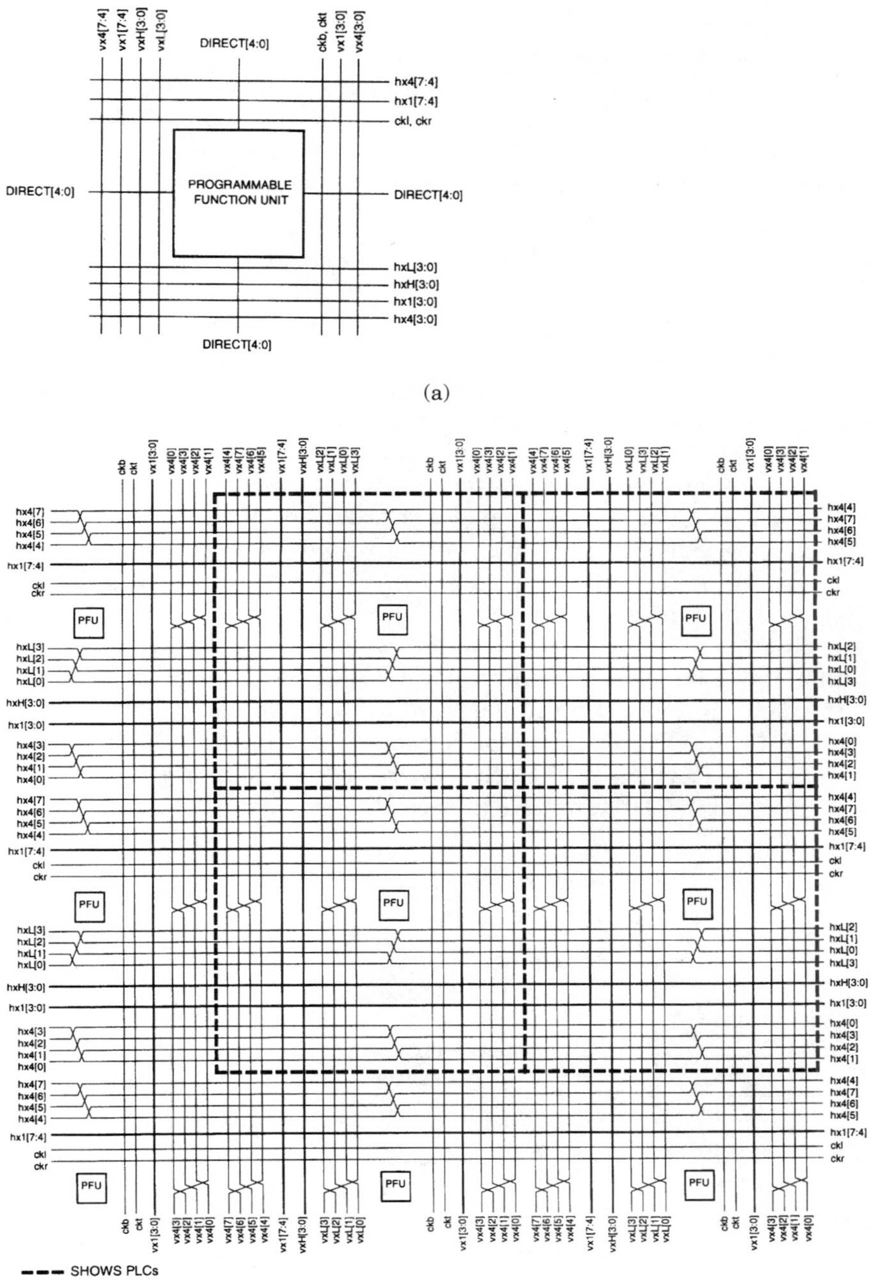

(b)

Figure 3.18 (a) Schematic of inter-PLC routing within one PLC and (b) global view of inter-PLC routing resources across multiple PLCs *(from Ref. 19 with permission of Lucent Technologies).*

allows direct routing of up to five signals without using the inter-PLC routing. The R-nodes occur in groups of four and differ in the numbers of PLC spanned. The x1 R-nodes span one PLC, the x4 R-nodes span four PLCs, the xH R-nodes span one-half the width (height) of the PLC array, and the xL R-nodes span the width (height) of the PLC array. All types of R-nodes run in both horizontal and vertical directions.

- *Programmable input/output cells (PICs) and routing resources.* The programmable input/output cells (PICs) are located along the periphery of each device, interface to the four bonding pads, and contain the necessary routing resources to provide an interface between I/O pads and the PLCs. Each PIC consists of input buffers, output buffers, and routing resources. The PICs output drivers have programmable drive capability and slew rates. The PIC routing is designed to efficiently route four-bit-wide buses. The connections between PLCs and the I/O pad are provided by two basic types of routing resources. These are routing resources internal to the PIC and routing resources used for PIC-PLC connection. Figure 3.19 shows a high-level view of these PIC routing resources.

  There is no direct connection between the inter-PIC R-nodes and the PLC R-nodes. All connections to/from the PLC must be made through the connecting R-nodes, which are perpendicular to the R-nodes in the PIC. Twenty-nine R-nodes in the PLC can be connected to the 15 R-nodes in the PIC. There are eight bidirectional (four TRIDI/four TRIDIH) buffers in each PIC.

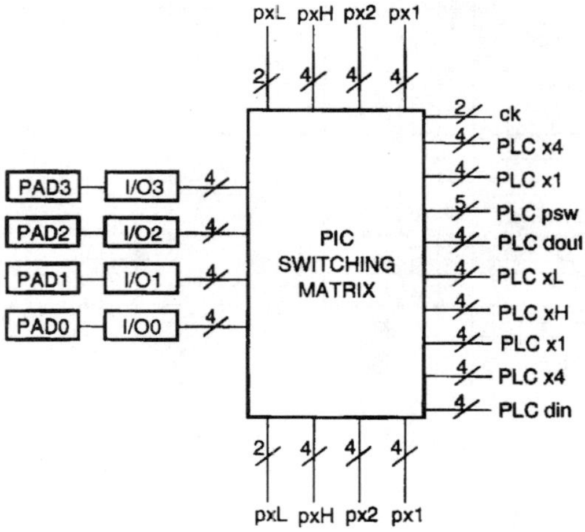

Figure 3.19 A high-level view of the PIC routing resources *(from Ref. 19 with permission of Lucent Technologies).*

- *Interquad routing.* In the ORCA 2C family, the PLC array is split into four equal quadrants, and interquad routing is used to route signals between the quadrants—especially to the quadrant in the opposite corner. The two types of interquad blocks are called *vertical quad blocks* (vlq), which run between quadrants on the left and right, and *horizontal interquad blocks* (hlq), which run between top and bottom quadrants. Both hlq and vlq blocks have the same logic resources.
- *Clock distribution network.* The ORCA OR2C family clock distribution scheme uses primary and secondary clocks. For the primary clock distribution system, if the clock signal is from an I/O pad, it can be driven into a clock R-node. The clock R-nodes do not provide clock signals directly to the PFU; they act as clock spines from which the clocks are branched to xL R-nodes. The xL R-nodes then feed the clocks to PFUs. A multiplexer in each PLC is used to transition from the clock spine to the branch. There are times when a primary clock is either not available or not a desirable choice, and a secondary clock is needed.

  A routing architecture enhancement has been made in the ORCA OR2C40, where each quadrant of the device is split into smaller arrays of PLCs called subquads. Each of these subquads is made up of a 4×4 array of PLCs (for a total of 16 per subquadrant), except at the outer edges of the array, which have less than 16 PLCs per subquad. The new routing resources, called subquad R-nodes have been added between adjacent pairs of subquads. Figure 3.20 shows block diagram of the subquad blocks containing a 4×4 array of PLCs, the interquad routing R-nodes, and the subquad routing R-nodes. All of the inter-PLC routing resources discussed for ORCA 2C family, continue to be routed between a PLC and its adjacent PLC, even if the two adjacent PLCs are in different subquad blocks.

ORCA OR2CA/OR2TA families. The ORCA OR2CA/OR2TA families are the new generation of FPGAs, based on the older 2C/2T families design and architecture migration to advanced 0.35/0.30 μm CMOS process.[20] These two processes are individually optimized for 5 V (2CA family) and low-power 3.3 V (2TA family) operations. These are some of the new architectural enhancements to improve performance and functionality:[21]

- Implementation for support of parallel multiplier
- Efficient support of synchronous single-port and dual-port memories
- Incorporation of capability for creation of large memory, using port-enable control and internal tristate buffers

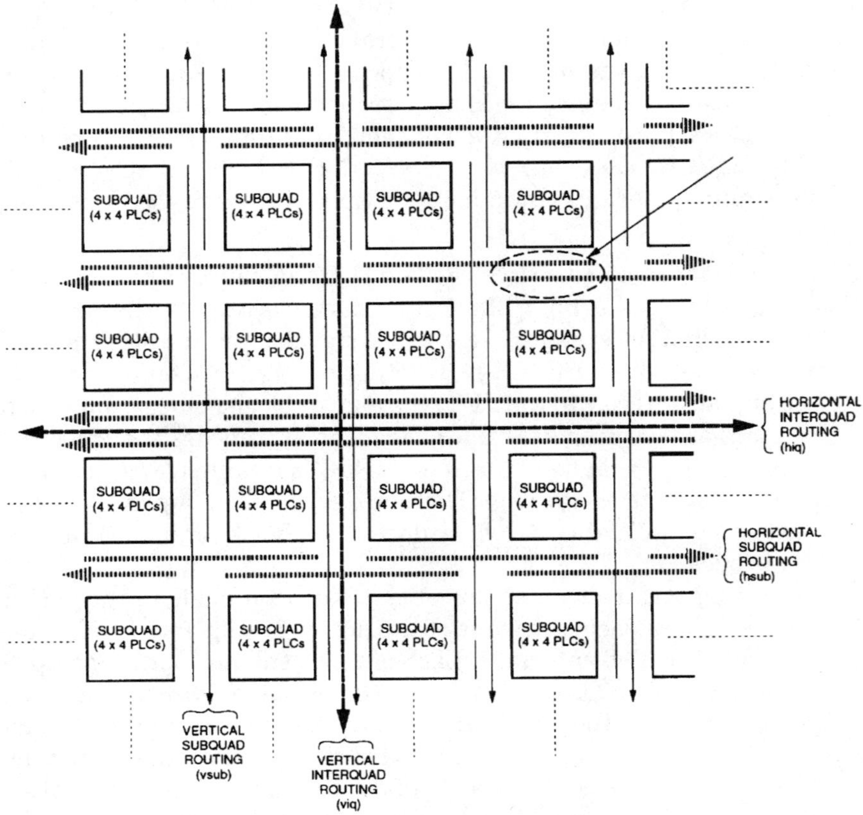

Figure 3.20 ORCA OR2C40 subquad block containing a 4 × 4 array of PLCs and subquad routing. The dotted circle indicates that additional details of horizontal interquad routing are available in Ref. 21. *(From Ref. 21 with permission of Lucent Technologies.)*

- Provide system integration with 5 V tolerant I/O buffers for interfacing with 3.3 V 2TA family

The multipliers are commonly used in DSP applications, and 2CA/2TA families provide a solution for implementing fast parallel multipliers by using ORCA logic blocks called *programmable function units* (PFUs). Figure 3.21*a* shows the PFU configured as 4 × 1 multiplier with X_{0-3} and Y_n as the unsigned 4-bit and 1-bit data, P_{in0}–P_{in3} and P_{out0}–P_{out3} as the input and output partial products, and C_{in} and C_{out} as the carry-in and carry-out signals. The PFU contains the four AND gates and the four-bit adder that are needed for the 4 × 1 multiplier. Since the fast-carry routing is used for much of the critical path, this means improved performance. Each PFU multiplier can be stacked together to form deeper and/or wider multipliers.

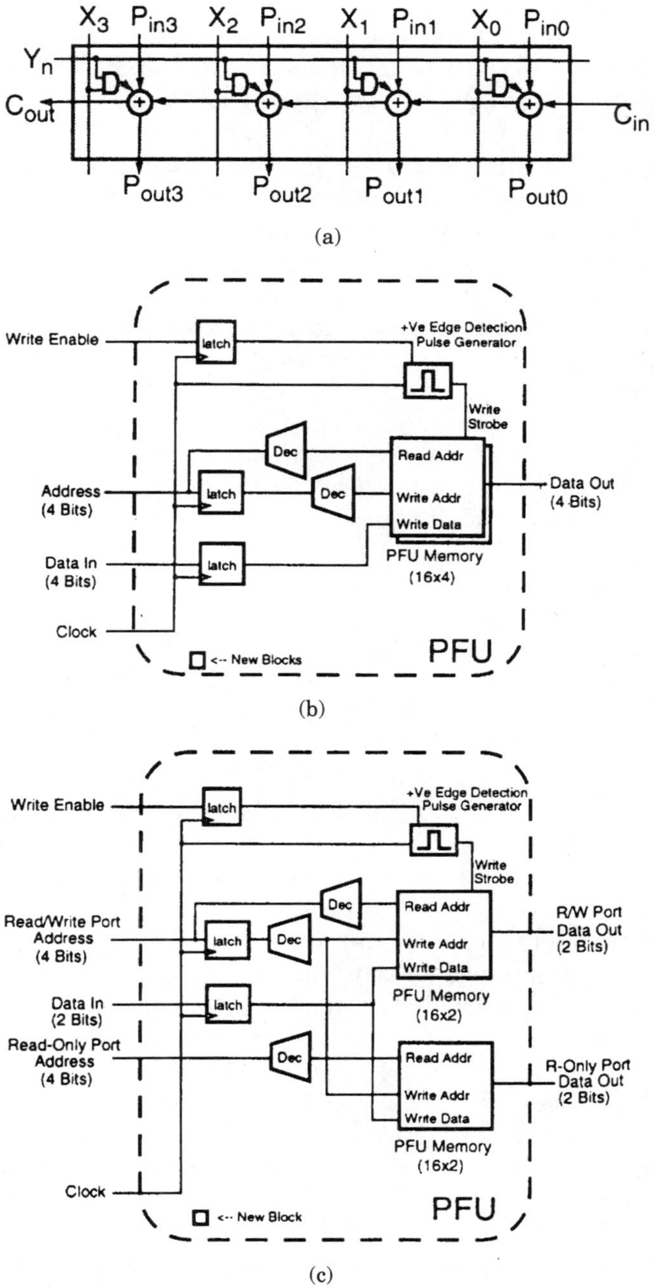

Figure 3.21 ORCA OR2CA/OR2TA families programmable functional unit (PFU) configured as (a) a 4 × 1 multiplier, (b) synchronous 16 × 4 memory, and (c) synchronous 16 × 2 dual-port memory *(from Ref. 21 with permission of Lucent Technologies).*

Most of the advanced digital systems require some form of memory, such as small registers for state machine designs, FIFOs for communication applications, and large RAMs for general data storage. To accommodate these different requirements, the logic blocks of some SRAM-based FPGAs can be converted into distributed asynchronous memories and combined appropriately for implementation of various memory requirements. Figure 3.21*b* shows OR2CA/OR2TA PFU support for implementation of 16 × 4 synchronous memory operations.

All of the incoming address, data, and write enable signals are latched by the clock while memory read operations remain asynchronous. Then, if the write enable is active, an internal write strobe is generated from the same clock edge to write the latched data into the latched address location. For the next half-cycle, all latches become transparent to sample the incoming signals. This design simplifies memory timing requirements, since all setup/hold times are now referred to one clock edge. This makes it easier for the output data to be latched by the configurable PFU flip-flops to meet the timing requirements of the next state.

Figure 3.21*c* illustrates the implementation of a 16 × 2 dual-port memory from the available 16 × 4 single-port memory. In this mode, the 16 × 4 single-port memory is divided into two banks. The write address and data buses of the two banks are connected together so that any arriving data is written into both the banks. The same address bus is also connected to the read address of the first bank, as in the single-port mode. However, another address bus is connected to the read address of the second bank so that it can be accessed independently. Therefore, each PFU can be constructed as a 16 × 2 dual-port memory, with one port for both read and write operations and the other port for read only.

In case memory depth of more than 16-bit words is required, multiple PFU memories can be grouped for implementation. As an example, for implementation of a 64 × 4 memory, four 16 × 4 PFU memories are needed.

ORCA OR3CA/OR3TA families. Lucent Technologies' latest introduction is the 3C series of SRAM-based FPGAs architecture, which is being merged with standard cell blocks to achieve greater system functionality and higher levels of performance.[22] The combined technology allows designers to define key functions in the standard cell blocks early and use programmable logic for the implementation of last-minute design changes. The OR3C/OR3T families are basically an extension of the OR2C/OR2T families, with several enhancements that allow the chips to operate with system clock rates from 10 to 80 MHz, and internal blocks of RAM to be accessed at rates over 100 MHz. To ease the system interface requirements, the chips include a dual-use microprocessor interface that can be employed to perform configuration, readback, device control, and device status.

A major improvement in the internal architecture is the addition of new blocks called *supplemental logic and interconnect cells* (SLICs). These blocks have a complexity roughly equivalent to a 22V10 PAL (and look very much like PALs), which gives OR3C/OR3T series a combined FPGA and CPLD architecture. These SLICs can implement functions that FPGAs can not perform efficiently, such as the implementation of 10-bit decode, 10-bit AND/OR functions, and so forth. Another improvement is in the clock routing to boost the speed. Its system clock allows any input pin to serve as a clock, and express clock lines have been added to the OR3C/OR3T array. The output of one look-up table (LUT) can be routed to another LUT in the same block without going to external routing pool, which provides 30 percent speed improvement in critical paths with cascaded multiple LUTs. There is also some diagonal routing capability provided for the nearest-neighbor connections.

ORCA OR3C/OR3T series FPGAs consist of three basic elements: programmable logic cells (PLCs), programmable input/output cells (PICs), and system-level features. An array of PLCs is surrounded by PICs. Each PLC contains a programmable function unit (PFU), a SLIC, local routing resources, and configuration RAM. Figure 3.22 shows simplified block diagrams of a PFU and a SLIC.

Each PFU contains eight 4-input (16-bit) look-up tables (LUTs), eight latches/flip-flops (FFs), and one additional flip-flop. The LUTs may be used individually or combined to produce 5-input LUTs, or cascaded to achieve complex functions of up to 21 inputs using the new soft-wired LUT connections. Each PFU may also be configured as a synchronous 32 × 4 single- or dual-port RAM or ROM. The FFs (or latches) may obtain input from LUT outputs or directly from invertible PFU inputs, or they can be tied high or low.

Most of the FPGA logic operations are performed in the PFU, whereas the decoders, PAL-like functions, and three-state buffering can be performed in the SLIC. The SLIC is connected to PLC routing resources and the outputs of the PFU. It contains three-state, bidirectional buffers and logic to perform up to a ten-bit AND function for decoding, or an AND-OR-INVERT to perform PAL-like functions. The PICs can be used to register signals, perform input demultiplexing and output multiplexing, and other functions on two output signals.

The OR3C/OR3T family is based on 0.35 μm, 4-level metal technology and ranges in density from devices such as OR3C/OR3T30, with 19k to 44k usable gates, 224 user I/Os, and 14 × 14 array size, to the OR3T225, with 140k to 320k usable gates, 608 user I/Os and 38 × 38 array size. It should be noted that the lower number specified in us-

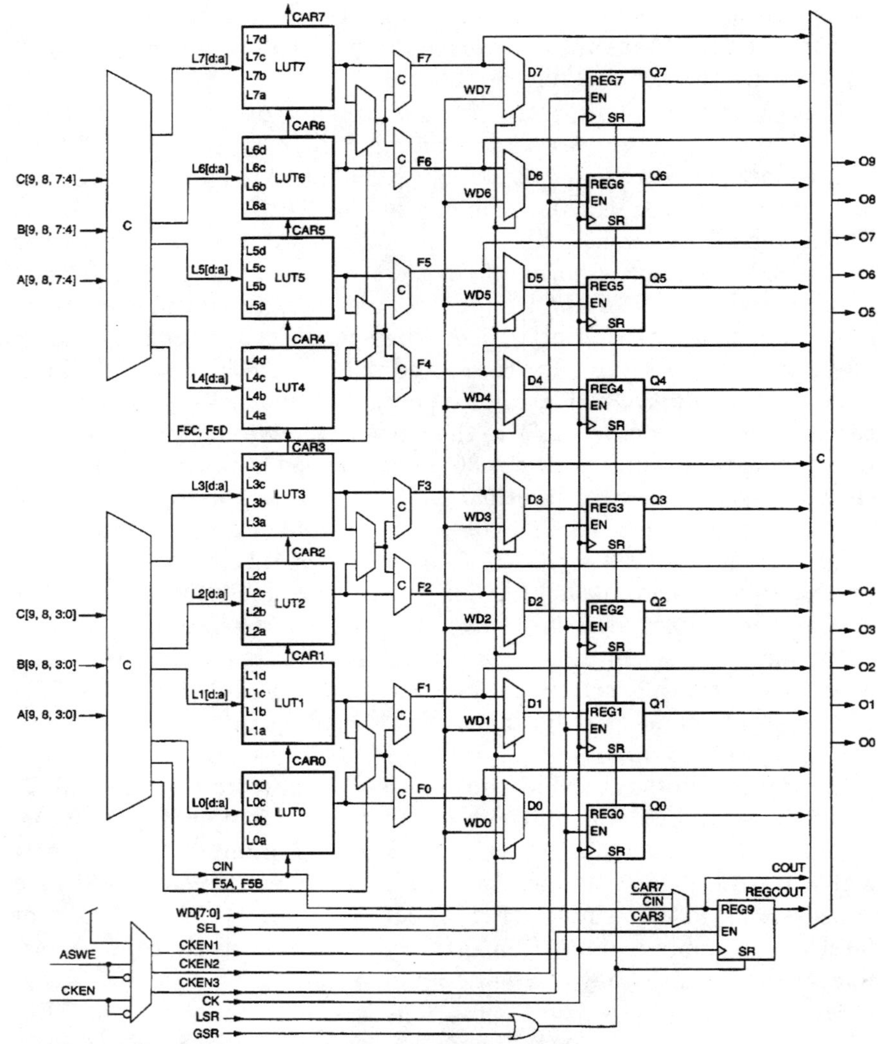

Note: LSR can be disabled on a per-nibble basis, and GSR can be disabled per PFU.

Figure 3.22a ORCA OR3C/OR3T family block diagram, programmable function unit (OFU) *(continues).*

able gate range is based on logic-only designs and assumes 96 gates per PFU, whereas the higher gate utilization number assumes 30 percent of the design to be RAM-based, and PFUs used as RAM are counted at 4 gates/bit.

ORCA FPGA design and development system. Lucent Technologies supports a number of different development system products, including

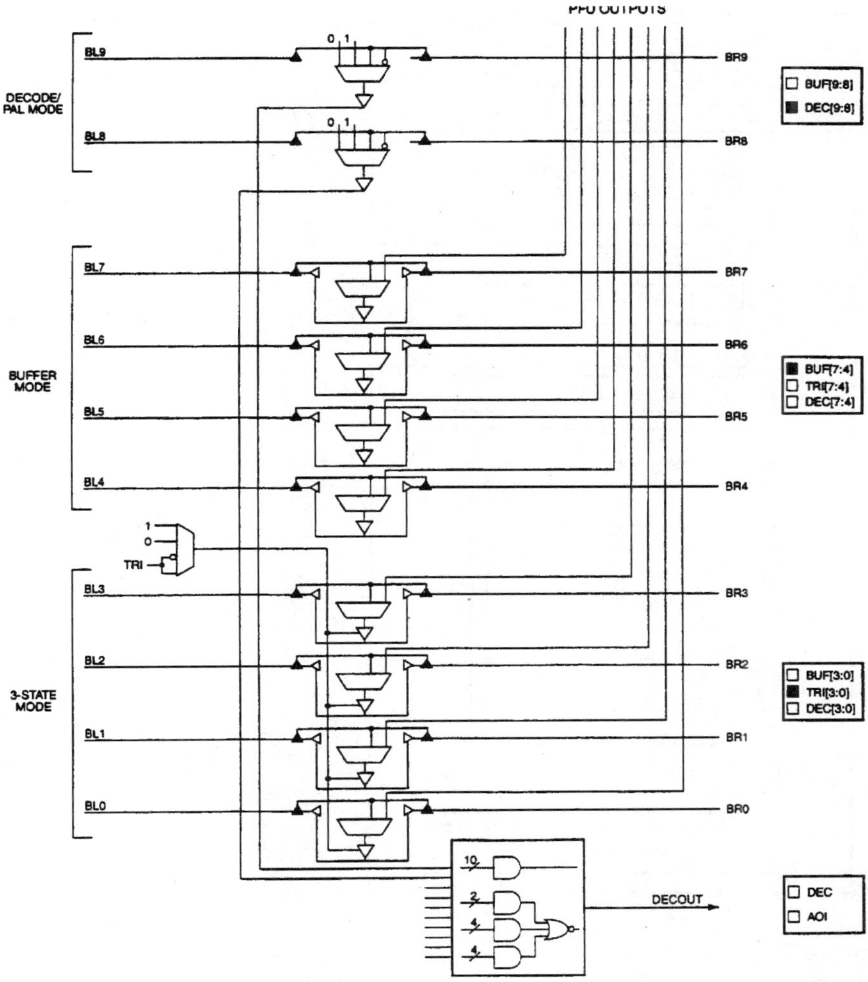

Figure 3.22b ORCA OR3C/OR3T supplemental logic and interconnect cells (SLICs) *(from Ref. 22 with permission of Lucent Technologies).*

popular CAE tools, libraries, and interfaces, which are optimized for the ORCA family architectures. As the designs increase in complexity, the use of high-level description languages (VHDL or Verilog HDL) and synthesis tools can increase productivity. A proprietary tool, called *synthesis compiler for user programmable arrays* (SCUBA), includes libraries, design flows, and algorithms developed for efficient synthesis of FPGA architectures.

Figure 3.23 shows the flowchart of ORCA Foundry development system. After a design is complete, it must be implemented into a selected architecture, which usually requires the three major steps of mapping,

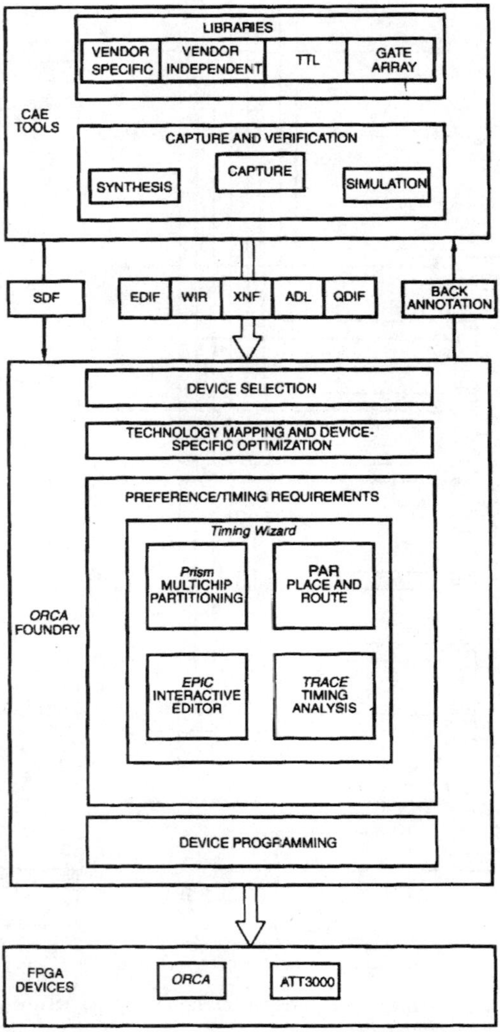

Figure 3.23 ORCA Foundry development system (from Ref. 21 with permission of Lucent Technologies).

placement, and routing. ORCA Foundry can be used to perform these three functions, as well as static timing analysis and for generating the bit stream used for device configuration. Once the design has been implemented, a bit stream of the file that contains the programming information for the RAM cells can be created and downloaded into the FPGA. Once downloaded, it causes the FPGA to perform the desired function.

The design verification and post-layout timing simulation can be performed using any of the simulators that were used for functional simulation. The ORCA Foundry development system provides an interface to these third-party simulators, which allows timing information to be transferred to the simulator after the map, place, and routing steps have been completed. The timing simulation usually requires the same input stimulus as the functional simulation and is used for debugging timing implementations that are too slow, in race conditions, or have setup/hold timing violations.

3.3.5 Motorola programmable array (MPA) family

MPA 1000 family architecture. The Motorola programmable array (MPA) 1000 is a family of reprogrammable SRAM based devices manufactured on a standard submicron, three-layer metal CMOS process with logic capacity ranging from 3500 to more than 22,000 equivalent FPGA gates.[23] The logic resources in the MPA 1000 family are fine grain, with each logic cell holding a single gate or a storage element, which allows the logic block size variation in accordance with the specific function being implemented. The MPA family interconnect structure is partitioned into four levels: (1) device, (2) quadrant, (3) zone, and (4) local.

Figure 3.24*a* shows the architectural overview of MPA 1000 family. At the highest level of hierarchy, the logic array is divided into four quadrants, each consisting of 2×2 (MPA 1016), 3×3 (MPA 1036), 4×4 (MPA 1064), and 5×5 (MPA 1100) zones, as shown in Fig. 3.24*b*. Each zone consists of 10×10 block of logic cells surrounded by port cells at its periphery. These port cells provide programmable connections between the zones and global resources to the global buses. The interquadrant switches provide programmable connection points between the quadrants. The I/O cells are present around the chip periphery, ranging from 80 for MPA 1016 to 200 for MPA 1100.

Figure 3.24*c* shows the core cell structure. Each of the two logic inputs to the core cells originates from one of seven sources, of which five of the sources comes from local-level and two from zone-level interconnect. The output of each core cell may be connected to eight other core cells using local interconnect and may be connected to four zone-level buses. The core cell may be used to repower a signal, provide additional routing, and allow a reprogrammable means of adjusting the routing resources to accommodate individual design requirements. The core cell has three states: repowering buffer, primary function, and secondary function. All core cells contain the ability to invert the inputs and perform a two-input AND gate function with programmable input inversion. The core cells are multifunctional, and a group of

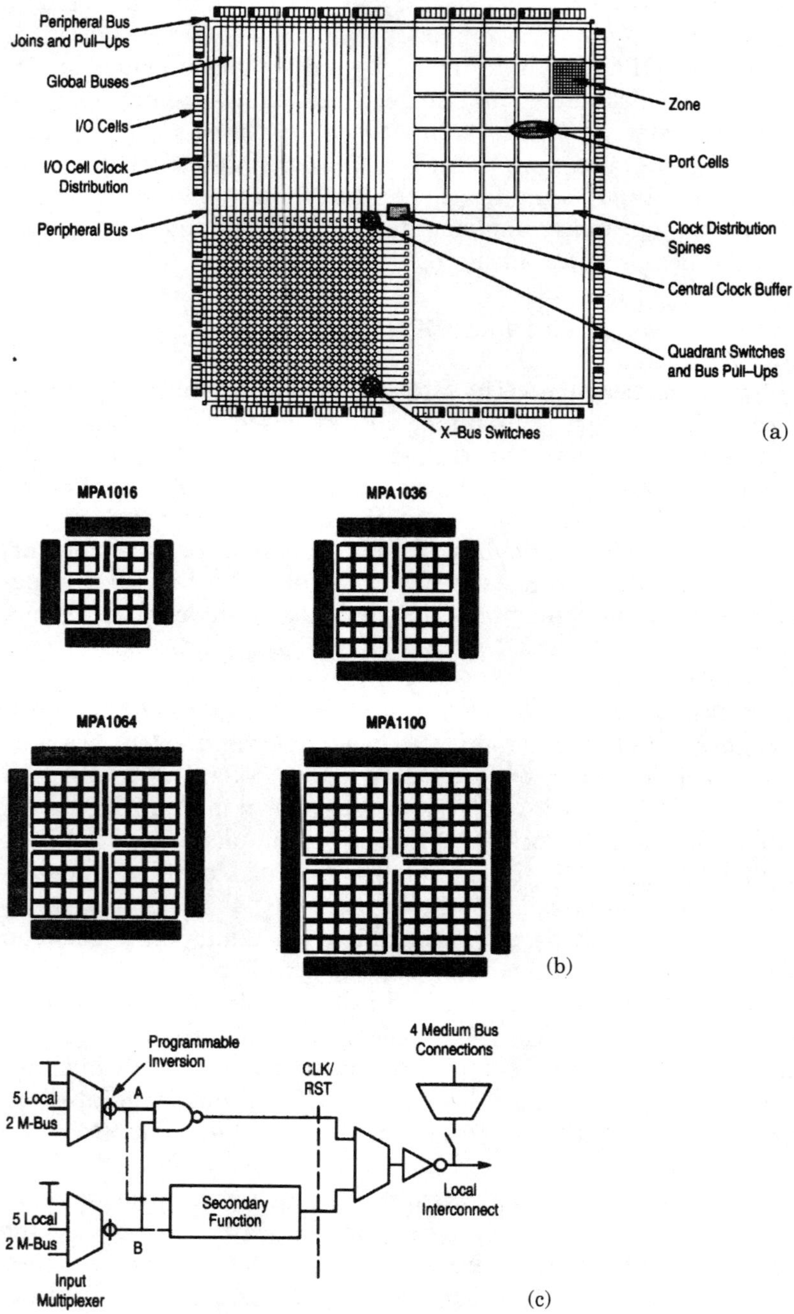

Figure 3.24 MPA 1000 family: (a) architectural overview for MPA 1016, MPA 1036, MPA 1064, and MPA 1100, (b) logic array/zones, and (c) core cell structure *(from Ref. 23 with permission of Motorola Corp.).*

four cells is defined as a *tile*. Each of the core cells within a tile has one of four secondary functions: XOR (two each), register (DFF or latch with CK enable, and set or reset), and wired-OR. Additional circuitry allows core cells to be grouped together to form higher-level functions such as a full adder, two-input MUX, T flip-flop, or JK flip-flop.

Motorola MPA 1000 family routing resources fall into three hierarchical categories: local interconnect, medium bus, and global bus. Additional interconnect resources include special 'X' buses and wired-OR buses. The port cell is the major connection exchange at the zone boundary, providing connection between all types of buses except local interconnect. Typical uses are connection of the medium buses of neighboring zones and connection of the zone's medium buses to the global routing. The vertical port cell also delivers clocks and resets into each zone after receiving them from the dedicated primary clock tree. The routing sources and their major functions, including port cells, local interconnect, medium buses, global buses, and X-buses, are briefly described and shown in Fig. 3.25.

The MPA 1000 clock distribution is implemented through a dedicated network consisting of eight dedicated clock input lines connected to the I/O cells on each device edge, a central clock buffer, a distribution comb structure, zone corner clock selection cells, and zone port cells along the top of each zone. The zone corner cell selects two of the eight lines for zone clocks and two of the eight lines for zone reset. Zone registers are connected to these clocks and reset signals through the top row of the port cells. The comb extends into the I/O cells via a similar clock selection cell, which is attached to each group of five I/O cells. This group is called an I/O zone.

The configuration logic is responsible for loading the external configuration data into the device. There are two basic modes: (a) micro mode, and (b) boot from ROM (BFR) mode. The micro mode allows the devices to be configured from a microprocessor or a similar system through a conventional peripheral interface. The BFR method, which is commonly used, allows loading up of configuration information through the serial EPROM.

MPA 1000 design and development system. The MPA design system is a bridge between the design capture and simulation environment provided by third-party vendors (e.g., Synopsys, Viewlogic, Exemplar, and so on) and Motorola FPGAs. The designs can be captured using schematics, high-level languages such as VHDL and Verilog-HDL, or a combination of these entry methods using commercially available design capture and logic synthesis software and the appropriate interface kit. Alternatively, existing designs can be retargeted from other PLDs to the MPA device. The design importation begins with a netlist

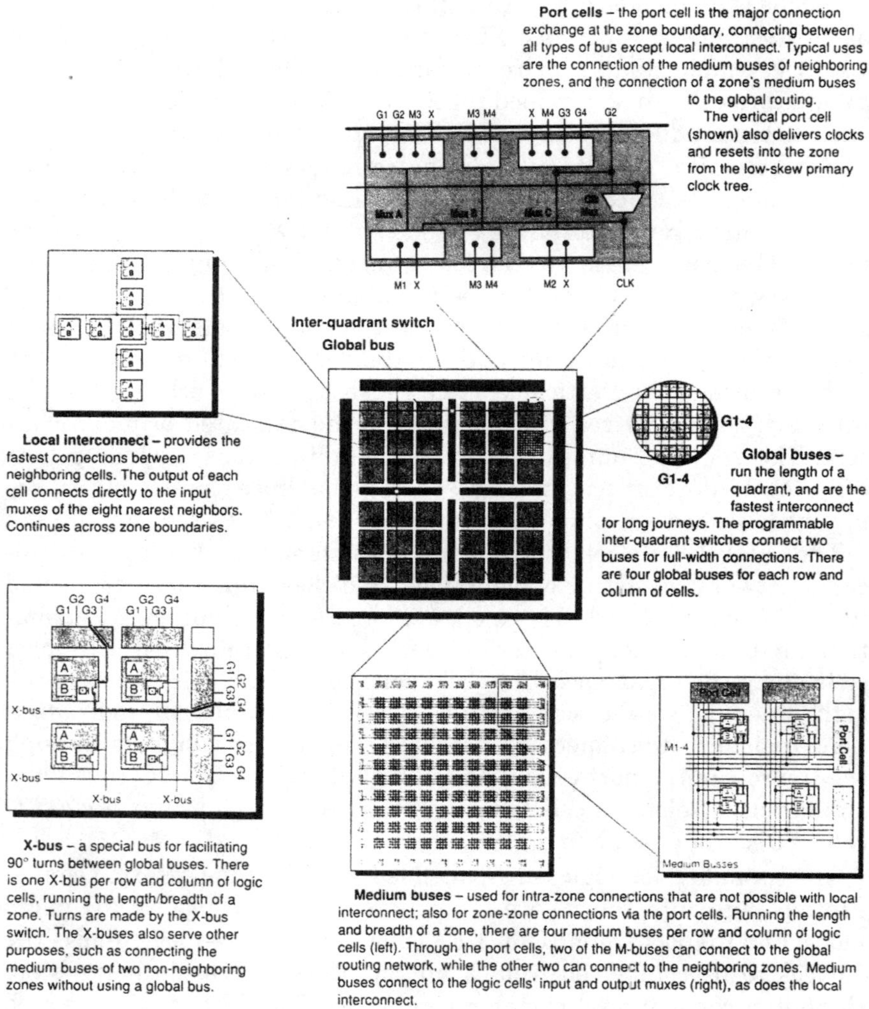

Figure 3.25 MPA 1000 family routing resources including port cells, local interconnect, medium buses, global buses, and X-buses *(from Ref. 23 with permission of Motorola Inc.).*

and an optimal clock specification file. The clock specification file provides a mechanism for the user design capture tools to document system-level timing requirements.

A design is automatically analyzed, optimized, transformed into MPA cells, partitioned, placed, and routed based on timing constraints for every path in the design. The Autolayout process makes use of the hierarchical organization of the MPA device for an implementation that meets timing requirements.

3.3.6 QuickLogic FPGAs

The architecture of QuickLogic pASIC families of FPGA is based on amorphous silicon antifuse technology called the ViaLink as the programmable element (see Sec. 3.2 for ViaLink technology discussion) and an array of flexible logic cells that have been optimized for implementation of high-speed arithmetic, data path, state machine, random, and glue logic functions.[16] The pASIC 1, pASIC 2, and recently introduced pASIC 3 families of FPGAs are briefly discussed below.

pASIC 1 family. The pASIC 1 device architecture consists of an array of user-configurable logic building blocks called *logic cells,* which are set up in a grid of metal wiring channels. The array contains internal logic cells, I/O cells, and dual-layer vertical and horizontal metal routing channels. The output of any cell may be programmed to connect to the input of any other cell through the ViaLink elements located at the wire intersections. The single line between logic cell represents channels containing up to 22 wires. Each of the internal logic cells has the logic capacity of 30 "gate-array" gates. The pASIC 1 family ranges in density from 2000 to 14,000 usable PLD gates devices, such as the QL8x12B, with 96 logic cells and 160 flip-flops maximum and the QL24x32B, with 768 logic cells and 948 flip-flops maximum.

Figure 3.26 shows the pASIC 1 internal logic cell, which is a general building block consisting of two six-input AND gates, four two-input AND gates, three 2-to-1 multiplexers, and a D flip-flop. In addition to the dedicated flip-flop, logic gates in each cell can be configured to provide two latches. This logic cell offers up to 14-input wide gating functions. It allows implementation of all possible Boolean functions of up to three variables as well as many other functions of up to 14 variables. The multiplexer output feeds the D-type flip-flop, which can also be configured to provide J-K, S-R, or T-type functions, as well as count with carry-in. Two independent SET and RESET inputs can be used to asynchronously control the output condition.

For pASIC 1 devices, three types of I/O structures are provided to configure buffering functions at the external pads: (1) bidirectional I/O cell, (2) dedicated input (I cell), and (3) clock input (I/CLK) cell. Multiple logic cells can be programmed to form a complex logic function by interconnection through the routing channels.

The routing resources include four types of signal wires: (1) segmented wires, (2) express wires, (3) clock wires, and (4) quad wires [for QL16x24B and QL24x32B devices only]. The segmented wires are used mostly for local interconnections and have a ViaLink element, known as *cross link,* at every crossover point. They may also be connected to the segmented wires of cells above and below through the Vi-

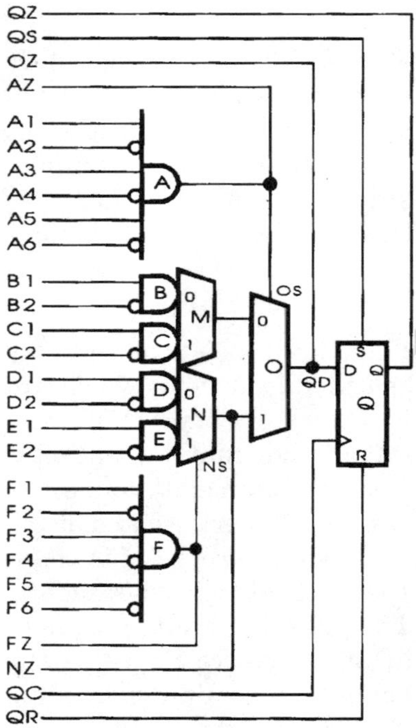

Figure 3.26 QuickLogic pASIC 1 family internal logic cell block diagram *(from Ref. 16 with permission of QuickLogic Corp.).*

aLink elements called *pass links*. The express lines are similar to segmented wires, except they are not divided by the pass links. Express wires may also be used to deliver the clock signals into the multiplexer region of the cell for combinatorial gating.

The dedicated clock wires are lightly loaded with only three links per cell to distribute high-speed clock edges to the flip-flop CLK, SET, and RESET pins. The quad wires are similar to segmented wires, since they are also used for local interconnect but, instead of having pass links above and below each cell, they have pass links every fourth logic cell. The horizontal wiring channels, called rows, provide connections via cross links to other columns of logic cells and the chip periphery. The programming of appropriate ViaLink elements allows electrical connection to be made from any logic cell output to the input of any other logic or I/O cell.

pASIC 2 family. In the pASIC 2 family, the densities range from 5000 to 16,000 usable PLD gates devices, such as the QL2003, with 192 logic cells and 408 flip-flops maximum, to the QL2009, with 672 logic cells and 1236 flip-flops maximum.[16] The pASIC 2 family 0.65 μm triple-layer metal process differs from pASIC 1 two-layer metal process, since the routing tracks and amorphous ViaLink antifuses are placed between the second and third layers of metal, which allows considerable reduction in die size for comparable densities (see Fig. 3.5*a*). The routing and capacitance is further reduced through a change in the programming link accessing circuitry.

The pASIC 2 logic cell consists of two six-input AND gates, four two-input AND gates, six 2-to-1 multiplexers, and one D flip-flop with asynchronous preset and clear controls. The total fan-in of the cell is 29 (including register control lines), which allows implementation of up to five independent fine-grained functions, one high-speed 16-input function or any intermediate combination. Complex functions such as counters and state machines can be implemented within a single level of logic, allowing them to run as fast as 150 MHz. Smaller functions such as combinatorial logic synthesized from the HDL may be mapped in parallel into independent fragments. The pASIC 2 family includes I/O cells with individually controlled output enable signals, allowing them to have input, output or bidirectional operation.

The pASIC 2 family features full JTAG IEEE 1149.1 scan path testability and is offered in both 5.0 V and 3.3 V supply versions. These devices provide a higher I/O to logic and register ratio than the pASIC 1 family, which allows more efficient implementation of high I/O applications such as the bus interfaces.

pASIC 3 family. The latest QuickLogic introduction is pASIC 3 family in 0.35 μm and four-layer metal process with device performance levels approaching those normally found in MPGAs. This new pASIC 3 family includes devices with logic capacities ranging from 12,000 to 100,000 usable PLD gates and up to 15,000 bits of embedded RAM. The 3.3 V devices are capable of implementing 33 MHz PCI 2.1 master and slave applications with zero-wait states.

The pASIC 3 uses variable-grained logic cell architecture in which a single logic cell can be configured with up to 16 inputs for implementation of wide combinatorial functions, or the cell can be configured as up to 5 completely independent logic elements for higher device utilization when implementing multiple small logic structures. This flexibility is especially effective for HDL based designs and logic synthesis tools that prefer small logic elements, typical of gate array structures.

An enhancement being planned for pASIC 3 family in mid-1998 is to offer configurable modules of embedded RAMs for higher performance and system-level integration.

QuickLogic design and development tools. Software support for pASIC 1 and pASIC 2 families is available through two basic environments. The QuickWorks suite provides a complete FPGA software solution from design entry and logic synthesis to place-and-route simulation. The QuickWorks™ Toolkit includes VHDL/Verilog synthesis capability and other features such as HDL editing, comprehensive Verilog simulation plus QuickLogic's SpDE software. The QuickTools is targeted toward the designers using a third-party CAE environment such as Synopsys, Mentor Graphics, Cadence, Viewlogic, and so on, for design entry, synthesis, or simulation. The QuickTools reads Verilog, VHDL, and EDIF netlists and provides design optimization, place and route, timing analysis, and back-annotation support.

The design support for pASIC 3 family is available through new releases of QuickWorks and QuickTools design software packages. The QuickChip third-party design packages also include support for pASIC 3 devices.

3.3.7 Xilinx FPGAs

Logic cell array (LCA) families general architecture. Xilinx user programmable gate arrays consists of three major configurable elements: configurable logic blocks (CLBs), input/output blocks (IOBs), and interconnections.[24] The CLBs provide the functional elements for user's logic implementation and can be configured as RAM or ROM also. The IOBs provide the interface between the package pins and internal signal lines. The programmable interconnect resources provide routing paths to connect the inputs and outputs of the CLBs and IOBs into appropriate networks. The customized design configuration is established by programming internal static memory cells, which determine the logic functions and interconnections implemented in the LCA device.

The first generation of Xilinx devices in the XC2000 family featured CLBs consisting of a combinatorial function generator capable of implementing four-input Boolean logic function and a single storage element. The second generation of Xilinx devices in the XC3000 family provided several enhancements over the XC2000 family, such as expanding CLB functionality to implement wider Boolean logic functions and including a second flip-flop in each logic block. Xilinx is not recommending XC2000 and XC3000 families for new designs.

The third generation of Xilinx the basic XC4000 family extends the LCA architecture by providing more powerful logic blocks, enhanced IOBs, and interconnection resources. XC4000E series is industry standard 5 V devices based on 0.5 µm, triple-layer metal (TLM) CMOS process with typical gate range up to 45,000 gates. Two versions of this family are currently recommended for new designs by Xilinx, as follows:

1. *XC4000XL.* The XC4000XL family has more routing resources than the older XC4000/E families and is therefore much easier to route, and it provides higher performance. The usable gates range from 1500 to 180,000 gates (for device types XC4002XL to XC4085XL) and the number of CLBs from 64 to 3136.
2. *XC4000XV.* This is a 2.5 V family recently introduced by Xilinx, based on a 0.25 μm, five-layer metal process using stacked vias. This family will include device offering from XC40125XV to the XC40250XV with usable gate counts ranging from 265k to 500k gates. The internal operation of this family is 2.5 V, but it uses a separate 3.3 V supply for I/Os and is 5 V compliant. The XC4000XV has more routing resources than the XC4000XL family.

The LCA XC6000 family is an enhanced version for microprocessor interface applications and fast configurations. The devices in this family range from XC6209, with typical usable gate range of 9k to 13k, to XC6264, with 64k to 100k usable gate range. This family, used for reconfigurable logic applications, is discussed in Chap. 6.

The latest introduction by Xilinx is the next-generation Virtex FPGA family that will be developed through 1999 and beyond. This new family is based on a 0.25/0.18 μm, five-layer metal process that can provide device density to 400,000 logic gates (or 32,000 logic cells) and higher, with system clock speeds exceeding 100 MHz. This family architecture, developed with logic synthesis users in mind, uses simplified logic mapping and greater routing resources. It uses three levels of memory hierarchy: CLB-based memory, block embedded memory, and off-chip memory access. It has configurable I/Os for mixed-signal applications and PCI compliance. Virtex FPGA family is targeted to IP core solutions and is discussed in more detail in Chap. 8.

XC 4000 family. A number of architectural enhancements have been made in the Xilinx XC4000 family, especially in the configurable logic block (CLB), which is surrounded by a versatile set of routing resources, resulting in more "effective gates per CLB".[24] Three major elements of the XC4000 architecture are CLBs, I/O blocks (IOBs), and programmable interconnect. These are described below, followed by a summary of enhancements made in Xilinx third-generation XC4000 family, relative to the first- and second-generation families.

- *Configurable logic blocks (CLBs).* Figure 3.27*a* shows the block diagram of the XC4000 CLB. Each CLB consists of a pair of flip-flops and two independent four-input function generators. Thirteen CLB inputs and four CLB outputs provide access to the function generators and flip-flops. These are more than double the number of I/Os

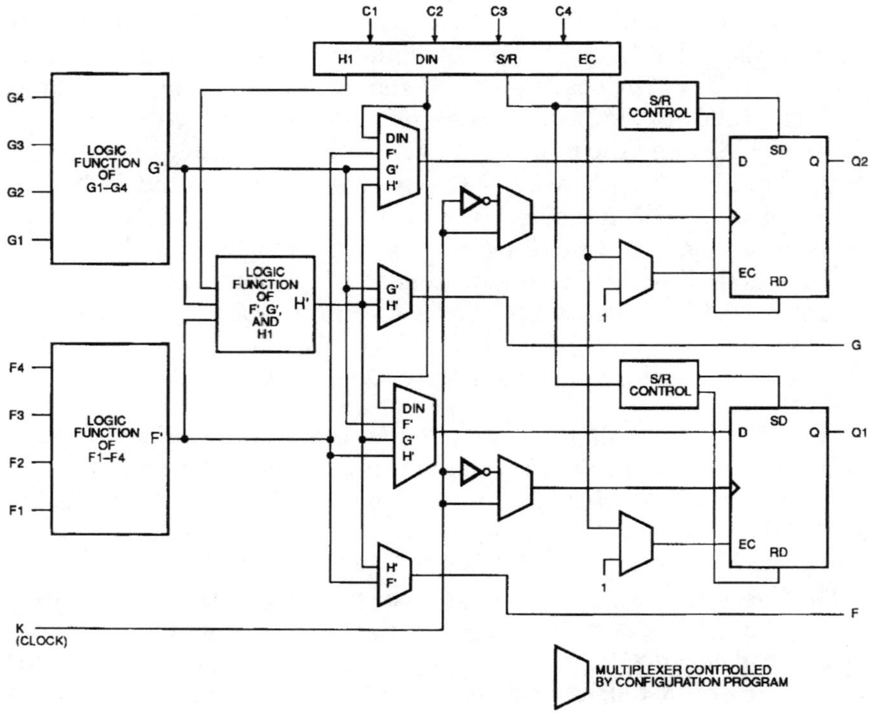

Figure 3.27a Xilinx XC4000 block diagram, configurable logic block (CLB) *(continues)*.

available on the XC3000 family and are connected to the programmable interconnect resources outside the block. Four independent inputs are provided to each of the two function generators (F1–F4 and G1–G4). These function generators, whose outputs are labeled F and G, are implemented as memory LUTs, and each is capable of implementing any arbitrarily defined Boolean function of their four inputs. A third function generator, H´, can implement any Boolean function of its three inputs: F´, G´, and a third input from outside the block (H1). The signals from the function generators can exit the CLB on two outputs: F´ or H´ can be connected to the F output, and G´ or H´ can be connected to the G output. Therefore, a CLB can be used to implement any of the following configurations: any two independent functions of up to four variables, any single function of up to five variables, any function of four variables together with some functions of five variables, or even some functions of up to nine variables.

The two storage elements in the CLB are edge-triggered D-type flip-flops with common clock (K) and enable clock (EC) inputs. A

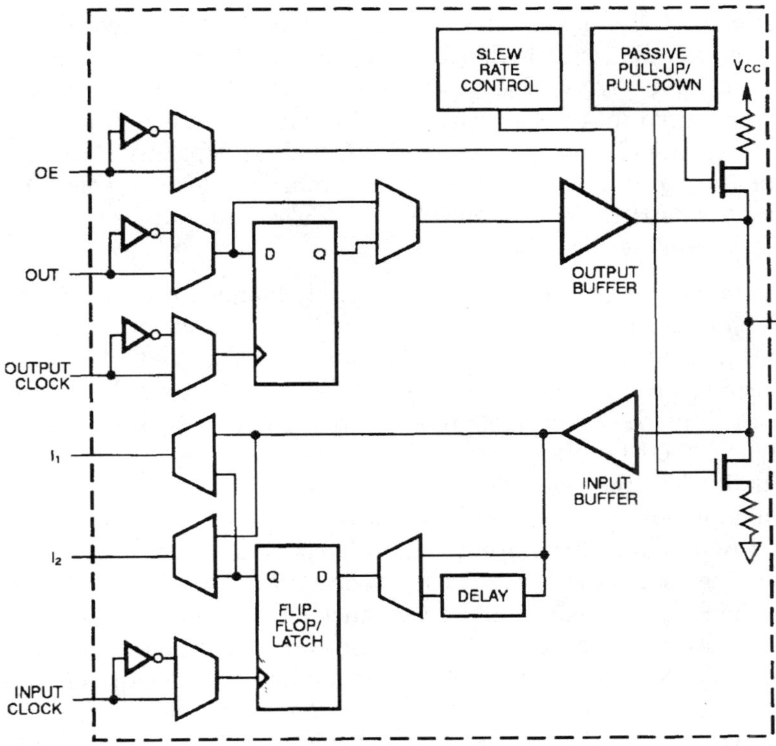

Figure 3.27b Xilinx XC4000 block diagram, I/O block *(from Ref. 24 with permission of ©Xilinx Corp.).*

third common input (S/R) can be programmed as either an asynchronous set or reset signal independently for each of the two registers; this input can also be disabled for either flip-flop. The source of the flip-flop data input is programmable and is driven either by the functions F, G, and H, or the direct in (DIN) block input. The flip-flops drive the Q1 and Q2 CLB outputs. Multiplexers in the CLB map the four control inputs, labeled C1 through C4, into four control signals (H1, DIN, S/R, and EC) in any arbitrary manner. Also, each CLB F and G function generator contains the dedicated arithmetic logic for generation of carry and borrow signals to increase performance of the adders, subtractors, accumulators, and so on.

- *Input-output blocks (IOBs).* The user-configurable I/O blocks provide the interface between external package pins and the internal logic, as shown in Fig. 3.27*b*. Each IOB controls one package pin and can be configured for input, output, or bidirectional signals. Two paths labeled I_1 and I_2 bring input signals into the array. Inputs are routed to an input register that can be programmed as either an edge-triggered flip-flop or a level-sensitive transparent latch. The I_1 and I_2 signals that exit the block can each carry either the direct or registered input signal. The output signals can pass directly to the pad or be stored in an edge-triggered flip-flop. There are number of other programmable options in the IOB, such as separate clock signals for the input and output registers. Embedded logic in the IOBs contains test structures compatible with IEEE Standard 1149.1 for boundary scan testing.

- *Programmable interconnect.* All internal connections are made with the metal segments and programmable switching points for implementation of desired routing. In XC4000 family, the CLB inputs and outputs are symmetrically distributed on all four sides of the block to provide additional routing flexibility as compared to XC2000/XC3000 families. There are three main types of interconnects defined by the relative length of their segments: single-length lines, double-length lines, and long lines.

  The single-length lines are a grid of horizontal and vertical lines intersecting at a switch matrix between each block. Figure 3.28*a* shows the single-length interconnect lines surrounding one CLB in the array. Each switch matrix consists of programmable n-channel pass transistors used to establish connections between the single-length lines. The single-length lines are normally used to conduct signals within a localized area and to provide the branching for nets with the fanout greater than one.

  The double-length lines consist of a grid of metal segments, which are twice as long as the single-length lines, i.e., a double-length line runs past two CLBs before entering a switch matrix. The double-length lines provide the most efficient implementation of intermediate length, point-to-point interconnections.

  The long lines form a grid of metal interconnect segments that run the entire length or width of the array, as shown in Fig. 3.28*b*. Additional long lines in the vertical plane can be driven by special global buffers, which are designed to distribute the clocks and other high-fanout control signals throughout the array. Long lines are intended for high-fanout, time-critical signal nets. The communication between long lines and single-length lines is controlled by programmable interconnection points at the line intersection.

The following are some of the enhancements made in Xilinx third-generation XC4000 family, relative to the first- and second-generation families:

- *Speed/performance improvement.* XC4000 is specified for a worst-case maximum clock rate of 60 to 70 MHz, which results from three

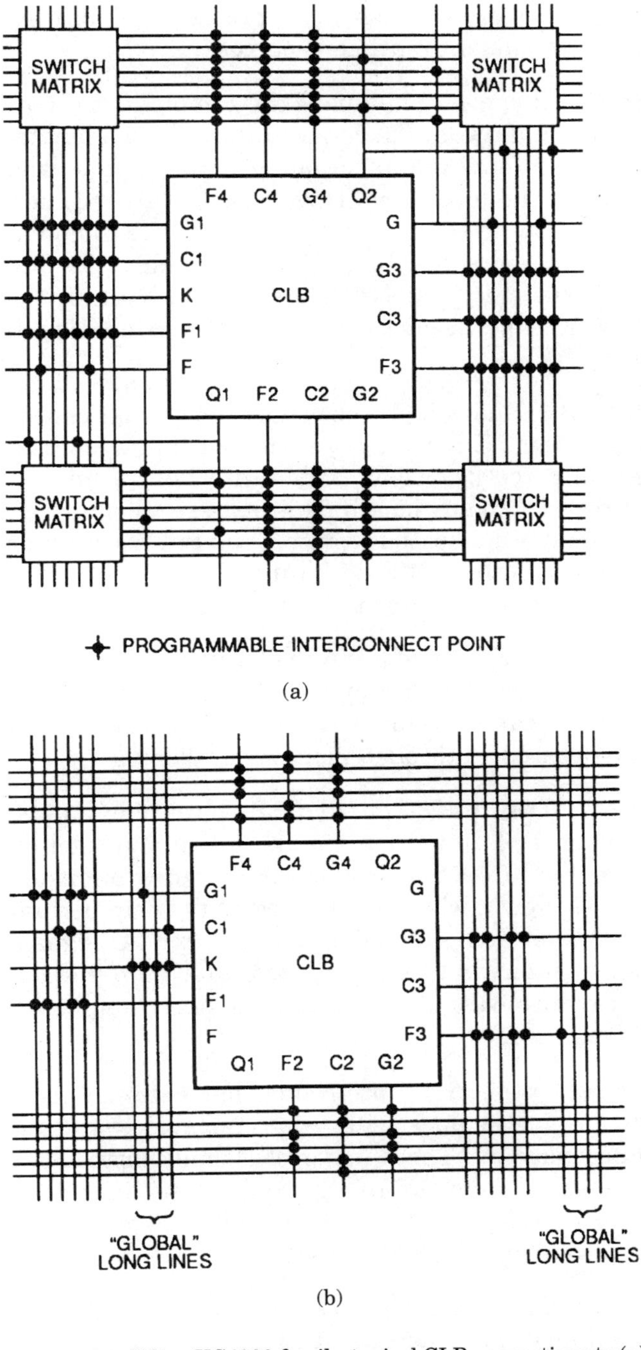

Figure 3.28 Xilinx XC4000 family, typical CLB connections to (a) adjacent single-length lines and (b) long line routing resources *(from Ref. 24 with permission of ©Xilinx Corp.).*

basic improvements in the areas of advanced processing, improved architecture, and more abundant routing resources.

- *More outputs.* The CLB can not only pass the combinatorial output(s) to the interconnect network but can also store the combinatorial results(s) or other incoming data in one or two flip-flops and connect their outputs to the interconnect network. In earlier-generation families, a choice had to be made—either output the combinatorial function or the stored value. In the XC4000 family, the flip-flops can be used as registers or shift registers without blocking the function generators from performing a different (even unrelated) job.

- *Fast carry.* Each CLB includes high-speed carry logic that can be activated by configuration. For example, each four-input function generator can be configured as two-bit adder with built-in hidden carry that can be expanded to any length. This fast-carry circuitry is equally effective in counter applications, since it can put two counter bits into each CLB and run them at a clock rate of up to 50 MHz for 16-bit counters, whether the counters are loadable or not.

- *Pipelining.* The abundance of flip-flops in the CLBs provides a tool for implementing higher-performance pipelined designs in which a function is divided into smaller subfunctions that can be executed in parallel, with results passed on through pipeline flip-flops.

- *Faster wide decoding.* The XC4000 family CLBs have nine inputs, which means any decoder of up to nine inputs can be implemented quite efficiently. Also, the XC4000 family has 16 fast programmable decoders, each with 40 inputs, which are located at the chip periphery with four decoders on each chip edge. They accept I/O signals and internal signals as input and generate a dedicated output. Each decoder AND gate can also be split in two when a larger number of narrow decoders is required.

- *Abundant routing resources.* The interconnection between logic blocks is made by the metal lines with programmable switching points and switching matrixes. The number of globally distributed signals has been increased from two to eight, and these lines have access to any clock or logic input. The XC4000 family has more than double the routing resources of XC3000 family.

- *On-chip memory.* The XC4000 family provides on-chip static memory resources. An optional mode for each CLB makes the memory LUTs in the function generators usable as either a 16×2 or 32×1 bit array of read/write memory cells. The configuration of CLB function generators as read/write memory does not affect the functionality of the other portions of CLB, except for the redefinition of control signals. These distributed RAMs have the same access time of about 5 ns for the logic delay and write time of about 10 ns.

Xilinx FPGA design and development system. The Xilinx Automated CAE Tools (XACT) set provides the design environment for the Xilinx LCA series of FPGAs. All Xilinx development system software is integrated under the Xilinx Design Manager (XDM™), which provides the designers with a common user interface, regardless of their choices of design entry and verification tools. Xilinx XACTstep™ development system supports popular design environments such as Viewlogic Systems, Mentor Graphics, Synopsys, OrCAD, Xilinx-ABEL (state machine module generator), and LogiBLOX™. The LogiBLOX is a module generator for small logic functions (e.g., counters, registers, adders/subtractors, and so on). It features tight integration with schematic capture packages and is included with all Xilinx development systems.

The XACTstep development system supports both simulation and in-circuit debugging techniques. For simulation, the system extracts the post-layout timing information from the design database. This data can then be sent to the simulator to verify timing-critical portions using XDELAY, the Xilinx static timing analyzer, and then back-annotation for debugging effort. The transfer into and out of XACTstep development system is provided through industry standard EDIF/SDF/VITAL formats. Xilinx also has a graphical *floorplanner* for more complex designs (>50,000 gates).

Xilinx XACTstep software solutions are available in two product series, as follows:

- *Foundation Series Software*™. This fully integrated tool set provides the designers with an easy access to design entry, synthesis, implementation, and simulation tools. The Foundation Series features support for standard HDL based designs, including VHDL, and ABEL. Figure 3.29 shows the block diagram for Foundation Series design flow.
- *Alliance Series Software*™. This is targeted for the designers wishing to integrate Xilinx implementation software into their existing design methodologies. The designer can choose from a wide variety of integrated design solutions, including schematic capture, synthesis, and PLD language-based design tools and platforms. It includes libraries and design kits in Alliance Series configurations for several EDA environments.

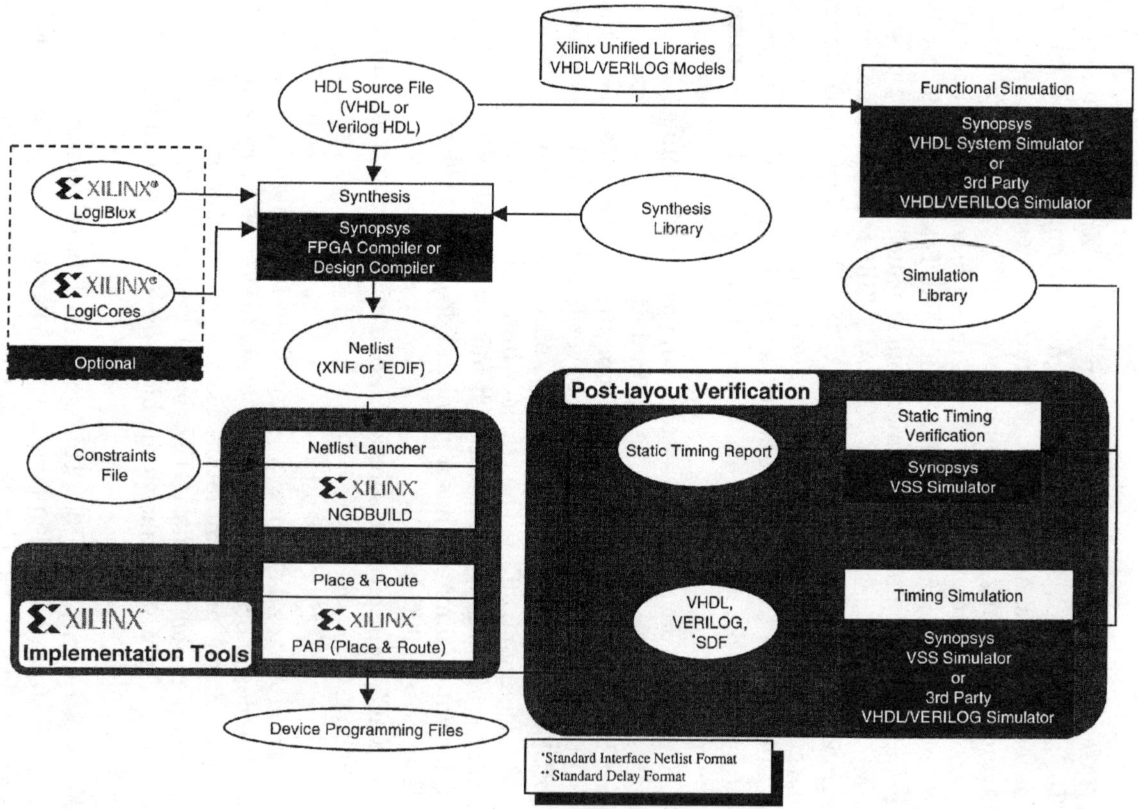

Figure 3.29 Block diagram of Foundation Series Software™ design flow *(from Ref. 24 with permission of ©Xilinx, Inc.).*

References

1. Jonathan Rose, et al., "Architecture of field programmable gate arrays," *Proc. of the IEEE,* vol. 81, no. 7, July 1993.
2. Jonathan Greene, et al., "Antifuse field programmable gate arrays," *Proc. IEEE,* vol. 81, no. 7, July 1993.
3. M. Chrzanowska-Jeske, "Architecture and synthesis issues in FPGAs," copy of paper received at CAD Tools Developers Conference.
4. M. Pedram, et al., "Design and analysis of segmented routing channels for row-based FPGAs," *IEEE Trans. On Computer Aided Design of Integrated Circuits and Systems,* vol. 13, no. 12, Dec. 1994, pp. 1470–1479.
5. A. El Gamal, et al., "Segmented channel routing is nearly as efficient as channel routing (and just as hard)," in *Proc. Conf. Advanced Research in VLSI,* Santa Cruz, CA March 1991.
6. Stephen Brown, et al., "Minimizing FPGA interconnect delay," *IEEE Design and Test of Computers,* Winter 1996 Issue, pp. 16–23.
7. S. Brown, et al., "Segmented routing for speed-performance and routability in field programmable gate arrays," *J. of VLSI Design,* vol. 4, no. 4, 1996, pp. 275–291.
8. Satwant Singh, "The effect of logic block architecture on FPGA performance," *IEEE Journal of Solid State Circuits,* vol. 27, no. 3, March 1992, pp. 281–287.
9. Dave Bursky, "Efficient RAM-based FPGAs ease system design," *Electronic Design,* Jan. 22, 1996, pp. 53–62.
10. Steven J. E. Wilton, et al., "Memory/Logic interconnect flexibility in FPGAs with large embedded arrays," *Proc. IEEE 1996 Custom Integrated Circuits Conference,* pp. 8.2.1–8.2.4.
11. Samir J. Sathe, et al., "Comparison of commercially available FPGA architectures," *Proc. IEEE 1994 Custom IC Conference,* pp. 383–385.
12. C. Hu, "Interconnect devices for field programmable gate arrays," *Proc. of IEEE International Electron Devices Meeting,* Dec. 1992, pp. 591–594.
13. David Marple and Larry Cooke, "Programming antifuses in Crosspoint FPGAs," *Proc. IEEE 1994 Custom IC Conference,* pp. 185–188.
14. K. El-Ayat, et al., "A CMOS electrically configurable gate array," *J. of Solid State Circuits,* 1989, 24(3): 752–762.
15. *Actel FPGA Databook and Design Guide,* 1996.
16. *QuickLogic Data Book,* 1996/1997.
17. *Atmel Configurable Logic Data Book,* August 1995.
18. Gatefield FPGA Data Sheets supplied by Gatefield Corporation.
19. Lucent Technologies *Field Programmable Gate Arrays Data Book,* April 1995 and data sheets supplied by Lucent Technologies.
20. ORCA OR2CxxA (5.0V) and OR2TxxA (3.3V) Series FPGA Data Sheets, August 1996 (Lucent Technologies).
21. T. Ngai, et al., "A new generation of ORCA FPGA with enhanced features and performance," *Proc. IEEE 1996 Custom IC Conference,* pp. 247–250.
22. Lucent Technologies ORCA OR3Cxxx/OR3Txxx Series FPGAs Advance Product Brief, October 1996.
23. *Motorola Programmable Array (MPA) Data Book* 1996 and data sheets supplied by Motorola Corp.
24. Xilinx, *The Programmable Logic Data Book,* 9/96, and updated figures with related text, supplied courtesy of ®Xilinx, Inc.

Chapter

4

Programmable Logic Device Design

4.1 Simple Combinatorial and Registered Logic Design Functions

4.1.1 Combinatorial logic design functions

In general, combinatorial logic design applications can be segmented into five major categories, as follows:

Encoders and decoders. These are the two most widely used functions in digital design. In case of encoder, a binary code of n bits can be used to represent $2n$ distinct pieces of coded data. A simple combinatorial encoder is a circuit that generates n bits of output information based on one of the $2n$ unique data combinations presented at its input. Figure 4.1 shows the block diagram of an encoder. The encoder design methodology is based on the Boolean equations derived from the truth table shown in Table 4.1*a*. The Boolean equations that can be optimized using Karnaugh maps or the software minimizer are shown below.

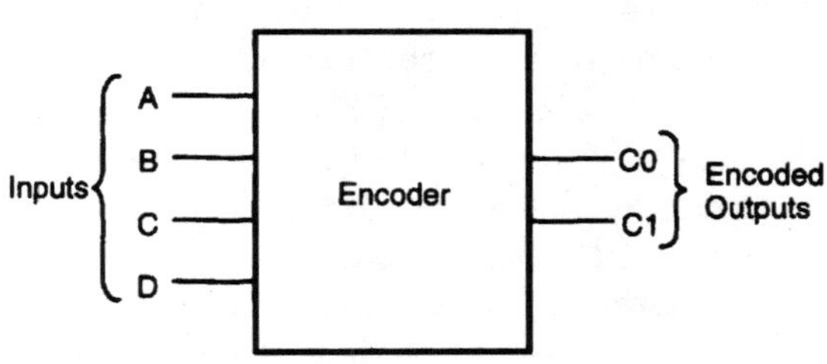

Figure 4.1 Block diagram of an encoder.

TABLE 4.1 Truth Tables

(a) A Typical Encoder

Inputs				Outputs	
A	B	C	D	C0	C1
1	0	0	0	L	L
0	1	0	0	L	H
0	0	1	0	H	L
0	0	0	1	H	H

(b) A Priority Encoder

Inputs				Outputs	
A	B	C	D	C0	C1
1	0	0	0	L	L
0	1	0	0	L	H
0	0	1	0	H	L
0	0	0	1	H	H
X	1	0	0	L	H
X	X	1	0	H	L Priority assignments
X	X	X	1	H	H

$$C1 = \overline{A} * B * \overline{C} * \overline{D} \quad + \quad \overline{A} * \overline{B} * \overline{C} * D$$

$$C2 = \overline{A} * \overline{B} * C * \overline{D} \quad + \quad \overline{A} * \overline{B} * \overline{C} * D$$

A priority encoder is a combinatorial circuit block similar to a general encoder except that its inputs have an assigned priority. These priority encoders are often used in applications such as state machines to detect the occurrence of highest priority event, microprocessor interrupt controllers, and bus control arbitration schemes. Table 4.1*b* shows the truth table for a four input signal (A, B, C, and D) priority encoder. These inputs are to be encoded as LL, LH, HL, and HH outputs, given that D has priority over C, C over B, and B over A. The Boolean equations for this priority encoder can be derived from the truth table in the same way as for the general encoder and then minimized.

A decoder performs the reverse function of an encoder by converting an n-bit code to one of its 2^n unique representations. It is a combinatorial circuit designed such that only one of its several outputs will be asserted at any time, based upon the unique input codes. A decoder can have as many outputs as there are possible binary input selection combinations. However, only one output is asserted at a time. A decoder (also known as a *demultiplexer*) can be considered as a logic function which, depending on the select code applied, connects one data input to the selected output. Table 4.2 shows the truth table

TABLE 4.2 Truth Table for an Active-Low 4-to-16 Decoder

Input select lines				Output lines															
A	B	C	D	Q0	Q1	Q2	Q3	Q4	Q5	Q6	Q7	Q8	Q9	Q10	Q11	Q12	Q13	Q14	Q15
0	0	0	0	0	1	1	1	1	1	1	1	1	1	1	1	1	1	1	1
0	0	0	1	1	0	1	1	1	1	1	1	1	1	1	1	1	1	1	1
0	0	1	0	1	1	0	1	1	1	1	1	1	1	1	1	1	1	1	1
0	0	1	1	1	1	1	0	1	1	1	1	1	1	1	1	1	1	1	1
0	1	0	0	1	1	1	1	0	1	1	1	1	1	1	1	1	1	1	1
0	1	0	1	1	1	1	1	1	0	1	1	1	1	1	1	1	1	1	1
0	1	1	0	1	1	1	1	1	1	0	1	1	1	1	1	1	1	1	1
0	1	1	1	1	1	1	1	1	1	1	0	1	1	1	1	1	1	1	1
1	0	0	0	1	1	1	1	1	1	1	1	0	1	1	1	1	1	1	1
1	0	0	1	1	1	1	1	1	1	1	1	1	0	1	1	1	1	1	1
1	0	1	0	1	1	1	1	1	1	1	1	1	1	0	1	1	1	1	1
1	0	1	1	1	1	1	1	1	1	1	1	1	1	1	0	1	1	1	1
1	1	0	0	1	1	1	1	1	1	1	1	1	1	1	1	0	1	1	1
1	1	0	1	1	1	1	1	1	1	1	1	1	1	1	1	1	0	1	1
1	1	1	0	1	1	1	1	1	1	1	1	1	1	1	1	1	1	0	1
1	1	1	1	1	1	1	1	1	1	1	1	1	1	1	1	1	1	1	0

for an active-low 4-input to 16-output decoder, for which the Boolean equations can be derived. The most common decoder applications are the address decoders used in most microprocessors and bus interfaces.

The encoders typically require a large number of inputs and fewer outputs, whereas the decoders typically require a fewer inputs and large number of outputs. An important device selection consideration for encoders and decoders is the number of product terms required for a design. This often determines whether a design can fit successfully in a target PLD. A careful selection of code values and priority assignments in priority encoders can help reducing the required number of product terms.

Multiplexers. A multiplexer, also referred to as a *data selector,* is widely used in digital design. It is a special combinatorial circuit designed to gate one of several inputs to a single output. The input selected for connection to the output is controlled by a separate set of input select lines. Figure 4.2 show the general block diagram of a 2^n-to-1 multiplexer. Table 4.3 shows the truth table for a three-input to one-output (3:1) multiplexer, which uses two select signals A and B to select one of the three inputs data to be sent to the output. This truth table can be used to derive the Boolean equations for a multiplexer design implementation.

Typically, multiplexers require more inputs than outputs; therefore, target PLDs with a large number of inputs and I/Os are more useful. Several multiplexers are often used simultaneously to route multiple address and data bits, under the control of the same select lines.

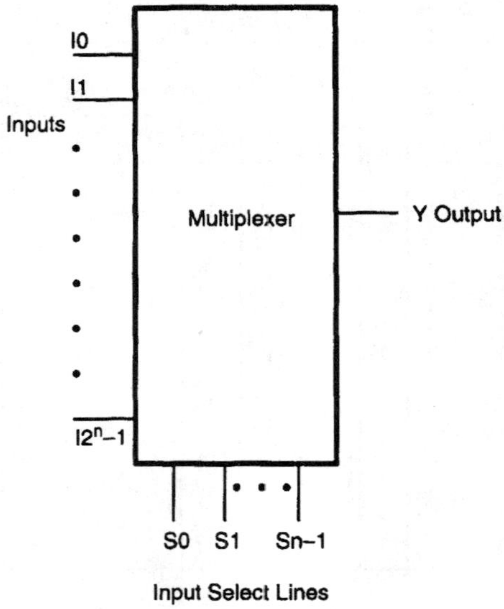

Figure 4.2 Block diagram of 2^n-to-1 multiplexer.

TABLE 4.3 Truth Table for a 3-to-1 Multiplexer

Select		Inputs			Output
B	A	I1C0	I1C1	I1C2	01Y
0	0	0	X	X	0
0	0	1	X	X	1
0	1	X	0	X	0
0	1	X	1	X	1
1	0	X	X	0	0
1	0	X	X	1	1

In such cases, where the number of device inputs and outputs exceeds device limits, multiple devices can be cascaded. Cascading can also used for large multiplexers that do not fit in a single device, provided the select bits are carefully selected for each PLD, to minimize the number of product terms.

Comparators. A comparator is a combinational circuit whose primary function is to compare the relative magnitude of two binary numbers. A basic comparator compares two numbers only for equality and generates the EQL signal (signifying A = B). An extension of this is the magnitude comparator, which also generates the LES signal (signifying A < B) and GTR signal (signifying A > B). Table 4.4 shows the

TABLE 4.4 Truth Table for a Comparator

Inputs				Outputs		
A		B		EQL	LES	GTR
A2	A1	B2	B1	A = B	A < B	A > B
0	0	0	0	1	0	0
0	0	0	1	0	1	0
0	0	1	0	0	1	0
0	0	1	1	0	1	0
0	1	0	0	0	0	1
0	1	0	1	1	0	0
0	1	1	0	0	1	0
0	1	1	1	0	1	0
1	0	0	0	0	0	1
1	0	1	0	1	0	0
1	0	1	1	0	1	0
1	1	0	0	0	0	1
1	1	0	1	0	0	1
1	1	1	0	0	0	1
1	1	1	1	1	0	0

truth table for a two-bit comparator from which the Boolean equations for three output signals EQL, LES, and GTR can be derived. These equations can be optimized to provide the sum-of-product forms for comparator design implementation using PAL devices.

The number of product terms needed is directly related to the number of bits compared. For LES and GTR functions, the number of product terms required depends upon the number of bits compared in two operands, as well as their value. The comparators require large number of product terms, which for n-bit comparison is given by $2^n - 1$. The value of the comparands themselves affect the number of product terms used. When the comparison is made with comparands, which are power-of-two numbers, the number of product terms can be substantially reduced. The Boolean equations show that the comparators require exclusive-OR functions. Therefore, they can be more efficiently implemented in target PLDs that offer exclusive-OR functions

A widely used application for LES and GTR Boolean equations derived for comparators is in the design of range decoders that compare a number (address) to a given range of comparand (addresses). To decode a range of three-bit numbers from B to A, one needs to compare another number X such that A > X > B, and derive equations for GTR (A > X) and LES (B < X) functions. Then, the two equations can be combined to give a range signal, which will be asserted only when A > X and X > B. The number of product terms required is fairly large and can easily exceed 100 for an 8-bit range comparator.

To fit the design on a PAL device, a commonly used technique is to select the address range defined by A and B, such that the range extends from B + 1 to A – 1, where A and B + 1 are the power-of-two numbers. Since the address space is aligned on the power-of-two boundaries, a number of bits of the address comparands will be zero. For Boolean equations implementation, this substantially reduces the number of product terms required.

Adders/arithmetic circuits. Digital systems are designed to perform a variety of arithmetic instructions on binary numerical data. A good example is the ALU (arithmetic logic unit), whose basic function is that of an adder performing addition on two binary numbers. A binary adder takes two inputs, adds them up, and generates the binary sum. The basic building block of any adding circuit is a full adder, which is a one-bit adder with carry-in and carry-out. Table 4.5 shows the truth table for a full adder. Then, the Boolean equations for outputs Y and C_{OUT} can be derived from the truth table and are as follows:

$$Y = A * \overline{B} * \overline{C}_{IN} + \overline{A} * \overline{B} * C_{IN} + A * B * C_{IN} + \overline{A} * B * \overline{C}_{IN}$$

$$C_{OUT} = A * C_{IN} + A * B + B * C_{IN}$$

TABLE 4.5 Truth Table for a Full Adder

Inputs			Outputs	
A	B	C_{IN}	Y	C_{OUT}
0	0	0	0	0
0	0	1	1	0
0	1	0	1	0
0	1	1	0	1
1	0	0	1	0
1	0	1	0	1
1	1	0	0	1
1	1	1	1	1

The full adders can be cascaded to form large binary adders so that each carry-out is directed to carry-in for the next stage. These adders are known as *ripple adders* and can be easily implemented using combinatorial PAL devices, which also provide internal feedback. The ripple adders have the advantage that they can be cascaded to any length. However, such large ripple adders may be inefficient, since the carry-out from the least significant (LSB) has to travel all the way to the most significant bit (MSB), which can take a long time. Therefore, adders with built-in carry-look-ahead circuitry that simultaneously generates the carry-in signal for all of the bits can be faster. However, the implementation of such a carry-look-ahead adder requires a large number of product terms. For example, a four-bit carry-look-ahead adder requires more than 36 product terms in the sum-of-products terms. The Boolean equations for a carry-look-ahead full adder are as follows:

$$Y_0 = A_0 \oplus B_0 \oplus C_{IN}$$

where the carry-out signal is given by

$$C_0 = A_0 * B_0 + (A_0 + B_0) * C_{IN}$$

Latches. A latch is basically a simple combinatorial circuit in which the output is a function of the inputs and feedback. One of the most common uses for a latch is as a temporary storage for data or addresses. PLD based latches are often used for address decoders to assert the decoded signal for long duration. The most commonly used latch is the D-type latch for which the truth table is shown in Table 4.6. This latch is in *transparent mode* when the control signal latch-enable (LEN) is high and the input signal (/D) is available at the outputs. Figure 4.3 shows the block diagram of a transparent latch. When the LEN signal is low (logic 0), the input data is latched on the outputs and is retained until LEN goes back high (logic 1). As an example, in a typical address decoder, the input will be a combination of various address signals that are decoded using range comparators. The Boolean equations for a latch can be derived directly from the truth table and are as follows:

$$\overline{Q} = \overline{D} * LEN + \overline{Q} * \overline{LEN}$$

TABLE 4.6 Truth Table for a Simple Latch

Inputs		Outputs
/D	LEN	/Q
0	1	0
1	1	1
X	0	/Q (previous)

4.1.2 Registered logic design functions

Section 4.1.1 discussed combinatorial logic design functions, whose outputs are totally independent of any system clock. This section dis-

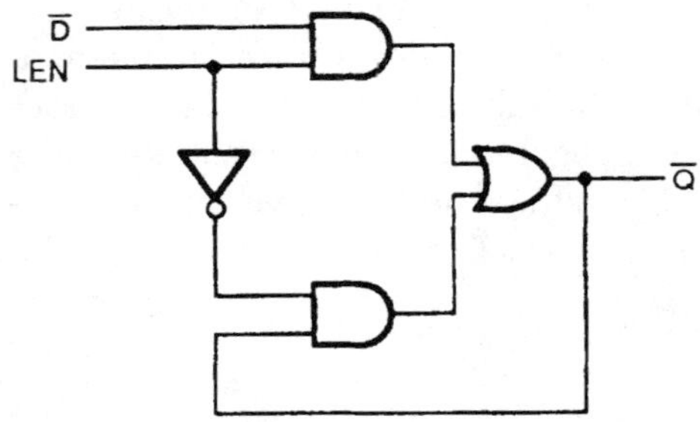

Figure 4.3 Block diagram of a transparent latch.

cusses registered logic functions, consisting of sequential circuits in which the outputs store their previous logic values until a new clock is applied. The storage elements for these sequential circuits are called *flip-flops,* and a bank of these flip-flops form *registers*, although individual flip-flops are also often referred to as registers. There are four basic types of flip-flops: S-R, J-K, T, and the D-type. Some PLDs provide the capability of configuring output registers as either D, T, J-K, or S-R. In certain designs, configurable flip-flops can reduce the number of product terms required. However, D-type flip-flops are the simplest to design with and will be used for discussion throughout this section.

A number of designs combine both registered and combinatorial functions. Also, there can be two distinct designs, one registered and one combinatorial combined on a single device for higher integration. There may be a design requirement where registered outputs need to be decoded using combinatorial logic. In many design software packages, the output registers are denoted by the ":=" assignment symbol, as compared to the "=" designation for a combinatorial output.

In registered logic functions, speed is a critical parameter, and the effects of clock as reflected in maximum frequency (f_{MAX}) specifications must be taken into consideration. The flexibility inherent in PLD design provides a choice of configurations from which different f_{MAX} parameters can be calculated. In general, there are three basic design approaches as follows:[1]

- In the first type of design, a PLD is used as a stand-alone registered design, as shown in Fig. 4.4*a*. In this configuration, to decide the next logic level of the registers, the present logic level needs to be available at the inputs of the registers before they are clocked. Therefore, the clock period is limited by the internal delay from the flip-flop outputs through the internal feedback and logic to the flip-flop inputs. This f_{MAX} is often designated as "f_{MAX} internal," and a simple counter is a good example of this type of design.

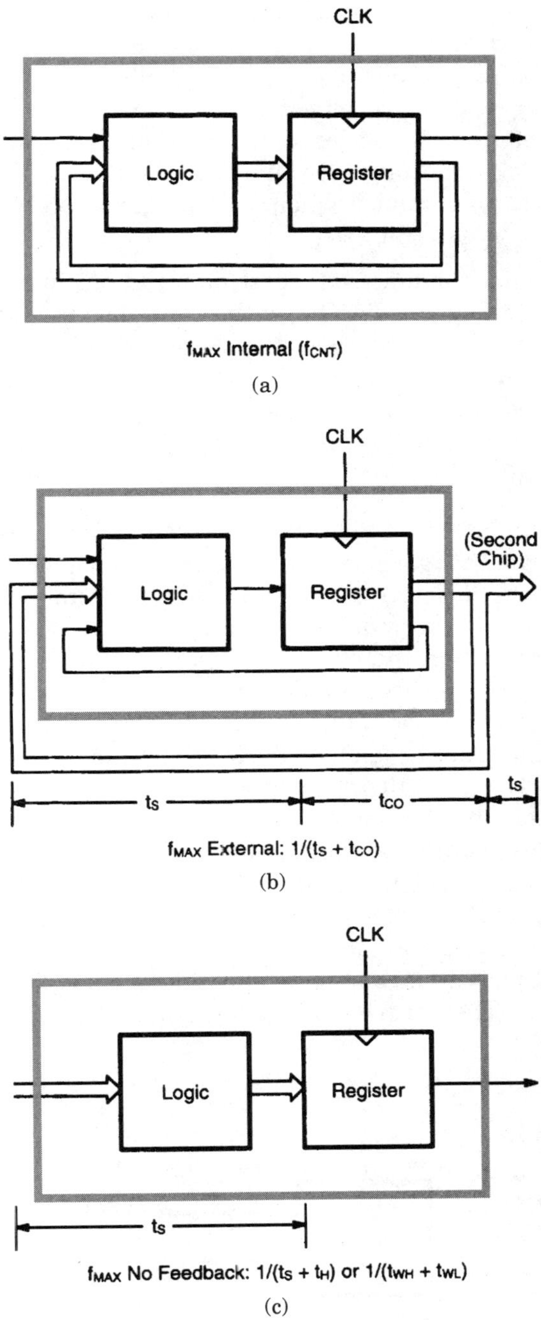

Figure 4.4 Three different f_{MAX} design configurations: (a) internal f_{MAX}, (b) external f_{MAX}, and (c) f_{MAX} with no feedback *(from Ref. 1)*.

- The second type of design is commonly used in system configurations with number of logic devices including PLDs that have a common clock. In this case, the registered outputs are sent off-chip, back to the device inputs of a second device, as shown in Fig. 4.4*b*. Therefore, the slowest path defining the period is the sum of the input setup time (t_s) and the clock-to-output (f_{co}) time. The reciprocal of (t_s + t_{co}) is f_{MAX}, which is the maximum frequency with external feedback and referred to as "f_{MAX} external."
- In the third type of design, a simple data path application, input data is presented to the flip-flop and clocked, without employing any feedback, as shown in Fig. 4.4*c*. In this case, the clock period is limited by the sum of data setup and data hold times ($t_s + t_H$). However, the minimum clock period ($t_{WH} + t_{WL}$) is usually a tighter limit. Therefore, the third f_{MAX} which is designated as "f_{MAX} no feedback" will be the lesser of ($t_s + t_H$) or $1/(t_{WH} + t_{WL})$.

The registered logic functions can be classified into two major categories: *synchronous* and *asynchronous* operations. In synchronous designs, the clock input of all the registers are tied together to a common clock. In comparison, for asynchronous designs, the flip-flop's clock inputs may not be tied together, and the clocks may be gated or even driven by other flip-flops. The synchronous design approach is commonly used for two major functions of data handling and control that require the use of counters and shift registers. A binary counter is the simplest form of a counter. However, there are other types of counters such as the modulo, Johnson, and Gray-code counters. The major differences between the various counters is the count sequence.

A counter is defined as a set of registers that cycles through a predetermined unvarying sequence. Figure 4.5 shows the general model of a synchronous counter, which has a common clock to all the flip-flops, and whose outputs are fed back to a combinatorial logic array called the *next-state (count) decoder*. The next count is generated by this logic

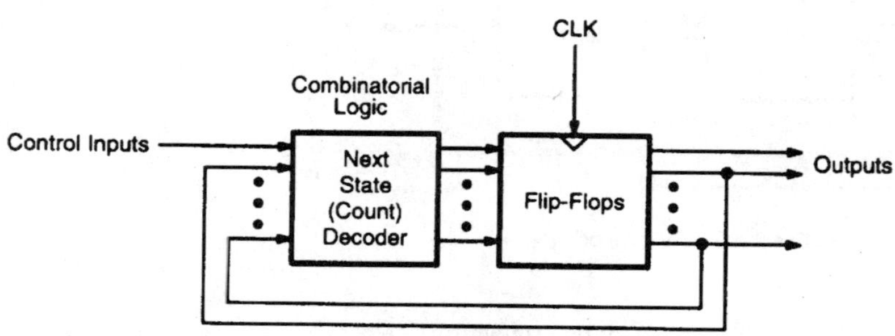

Figure 4.5 General model of a counter.

based on the present count and control inputs. The counters can be conveniently represented by bubble-and-arrow diagrams called the *state diagrams.* The various types of counters can be implemented by PLDs, using standard sum-of-products form of array, and are discussed below.

Binary counters. Figure 4.6 shows the state diagram of a 4-bit binary counter. Table 4.7 shows the truth table (also called the *transition table*) for such a counter. The Boolean equations for each bit can be derived directly from the truth table by collecting all the product terms where outputs are asserted high (ones). This will yield Boolean equations for devices with active-high outputs. These equations can be inverted for devices with the active-low outputs. The binary counter Boolean equations make use of exclusive-OR functions in the output. These XOR functions can be implemented in their sum-of-products

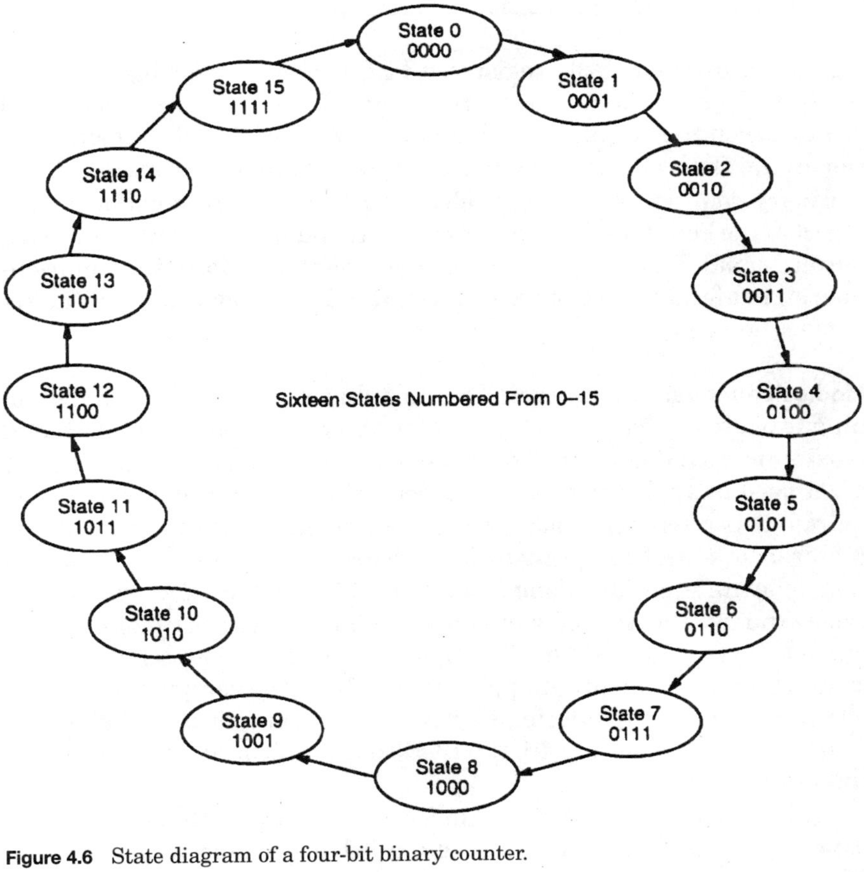

Figure 4.6 State diagram of a four-bit binary counter.

TABLE 4.7 Truth Table for a Four-Bit Binary Counter

Present state				Next state			
X3	X2	X1	X0	X3	X2	X1	X0
0	0	0	0	0	0	0	1
0	0	0	1	0	0	1	0
0	0	1	0	0	0	1	1
0	0	1	1	0	1	0	0
0	1	0	0	0	1	0	1
0	1	0	1	0	1	1	0
0	1	1	0	0	1	1	1
0	1	1	1	1	0	0	0
1	0	0	0	1	0	0	1
1	0	0	1	1	0	1	0
1	0	1	0	1	0	1	1
1	0	1	1	1	1	0	0
1	1	0	0	1	1	0	1
1	1	0	1	1	1	1	0
1	1	1	0	1	1	1	1
1	1	1	1	0	0	0	0

logic form in most of the registered PLDs. However, for big counters, many product terms are required. Some PLDs require a dedicated XOR gate on the outputs, which allows an AND-OR-XOR implementation of the Boolean logic and, therefore, fewer product terms.

Binary counters can be cascaded into two or more devices to construct large counters. The less significant counters generate an extra output signal. These signals are ANDed together to form the count-enable signal for a higher order (up) counter. For a down counter, the reverse scheme is used.

Modulo counters. In general, the modulus refers to the number of unique states traversed by a counter. A typical n-bit counter has a maximum modulus of $2n$. The non-power-of-two modulo counters are often used in applications where precise timing duration control is required. It is often necessary to introduce signal delays to meet timing requirements and compensate for possible timing skews or propagation delay differentials along two different signal paths. An example of this is the introduction of wait states to allow for access times of different memory elements. One example of a modulo counter is a BCD counter, which is useful in applications where the computer's outputs are generated using a decimal system. As compared to a 4-bit binary counter that can count to 16, the BCD counter terminates the count at the modulus of 10.

Table 4.8*a* shows the truth table for a BCD counter that counts from 0 to 9. Modulo counters can be designed in various ways. One ap-

TABLE 4.8 Truth Table for (a) a BCD Counter and (b) BCD Counter Illegal State Recovery to Count Zero

Present state					Next state			
Q3	Q2	Q1	Q0		Q3	Q2	Q1	Q0
				(a)				
0	0	0	0	0 ⇒ 1	0	0	0	1
0	0	0	1	1 ⇒ 2	0	0	1	0
0	0	1	0	2 ⇒ 3	0	0	1	1
0	0	1	1	3 ⇒ 4	0	1	0	0
0	1	0	0	4 ⇒ 5	0	1	0	1
0	1	0	1	5 ⇒ 6	0	1	1	0
0	1	1	0	6 ⇒ 7	0	1	1	1
0	1	1	1	7 ⇒ 8	1	0	0	0
1	0	0	0	8 ⇒ 9	1	0	0	1
1	0	0	1	9 ⇒ 0	0	0	0	0
				(b)				
1	0	1	0	10 ⇒ 0	0	0	0	0
1	0	1	1	11 ⇒ 0	0	0	0	0
1	1	0	0	12 ⇒ 0	0	0	0	0
1	1	0	1	13 ⇒ 0	0	0	0	0
1	1	1	0	14 ⇒ 0	0	0	0	0
1	1	1	1	15 ⇒ 0	0	0	0	0

proach is to derive the Boolean equations directly from the truth table for PLD implementation. For this BCD counter, the count values from 10 to 15 are illegal states, and the device can accidentally power up in one of these states. Therefore, for a good design, a mechanism must be built into the equations to allow it to recover back into a legal state. Table 4.8*b* shows the truth table for illegal state recovery to count zero. For active-low outputs, the Boolean equations can be directly derived from the truth table and optimized using Karnaugh maps or the software minimizer. Figure 4.7 shows the block diagram of a dual-BCD counter.

The technique used for counter design from zero to a fixed modulus can also be applied to a counter that counts down from a maximum power-of-two number to a fixed modulus, or even a counter that counts from one modulus to another. The registered PLDs used for modulo counters are similar to the ones selected for other counters. Since the counts used are binary, the PLDs with J-K, T-type flip-flops, or XOR gates help optimize the number of product terms used. The number of product terms required also depends on the modulus selected.

Another design factor to take into consideration for flip-flop selection is the illegal states. D-type flip-flops are generally more suitable

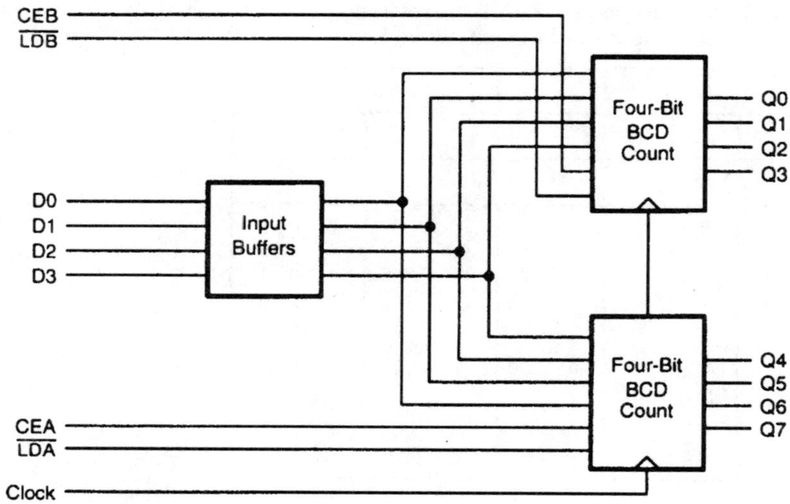

Figure 4.7 Block diagram of a dual-BCD counter.

for illegal state recovery than the J-K or T-type flip-flops. This is because, when no product terms are asserted, the D-type flip-flop resets to zero, whereas the designers using J-K or T-type flip-flops must design in illegal state recovery logic. For large modulo counters, the technique of generating Boolean equations from the truth table is tedious and impractical. An alternative design approach is to divide the large modulo counter into two smaller counters to simplify the design and help minimize the number of product terms.

Counters with encoding. The binary counter and BCD counters discussed so far generate binary output sequences. However, most peripherals require a predetermined sequence of control signals, which can be generated by decoding the binary sequence with combinatorial logic. Figure 4.8 shows the general model of a counter with combinatorial output decoding circuitry. This combinatorial circuit modifies the counter bits and generates the output signals in the sequence required for peripheral timing and control. However, since these circuits require extra combinatorial logic, they are not very efficient and are susceptible to hazards and output glitches.

It is possible to have different output codings for a four-bit counter. A commonly used technique for generating control signals is to implement the decoding circuitry before the registers. This will generate a counter with state values that are not a binary progression. Many different codes can be generated using such techniques. Some examples of these are Gray-code counters and Johnson counters discussed below.

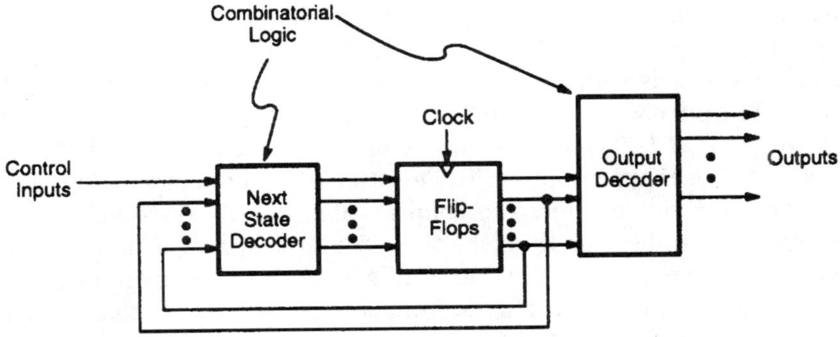

Figure 4.8 Block diagram of a counter with an output decoder.

- *Gray-code counters.* Gray-code counters are often used in digital designs for control timing functions. Their major advantage is the characteristic that only one output bit changes value for every clock cycle. These output signals can be easily decoded using a combinatorial decoder. Gray-code counters are used extensively as system clocks, since the different output bits provide different clock pulses without the risk of hazards. They are also used in specialized applications such as shaft encoders and real-time process controls. Table 4.9 shows the truth table for a four-bit Gray code counter. The Boolean equations can be directly derived from the truth table for PLD implementation.

TABLE 4.9 Truth Table for a Four-Bit Gray-Code Counter

Present state				Next state			
X3	X2	X1	X0	X3	X2	X1	X0
0	0	0	0	0	0	0	1
0	0	0	1	0	0	1	1
0	0	1	1	0	0	1	0
0	0	1	0	0	1	1	0
0	1	1	0	0	1	1	1
0	1	1	1	0	1	0	1
0	1	0	1	0	1	0	0
0	1	0	0	1	1	0	0
1	1	0	0	1	1	0	1
1	1	0	1	1	1	1	1
1	1	1	1	1	1	1	0
1	1	1	0	1	0	1	0
1	0	1	0	1	0	1	1
1	0	1	1	1	0	0	1
1	0	0	1	1	0	0	0
1	0	0	0	0	0	0	0

- *Johnson counters.* A Johnson counter is a part of a family of counters, known as *ring counters,* that are used for special applications where code symmetry is required. Ring counters are also used for timing purposes, since all the outputs are essentially a series of pulses. The code symmetry allows use of fewest possible product terms with a D-type register. Table 4.10 shows the truth table (legal states) for a five-bit Johnson counter, also known as *circular-shift counter.* As can be seen from the truth table, the counter first fills up with ones from left to right, then it fills up with zeros again. The output sequence shows that only one of Johnson counter bit changes for every clock period, like the Gray-code counter. A major advantage of Johnson counter is that it can be easily decoded with small two-input NAND gates and is suitable for high-speed applications.

  In general, an n-bit Johnson counter will produce a modulus of $2n$. The five-stage Johnson counter has 10 legal states and 22 illegal states. The implementation of Johnson counter is relatively simple and same regardless of the number of stages. Figure 4.9 shows the state diagram of a five-bit Johnson counter. When D-type flip-flops are used, the Q output of each flip-flop is connected to the D input of the following stage except for the Q output of the last stage, which is complemented and connected to the D input of the first stage. A major disadvantage of Johnson counter is the number of invalid (or illegal) states, which increases exponentially with the length of the counter. Johnson counters are quite susceptible to illegal states and can "hang up" easily. Therefore, they have to be designed taking into consideration illegal state recovery.

Shift registers. A shift register is a primary building block in digital system design and closely related to a ring counter. Its major use is for

TABLE 4.10 Five-Bit Johnson Counter Truth Table for Legal States

Present state					Next state				
Q4	Q3	Q2	Q1	Q0	Q4	Q3	Q2	Q1	Q0
0	0	0	0	0	1	0	0	0	0
0	0	0	0	1	0	0	0	0	0
0	0	0	1	1	0	0	0	0	1
0	0	1	1	1	0	0	0	1	1
0	1	1	1	1	0	0	1	1	1
1	1	1	1	1	0	1	1	1	1
1	1	1	1	0	1	1	1	1	1
1	1	1	0	0	1	1	1	1	0
1	1	0	0	0	1	1	1	0	0
1	0	0	0	0	1	1	0	0	0

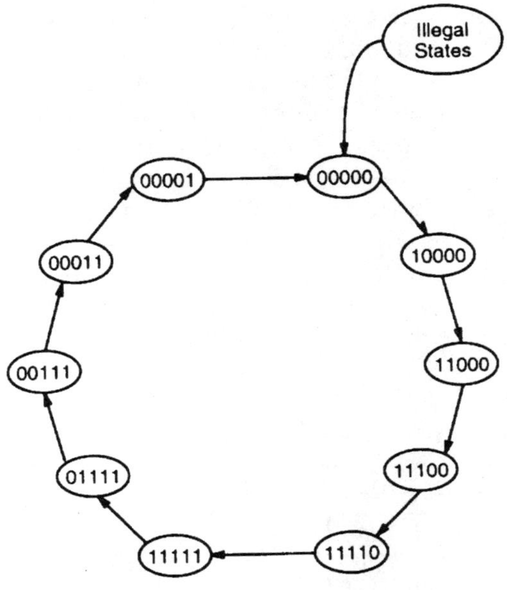

Figure 4.9 State diagram of a five-bit Johnson counter *(from Ref. 1).*

temporary data storage and data manipulation for advanced arithmetic and multiplication operations. The shift registers are frequently used in communications for converting parallel byte-wide data from the microprocessor to a serial data bit-stream for transmission. Figure 4.10 shows a block diagram of a shift register whose primary purpose is to shift data from one flip-flop to another. There are several types of shift registers, which are classified by the way in which the incoming data is received (parallel or serial) and how the outgoing data is transmitted (parallel or serial).

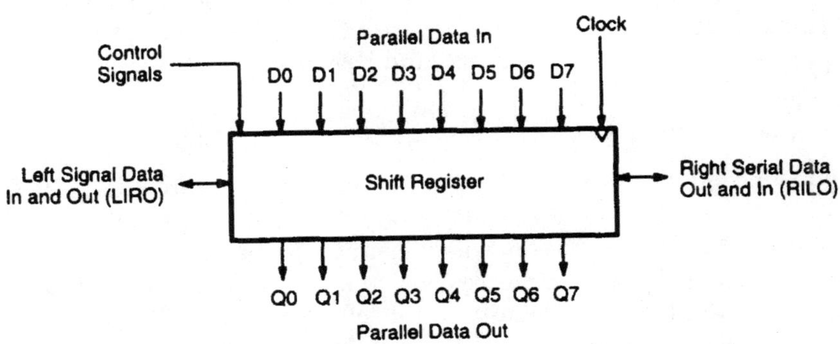

Figure 4.10 Block diagram of a shift register.

Table 4.11 shows the truth table for a universal shift register that provides both serial and parallel input and output functions. The data is shifted from one flip-flop to another in the left or right direction, depending upon the control signals I0 and I1. These inputs also control when the new parallel data is loaded onto the registers. During the left or right shift operations, serial data can be received and transmitted on serial pins LIRO and RILO. Since the flip-flops outputs appear on the output pins at all time, the parallel output data is always available. This universal shift register can be used for serial in/serial out, parallel in/parallel out, serial in/parallel out and parallel in/serial out. The Boolean logic equations can be directly derived from the truth table for PLD implementation.

TABLE 4.11 Truth Table for a Universal Shift Register

Q7	Q6	Q5	Q4	Q3	Q2	Q1	Q0	I1	I0	
Q7	Q6	Q5	Q4	Q3	Q2	Q1	Q0	0	0	;retain data
RILO	Q7	Q6	Q5	Q4	Q3	Q2	Q1	0	1	;shift right
Q6	Q5	Q4	Q3	Q2	Q1	Q0	LIRO	1	0	;shift left
D7	D6	D5	D4	D3	D2	D1	D0	1	1	;load data

In most data processing units, some form of data shifting or rotation is necessary. For applications such as floating-point arithmetic or string manipulation, ordinary shift registers are inefficient, since they require n clock cycles for an n-bit shift. A specialized shift register called a *barrel shifter* is used to shift (or rotate) data by any number of bits in a single clock cycle. In a typical computer system, the shifter is located at the output of the ALU. The storage registers on the output of the shifter are used in this architecture to pipeline the data operation and increasing throughput. The three-state buffer on the output registers helps provide interface to the data bus.

Figure 4.11 shows the block diagram of an eight-bit registered barrel shifter, and Table 4.12 shows its truth table. This registered barrel shifter requires a total of 14 inputs and 8 outputs.

4.1.3 Macros and libraries

In PLD design, the logic cell represents the basic building block. PLD manufacturers offer library cells that represent basic logic gates such as NAND/NOR, which are known as primitive cells (or just cells). The libraries also include hard macros, which are usually the simple functions that can fit into one logic cell and soft macros that fit into multiple logic cells. However, some of the vendors such as Actel have hard macros that require more than one logic module for implementation.

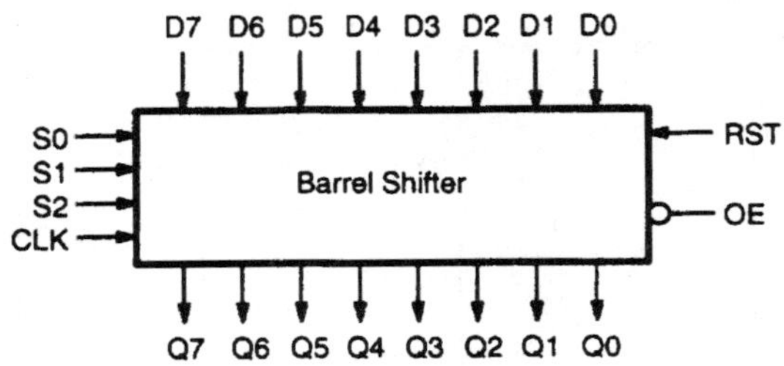

Figure 4.11 Block diagram of an eight-bit barrel shifter.

TABLE 4.12 Truth Table for an Eight-Bit Barrel Shifter

S2	S1	S0	Q7	Q6	Q5	Q4	Q3	Q2	Q1	Q0
0	0	0	D7	D6	D5	D4	D3	D2	D1	D0
0	0	1	D6	D5	D4	D3	D2	D1	D0	D7
0	1	0	D5	D4	D3	D2	D1	D0	D7	D6
0	1	1	D4	D3	D2	D1	D0	D7	D6	D5
1	0	0	D3	D2	D1	D0	D7	D6	D5	D4
1	0	1	D2	D1	D0	D7	D6	D5	D4	D3
1	1	0	D1	D0	D7	D6	D5	D4	D3	D2
1	1	1	D0	D7	D6	D5	D4	D3	D2	D1

In general, a hard macro includes placement information, so that even with a change in its position and orientation, the relative location of the transistors inside the macro, their layout and wiring interconnect remains fixed. In comparison, a soft macro contains only connection information between various programmable logic cells, and therefore, the placement and wiring for a soft macro can vary.

Both hard and soft macros offer a wide range of combinatorial and sequential logic functions and flexibility. Table 4.13 shows some examples of the combinatorial and sequential hard macros and soft macros available for Actel Integrator series of devices that include 1200XL and 3200DX families discussed in Sec. 3.3.1.[2] The table lists function, name of the macro, description, and number of sequential (S) and combinatorial (C) logic modules required for implementation.

4.2 Simple PLD Design Methodology

The implementation of a simple PLD design requires the creation of a design file that contains three types of information: (a) basic bookkeeping information, (b) design syntax, and (c) simulation syntax. Once the design file is complete, it is then assembled and simulated, and the resultant JEDEC file is downloaded to a device programmer

TABLE 4.13 Actel Integrator Series Hard and Soft Macros

Function	Macro	Description	Modules	
			S	C
		(a) Hard Macros, Combinatorial		
Combinatorial logic module	CM8	Combinational module (full 1200XL and 3200DX logic module)		1
Sequential logic module	DFM7A	Four-input D-type flip-flop with multiplexed data, active low clear, and active high clock	1	
	DFM7B	Four-input D-type flip-flop with multiplexed data, active low clear, and clock	1	
Adder	FA1A	One-bit adder, carry-in and carry-out active low, A-input active low		2
	FA1B	One-bit adder, carry-in and carry-out active low		2
	FA2A	Two-bit adder, carry-in and carry-out active low, A0 and A1 inputs active low		2
	HA1	Half-adder		2
	HA1A	Half-adder with active low A-input		2
	HA1B	Half-adder with active low carry-out and sum		2
	HA1C	Half-adder with active low carry-out		2
AND	AND2	Two-input AND		1
	AND2A	Two-input AND with active low A-input		1
	AND2B	Two-input AND with active low inputs		1
	AND3	Three-input AND		1
	AND3A	Three-input AND with active low A-input		1
	AND3B	Three input AND with active low A- and B-inputs		1
	AND3C	Three-input AND with active low inputs		1
	AND4	Four-input AND		1
	AND4A	Four-input AND with active low A-input		1
	AND4B	Four-input AND with active low A- and B-inputs		1
	AND4C	Four-input AND with active low A-, B-, and C-inputs		1

TABLE 4.13 *(continued)*

Function	Macro	Description	Modules	
			S	C
AND	AND4D	Four-input AND with active low inputs		2
	AND5B	Five-input AND with active low A- and B-inputs		1
AND-OR	AO1	Three-input AND-OR		1
	AO10	Five-input AND-OR-AND		1
	AO11	Three-input AND-OR		1
	AO1A	Three-input AND-OR with active low A-input		1
	AO1B	Three-input AND-OR with active low C-input		1
	AO1C	Three-input AND-OR with active low A- and C-inputs		1
	AO1D	Three-input AND-OR with active low A- and B-inputs		1
	AO1E	Three-input AND-OR with active low inputs		1
AND-XOR	AX1	Three-input AND-XOR with active low A-input		1
	AX1A	Three-input AND-XOR-INVERT with active low A-input		2
	AX1B	Three-input AND-XOR with active low A- and B-inputs		1
	AX1C	Three-input AND-XOR		1
Buffer	BUFF	Buffer with active high input and output		1
	BUFA	Buffer with active low input and output		1
		(b) Hard Macros, Sequential		
D-type	DF1	D-type flip-flop	1	
	DF1A	D-type flip-flop with active low output	1	
	DF1B	D-type flip-flop with active low clock	1	
	DF1C	D-type flip-flop with active low clock and output	1	
	DFC1	D-type flip-flop with active high clear	1	1

TABLE 4.13 *(continued)*

Function	Macro	Description	Modules	
			S	C
D-type	DFC1A	D-type flip-flop with active high clear and active low clock	1	1
	DFC1B	D-type flip-flop with active low clear	1	
	DFC1D	D-typc flip-flop with active low clear and clock	1	
	DFE	D-type flip-flop with active high enable	1	
	DFE1B	D-type flip-flop with active low enable	1	
	DFE1C	D-type flip-flop with active low enable and clock	1	
	DFE3A	D-type flip-flop with enable and active low clear	1	
	DFE3B	D-type flip-flop with enable and active low clear and clock	1	
	DFE3C	D-type flip-flop with active low enable and clear	1	
	DFE3D	D-type flip-flop with active low enable, clear, and clock	1	
	DFEA	D-type flip-flop with enable and active low clock	1	
	DFM	Two-input D-type flip-flop with multiplexed data	1	
	DFM1B	Two-input D-type flip-flop with multiplexed data and active low output	1	
	DFM1C	Two-input D-type flip-flop with multiplexed data and active low clock and output	1	
	DFM3	Two-input D-type flip-flop with multiplexed data and clear	1	1
	DFM3B	Two-input D-type flip-flop with multiplexed data and active low clear and clock	1	
	DFM3E	Two-input D-type flip-flop with multiplexed data, clear, and active low clock	1	1
	DFM4C	Two-input D-type flip-flop with multiplexed data and active low preset and output	1	
	DFM4D	Two-input D-type flip-flop with multiplexed data and active low preset, clock, and output	1	
	DFM6A	Four-input D-type flip-flop with multiplexed data, active low clear, and active high clock	1	
	DFM6B	Four-input D-type flip-flop with multiplexed data, active low clear, and clock	1	
	DFMA	Two-input D-type flip-flop with multiplexed data, active low clear, and clock	1	
	DFMB	Two-input D-type flip-flop with multiplexed data and active low clear	1	
	DFME1A	Two-input D-type flip-flop with multiplexed data and active low enable	1	
	DFP1	D-type flip-flop with active high preset		2
	DFP1A	D-type flip-flop with active high preset and active low clock		2
	DFP1B	D-type flip-flop with active low preset		2

TABLE 4.13 ***(continued)***

Function	Macro	Description	Modules S	Modules C
D-type	DFP1C	D-type flip-flop with active high preset and active low output	1	1
	DFP1D	D-type flip-flop with active low preset and clock		2
	DFP1E	D-type flip-flop with active low preset and output	1	
	DFP1F	D-type flip-flop with active high preset and active low clock and output	1	1
	DFP1G	D-type flip-flop with active low preset, clock, and output	1	
	DFPC	D-type flip-flop with active high preset, active low clear, and active high clock		2
	DFPCA	D-type flip-flop with active high preset and active low clear and clock		2
T-type	TF1A	T-type flip-flop with active low clear	1	
	TF1B	T-type flip-flop with active low clear and clock	1	

(c) Soft Macros

Function	Macro	Description	Max. logic levels	Modules S	Modules C
Adder	FADD10	Ten-bit adder	3		56
	FADD12	Twelve-bit adder	4		9
	FADD16	Sixteen-bit adder	5		97
	FADD8	Eight-bit adder	4		44
	FADD9	Nine-bit adder with active low carry-out	3		49
	VAD16C	Very fast 16-bit adder, no carry-in	3		91
	VADC16C	Very fast 16-bit adder with carry-in	3		97
Comparator	ICMP4	Four-bit identity comparator	2		5
	ICMP8	Eight-bit identity comparator	3		9
	MCMPC2	Two-bit magnitude comparator with enable	3		9

TABLE 4.13 ***(continued)***

			Max. logic levels	Modules S	Modules C
Comparator	MCMPC4	Four-bit magnitude comparator with enable	4		18
	MCMPC8	Eight-bit magnitude comparator with enable	6		36
Counter	CNT4A	Four-bit binary counter with load and clear	4	4	8
	CNT4B	Four-bit binary counter with load, clear, carry-in, carry-out	4	4	7
	FCTD16C	Fast 16-bit down counter, parallel loadable	2	19	33
	FCTD8A	Fast 8-bit down counter, parallel loadable	1	10	18
	FCTD8B	Fast 8-bit down counter, parallel loadable	1	9	13
	FCTU16C	Fast 16-bit up counter, parallel loadable	2	19	31
	FCTU8A	Fast 8-bit up counter, parallel loadable	1	10	17
	FCTU8B	Fast 8-bit up counter, parallel loadable	1	9	12
	UDCNT4A	Four-bit up/down counter with load, carry-in, and carry-out	5	4	13
	VCTD16C	Very fast 16-bit down counter, delay after load, registered control inputs	1	34	41
	VCTD2CP	Two-bit down counter, prescaler, delay after load, used to build VCTD counters	1	5	2
	VCTD2CU	Two-bit down counter, upper bits, delay after load, used to build VCTD counters	1	2	3
	VCTD4CL	Four-bit down counter, lower bits, delay after load, used to build VCTD counters	1	4	7
	VCTD4CM	Four-bit down counter, middle bits, delay after load, used to build VCTD counters	1	4	8
Decoder	DEC2X4	Two-to-four decoder	1		4
	DEC2X4A	Two-to-four decoder with active low outputs	1		4
	DEC3X8	Three-to-eight decoder	1		8
	DEC3X8A	Three-to-eight decoder with active low outputs	1		8
	DEC4X16A	Four-to-sixteen decoder with active low outputs	2		20
	DECE2X4	Two-to-four decoder with enable	1		4
	DECE2X4A	Two-to-four decoder with enable and active low outputs	1		4
	DECE3X8	Three-to-eight decoder with enable	2		11
	DECE3X8A	Three-to-eight decoder with enable and active low outputs	2		11
Latch	DLC8A	Octal latch with clear active low 8-bit data latch with low clear	1	8	

for configuration of the device. The two options available for the PLD designer are the Boolean logic equations and the state machine syntax. The Boolean equations are the only option for the combinatorial designs, and they can also be efficient for some registered designs. The Boolean equations can be derived from a combination of the functional description, the truth table, and/or the timing diagrams. The state machine approach is best suited for large registered control designs and can be derived from the functional description, state table, state diagram, and/or the timing diagram.

The Boolean logic equations are used to represent the sum-of-products logic form and are quite suitable for representing the two-level AND-OR structures available in most simple PLDs. A conventional approach to the design is to convert a design problem to its discrete SSI and MSI logic components for implementation in the PLDs. However, a direct implementation of a design in sum-of-products in a PLD can yield a more efficient and faster circuit. Boolean equations can be directly derived from the truth table, which is more often used in simple combinatorial designs, or from the timing diagrams used in registered control designs, as shown in Fig. 4.12*a*.[1] Figure 4.12*b* shows the state machine description approach.

The following are the programming languages commonly used for simple PLD designs:

- *ABEL.* Advanced Boolean Expression Language (ABEL) is the universal programming language developed by Data I/O Corporation.
- *CUPL.* Universal Compiler for Programmable Logic (Logical Devices trademark).
- *PALASM.* PAL Assembler language, originally developed by Advanced Micro Devices (AMD).

The design simulation is an integral part of the PLD design process, and the purpose is to exercise all inputs and test the response of system outputs to verify that they will function as desired. This is done through simulation test vectors, which identify any flaws in the design equations that could affect the logical operation of the programmed device. Therefore, the simulation vectors serve as a design debugging tool and eventually make up part of a larger set of test vectors called *functional test vectors*. These functional test vectors are used to test the programmed PLDs to identify any individual devices that may be defective. The simulation is included in the design file along with the logic equations. The simulation vectors (or events) can be directly derived from the truth table and the timing diagrams of the design. The logic level and functions of all signals can be expanded and expressed in a test vector form using software.

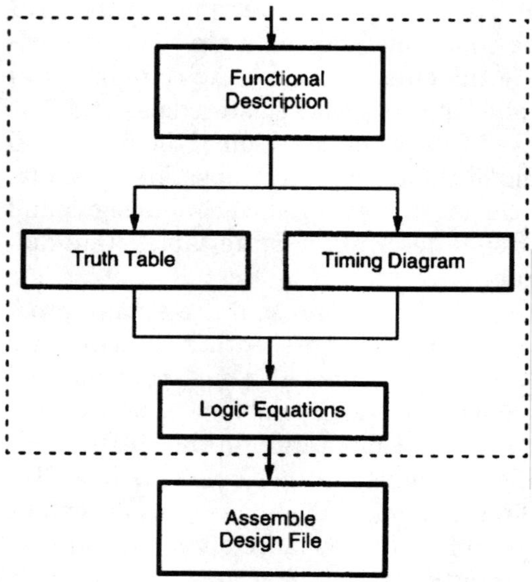

(a)

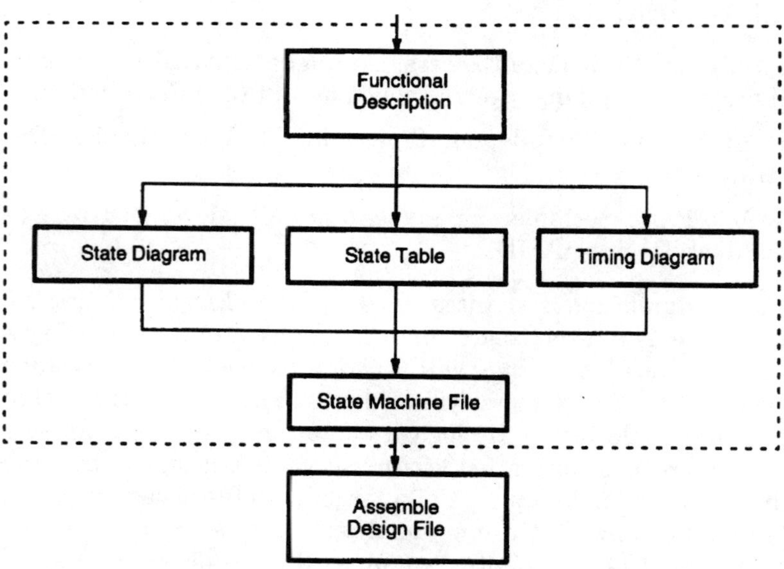

(b)

Figure 4.12 (a) Boolean logic equations approach and (b) state machine description *(from Ref. 1)*.

The simulator interprets the design file and generates the output logic levels and/or waveforms for design verification. Once the simulation is complete, the design file can be assembled to generate the JEDEC file. The procedure for assembly varies with different software packages. Once the design simulation is completed, the final step is device programming and testing. There are two types of programmers generally available: menu driven and device code based. The menu-driven programmer directly indicates the part type being programmed, whereas the latter type requires the user to enter the device code before programming. Once the JEDEC file has been downloaded, the device can be programmed and verified after the programming cycle. Figure 4.13 shows a typical PLD design, simulation, programming, and test flow chart.[1]

4.3 FPGA/CPLD Design Methodology

FPGA and complex PLD design methodology basically consists of following three steps: (1) design entry, (2) design implementation, and (3) design verification. Figure 4.14*a* shows the flow chart of basic FPGA and CPLD design methodology. These three major steps are discussed in the following sections.

4.3.1 Design entry

The design entry is the process of specifying logic functions needed for a given application. Schematic editors are the most popular tools for design entry. Their use for FPGA design requires both an appropriate library of logic symbols and a translator that can convert the schematic into a standard netlist format. FPGA symbol libraries and netlist interfaces are available for schematic editors from most popular third-party CAD tools suppliers such as Viewlogic Systems, OrCAD, Mentor Graphics, and Cadence Design Systems. These libraries include wide variety of commonly used logic functions that can be implemented in FPGA designs. Schematic entry tools are well suited to glue logic designs. However, as the designs grow in size and complexity, behavioral design approach that requires hardware description of the design in high-level languages, such as VHDL and Verilog HDL, are gaining popularity.

One of main attractions of the hardware description language (HDL) methodology is the design portability offered; i.e., a design described in HDL and use of logic synthesis compiler can help make the design migration to multiple FPGA/CPLD architectures easier. However, in reality, completely technology-independent FPGA designs with HDLs are difficult because of wide variations in coding styles and

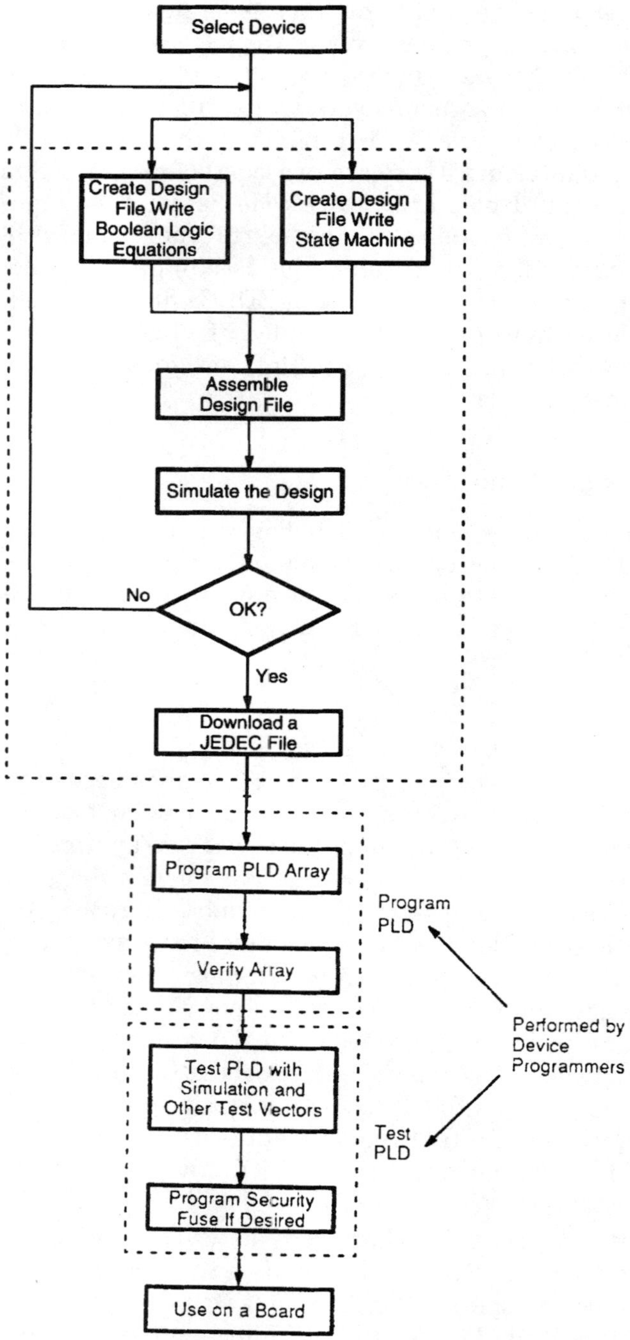

Figure 4.13 Typical PLD design, simulation, programming, and test flowchart *(from Ref. 1)*.

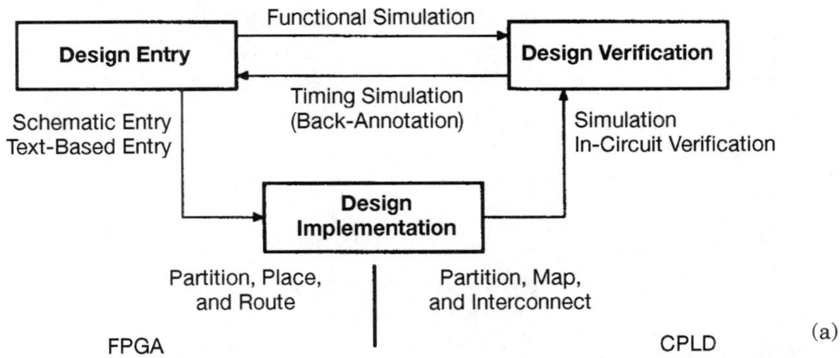

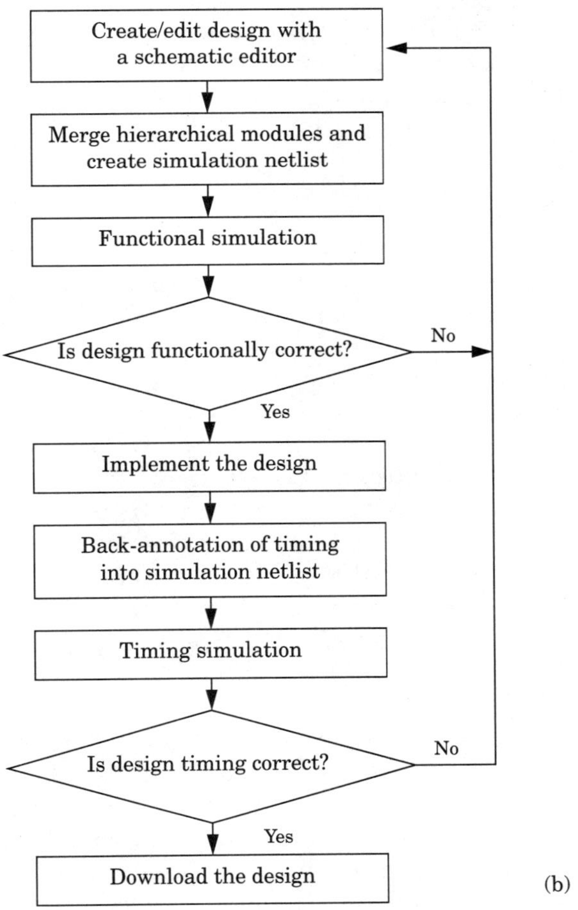

Figure 4.14 FPGA/CPLD (a) basic design methodology and (b) typical design flow steps *(from Ref. 3)*.

FPGA architectures. Therefore, this approach requires FPGA *fitters,* the device specific tools for logic optimization to fit the target FPGA architecture. Also, the migration of HDL-based designs is aided by standards such as *libraries of parameterized modules* (LPMs), a set of generic, parameterized logic functions. These high-level design languages and logic synthesis issues will be discussed in more detail in Sec. 4.4.

Most FPGA development systems support hierarchical design entry, i.e., the top-level drawings define the major functional blocks, whereas the lower-level descriptions define the logic for each block. These development systems can combine hierarchical elements that are specified with multiple design entry tools and support "mixed-mode" design entry, which allows the designers to intermix schematic, text, gate-level, and behavioral-level design descriptions. The major advantages of the hierarchical approach is that it allows large, complex designs to be broken down into smaller, more manageable pieces, which also eases debugging. Figure 4.14*b* shows the typical FPGA design flow steps.[3]

4.3.2 Design implementation

After the design entry, the physical implementation of design involves several steps, which include the translation of input files into the appropriate netlist format, merging the hierarchical design elements, deleting the unused logic, and mapping the design into FPGA's available logic resources. This is followed by the placement and routing phase, which involves determination of an optimal placement of the logic, and selection of the routing channels that connect the logic and I/O blocks. Design implementation tools, also known as *place-and-route* tools, apply a high degree of automation to optimize and map the design into the FPGA architecture. These tools tend to be unique to each FPGA architecture but should have a seamless interface with their design entry and verification tools. A number of sophisticated algorithms have been developed that automatically partition, place, and route a design within the target FPGA architecture.

The mapping partitions the design into chunks that can fit into the logic and I/O blocks of the target FPGA architecture. A partition advantage is gained by using a device viewer, a target tool for "peeking" inside the device architecture. This device, originally intended for post-route analysis, presents a good opportunity for determining the effects of various preroute implementation proposals. Once the designer becomes aware of the effects of mapping on placement and the effect of placement on routing, he can choose between a routing driven placement, a placement-driven mapping, and a mapping-driven implementation. The largest class of partitioning constraints and dependencies derive from the target device itself, and the major concerns include size, speed, and available resources. Therefore, the FPGA design success depends on the early recognition and consideration of the target device constraints and dependencies to achieve an optimal implementation.

The major objectives of *floorplanning* are to optimize the chip area and minimize the delay. For delay prediction, the designer needs to know the parasitics associated with the interconnect, i.e., the interconnect capacitance and resistance. The goal of a placement tool is to arrange all logic cells within the flexible blocks on a chip to (a) ensure that the router can complete the routing step, (b) minimize the total estimated interconnect length, and (c) minimize critical delays (i.e., meeting the timing requirements for critical nets). After completing the floorplanning, one can begin the placement of logic cells within the flexible blocks. The placement is more suited to automation than floorplanning.

A mapping algorithm for FPGAs must distinguish between the combinatorial logic gates and other functions such as flip-flops and I/O pads. For example, many FPGAs discussed in Sec. 3.3 use look-up tables (LUTs) to implement combinatorial logic, and a number of algorithms have been developed to map logic into the LUTs. Basically, these algorithms optimize area by minimizing the total number of LUTs, or they optimize speed by minimizing the number of levels of LUTs needed for logic implementation. These algorithms differ in their approach for logic partitioning, but all of them involve a matching step that determines whether a given subset of logic can fit into a single logic block. Some algorithms also take routability into consideration.

The placement step takes the defined logic and I/O blocks from the technology mapper for assignment to physical locations in the target FPGA. Many of the placement algorithms used for the masked programmed gate arrays can also be used for FPGAs, although with some modifications. In these algorithms, placement is optimized by means of cost functions, which are based on variety of criteria such as performance and timing. The timing-driven placement tools also incorporate performance requirements into the cost functions of the placement algorithms. Most FPGA placement tools are based on some combination of following three algorithmic approaches: (1) mincut, (2) simulated annealing, and (3) general force-directed relaxation (GFDR) algorithms.

The mincut placement algorithm[4] begins by dividing the design into two clusters and placing half of the blocks on either side of a cut line. Then, the elements are moved across the line and swapped to reduce cost, since the major criterion is the number of interconnects crossing the cut line to keep the connected blocks close together. The algorithm keeps on dividing each cluster into smaller clusters until each represents one logic block.

The simulated annealing[5] is a more complex algorithm, which is modeled on the process of crystallization of a solid from its molten form. In this algorithm, the elements of a randomly chosen placement are swapped such that, if the new placement results in an overall lower cost, it is accepted. If the placement results in a higher cost, there is a finite probability of acceptance of a bad move, which is proportional to the "temperature" during the "annealing" process. The temperatures are high at the beginning of the process and are gradually reduced. Since the simulated annealing requires so many iterations, it is critical that the placement objectives should be easy and faster to calculate. The simulated annealing requires the use of a slow cooling schedule, which means long CPU times. As a general rule, simulated annealing is capable of producing better results at the expense of long computer run times.

In the GFDR[6] algorithm, the nets between blocks are based on the physical model of a spring, whose strength is proportional to the criticality of the net and other cost considerations. The placement process involves swapping the location of two blocks or cycling the location of several blocks. The prime candidate destination for a given block is the *center of gravity* of those blocks connected to it.

Most FPGA routers' optimization process allows removal of an already established route for rerouting to accommodate another signal. The timing-driven routers calculate interconnect delays while they are routing and also include performance information in the decision process. The net criticality can improve layout results, although its effects are difficult for the user to control. An improvement in the timing of a critical net may cause other connections to become critical, and the results of second iteration may not be any better than the first attempt. Typically, several iterations of the place-and-route implementation software are required, and the results are analyzed. A design performance often depends on the speed of entire interconnect path traversing several nets. There are tools available, such as XACT-Performance from Xilinx and Timing Wizard from NeoCAD, that enable users to specify timing required along the entire path in an FPGA design.

In the placement process, estimated logic and routing delays help determine an optimal placement for each logic block in the FPGA. Once the placement is fixed, an automated router selects interconnects for the blocks. The designer can enter the specifications that reflect actual performance required for design, and the implementation program uses them to guide the placement and routing process. For timing simulation, FPGA implementation tools calculate the post-layout timing, including the actual delay of routing paths. This information is annotated into gate-level libraries for simulators such as Viewlogic System's Viewsim and Mentor Logic's Quicksim. Many simulation interfaces also support back-annotation, i.e., the mapping of timing information back into the signal names and symbols of the original schematic, to help ease the debugging effort. Most libraries supported by FPGA vendors provide enough information to support board-level simulation. Also, FPGA behavioral models such as SmartModel library from Logic Modeling can be used with a variety of third-party simulators to support system level verification.

The static timing analysis technique is also becoming quite popular for analyzing fully synchronous designs. Static timing analyzers examine a design's logic and timing to calculate signal path performance, identify possible race conditions, and detect set-up and hold-time violations. The latest FPGA implementation tools also have features such as incremental design support that are helpful in debugging and for making minor changes that are needed near the end of design cycle. After the placement and routing, the final step consists of generating the binary configuration program, which is downloaded into the FPGA to implement the specified logic.

4.3.3 Design verification

Design verification is a method of determining that the behavior, performance, logical function, electrical characteristics, and circuit timing of the design implementation meet the intended specification requirements. The FPGA design validation is usually performed through in-circuit testing and software simulation. The FPGA system design verification plan should address following questions and issues:[7]

- Define requirements or features that should be tested and the test methodology. Should the testing be performed through inspection, simulation, or formal methods?
- In the design process, when can each requirement or feature be tested: during the performance or behavioral modeling, or during synthesis code development, and so on?
- What version of the design representation will be used to verify each requirement or feature tested—whether at the system behavioral level, at the synthesizable component level, or at the system gate level?
- What kind of functional testing will be used: exhaustive testing, selective testing, or random pattern testing?

The software simulators (event-driven or cycle-driven), and hardware simulators (e.g., accelerators, emulators), dynamically exercise the UUT and predict how the actual device should respond based on the patterns supplied by a test bench. Simulators are usually divided into following categories or simulation modes:

- Behavioral simulation
- Functional simulation
- Static timing analysis
- Gate-level simulation
- Switch-level simulation
- Transistor-level simulation

Behavioral simulation includes no timing information and only provides an indication of whether the design will work. Functional simulation ignores timing and includes unit-delay simulation, which sets the delay to a fixed value (e.g., 1 ns). Gate-level or logic simulation can be used to check the timing performance. Static timing analysis is the most widely used form of simulation. Switch-level simulation is required only to verify the behavior of circuits with nodes that are not driven, or circuits using logic that is not complementary. The most complex and time-consuming transistor-level simulation is rarely used. Therefore, two types of simulators are normally used for CPLD/FPGA design: (1) logic simulator and (2) timing analysis tools.

The logic simulator for prelayout behavioral, functional and timing simulation is used to flag (or detect) any design errors. At this stage, in a nondeterministic FPGA architecture, the logic path delays are only estimates, since the wiring interconnect delays will not be known until the physical design aspect of place and route is complete. Then, based on postlayout timing information, designers need to add or back-annotate the post layout netlist, followed by an iterative process of postlayout timing simulation and verification.

The second type of simulator most often used is the timing analysis tool, which eliminates the need for supplying input waveforms as stimulus. The timing analyzer checks for critical paths that have large delays (e.g., due to a high-fanout net) and limit the speed of operation. The designers can set certain delay restriction on a net or a path as a timing constraint which, if exceeded, will be flagged as a timing violation. Each time timing violations are fixed, others may appear, especially while using place-and-route software that uses random algorithms. However, there are more complex and expensive logic synthesizers available that can automate this iterative design process.

A limitation of the simulation-based verification method is the amount of simulation required for design verification. The design verification coverage is only as good as the test bench suite. The development of test bench suite is the implementation of the verification plan. The test suites are often generated with the help of automated test-suite development tools. This includes all the test benches developed as part of the verification process. These test benches used in conjunction with the simulation-based verification tools provide a method of dynamically exercising the hardware design. The test benches are targeted at the various phases of the design process or various parts of the design hierarchy. A specific test bench consists of a copy of the unit under test (UUT), a routine to generate UUT stimulus, and another routine to check that the test was completed successfully. The UUT stimulus can be created either by some routine executing an algorithm or by reading in patterns from an external file.

The formal method-based design verification tools basically compare the logic functions of one model against the logic functions of another. If the individual functions are identical, the models must be identical. Two different types of tools are used: equivalence checkers and model checkers. Equivalence checking is more commonly used in the industry. It attempts to reduce or eliminate the need for simulation of derived models such as synthesis or gate-level models, but not the original top-level model. If a synthesis model is statically equivalent to a gate-level model, there is no need to simulate the gate-level model to verify the functional accuracy. Thus, eliminating the need for extensive gate-level simulations can save time.

The model equivalence does not provide information about whether the synthesis model is functionally correct or about the timing characteristic of the derived model. A major deficiency of currently available equivalence checking tools is that they work efficiently only in cases where the two design representations are already similar; i.e., where each of the designs has an identical set of memory elements (a synthesis model and the derived gate-level model).

Recently introduced verification methods and tools, based on static analysis techniques, are becoming more popular. The manually created test benches are being supplemented by the tools that automatically generate test. Also, the event-driven simulation tools are being replaced by cycle simulators, the simulation is being supplanted by formal test methods, and simulation-based timing analysis is being replaced by static timing analysis.

4.3.4 Programmable logic design using timing-driven tools

The programmable logic design approach with timing driven tools requires a timing specification that can be used to create solid timing constraints.[8] The first stage of timing-driven design is using this high-level specification to come up with an architecture that will perform the function in a satisfactory amount of time. If there are no timing estimates available, many iterations may have to be performed through the entire design process to meet the timing constraints. A timing-driven design tool must have a way to estimate the effects of a design decision on the timing of the design. Many FPGA design tools provide macros that have predefined logic mapping. The data sheets for FPGAs give some typical logic propagation and interconnect delay information. The logic propagation delay based on a device fanout is relatively accurate. However, typical interconnect delays vary considerably, depending on the architecture. After mapping, but before place and route, fixed-routing delay architectures offer much tighter estimates on timing. FPGAs and CPLDs all offer more than one method of routing and have different delays associated with the different routing methods.

CPLDs, with their sum-of-products architectures, can usually fit a large amount of combinatorial logic in a single level, which simplifies their timing estimation. For CPLDs, the placement process is relatively easy, since the placement within certain zones will have fixed timing. Some CPLD architectures such as Lattice ispLSI families (see Sec. 2.3.5) have relatively small placement-dependent timing effects. There are other CPLDs in which the placement-dependent timing effects resemble those of small, multiple sum-of-products PLDs. The connections within one of the small PLD blocks will have one short interconnect delay, whereas the connections that span between the blocks will have a longer interconnect delay.

Timing models are simplified block diagrams provided by the CPLD manufacturers to illustrate the propagation delays through a specified product family. Since logic can be implemented using different routing paths, an approximate propagation delay can be estimated by tracing the actual routing paths. As an example, for Altera MAX 5000 and MAX 7000 EPLD designs using MAX+PLUS II, the actual routing paths used can be identified by examining the equations listed in MAX+PLUS II Report File (.rpt). Then, the appropriate microparameters can be added to calculate the propagation delay through the Altera device. Figure 4.15 shows the timing model for MAX 5000/7000 EPLD.[9]

The graphical structures that correspond to making all the connections for a net are commonly known as *trees*. There are special classes of trees called *Steiner trees* that minimize the total length of interconnect delay and are basis for many routing algorithms. In a timing-driven placement, one must estimate delay for every trial placement, possibly for hundreds of thousands of gates. The minimum-length Steiner tree does not necessarily correspond to the interconnect path

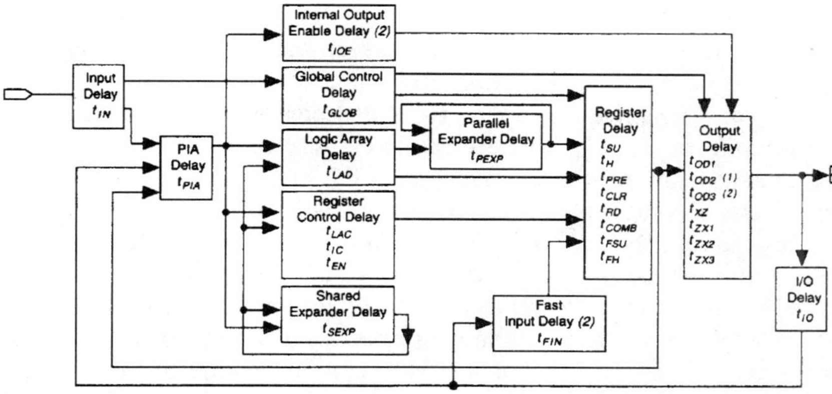

Notes:
(1) Not available in 44-pin devices.
(2) Only available in EPM7128E, EPM7160E, EPM7192E, and EPM7256E devices.

Figure 4.15 Altera MAX 5000/MAX 7000 EPLD timing model *(from Ref. 9).*

that minimizes delay. Therefore, to minimize delay path, one may have to route with non-Steiner trees.

The non-timing-driven tools attempt to connect all nets with the shortest possible delays while still fully routing the device. The routing algorithm and tools can automatically place and route with timing that meets the specification. However, if the desired goals are not achieved, there are two general approaches available, as follows: (a) prioritizing the nets to make critical nets with shorter delays, and (b) use interactive tools to alter the placement and routing of the design.

The prioritization of nets is usually performed by actually listing all the nets (or at least the critical nets) in the order the designer wants them to be routed or, as an alternate, to give some nets a criticality rating that places them in groups of different priority. The prioritization of nets works well in some cases for reducing the interconnect delays—especially when the percentage of nets that need to be very fast is relatively low. If one is already getting the shortest interconnect delay that the target FPGA offers, then no amount of prioritization is going to fix the problem. Then the only other options available are altering the logic design for greater speed or selecting another FPGA technology.

The timing-driven place-and-route tools avoid the deficiencies inherent in achieving timing goals by net prioritization. One deficiency is that the timing constraints are usually path based and seldom net based. Timing-driven place-and-route tools let the designers specify the delays based on signal paths and then create a placement and routing architecture that achieves the desired timing goals. Another deficiency with the net prioritization to meet timing constraints is that several iterations through the placement and routing steps are usually required, whereas the timing-driven tools allow the designers to specify system timing goals directly.

Timing-driven place-and-route tools use proprietary algorithms. A popular concept used in these place-and-route tools, as well as logic synthesis, is that of slack time. Typically, once the logic design is mapped into the logic blocks of the device architecture, the logic levels in each path and the associated logic propagation delays are known. A timing analyzer estimates the typical interconnect delay. Then, using the estimated timing information, the software can compute the slack time for each path, which is the maximum path timing constraint minus the typical path delay. The software can also compute the average slack time per net by dividing the path slack time by the number of logic levels. If the slack time is negative, the software can immediately flag the timing constraint as being tight, and no time is wasted trying to place and route a design to meet an impossible timing constraint. Therefore, timing-driven design tools at the system architecture level or the place-and-route level can use their timing estimation capabilities to flag timing problems so that they can be resolved earlier during design cycle.

4.4 High Level Language-Based Designs and Logic Synthesis

4.4.1 Hardware description languages: VHDL and Verilog

An HDL is a software programming language used for modeling the intended operation of a piece of hardware. In general, logic synthesis is the process of generating an optimized gate-level netlist for a logic design from some higher-level structural description (or language). There are two major types of hardware description languages in use. Single-module level HDLs are commonly used to describe small modules within a schematic based hierarchical design, e.g., ABEL®, CUPL®, and PALASM. These languages, and the earliest commercial synthesis tools, were developed in 1970s to support PAL types of devices.

The HDL constructs refer to three styles of defining functions: behavioral, data flow, and structural. The behavioral style defines the behavior of combinatorial logic circuits, such as the multiplexers and logic gates; the structural style defines the way for instances of previously compiled designs to be connected. The design language such as ABEL (Advanced Boolean Equation Language) provides high-level language constructs such as "if-then-else" for behavioral description, plus Boolean and arithmetic operations for data-flow description. It can also support limited code reuse by allowing instantiation of previously defined ABEL modules in other modules or ABEL design files.

In 1980s, *system-level HDLs* were developed that support more complex constructs and higher levels of abstraction than those offered by the single-module level HDLs. The most popular of these system-level HDLs are VHSIC hardware description language (VHDL) and Verilog-HDL. These HDLs are industry standards that are in compliance with IEEE standard 1076 (1993) for VHDL and IEEE 1364 for Verilog. The advantages of HDL-based designs include following:

- Higher productivity and yields
- Shorter development cycles, more product features, and reduced time to market
- Design portability and reuse, which results in increased flexibility for design modifications

VHDL is a programming language that has been designed and optimized for describing the behavior of digital circuits and systems. VHDL allows the behavior of complex electronic circuits to be captured into design system for automatic synthesis or for system simulation. VHDL has features that allow electrical aspects of circuit behavior such as functional operation, gate delays, and signal rise/fall times to be precisely described. The resulting VHDL simulation models can then be used as building blocks in larger circuits (e.g., by using schematics or block diagrams) for simulation purposes.

VHDL is similar to programming languages C and C++ and includes features useful for control and data representation. In addition, VHDL provides features that allow the description of concurrent events. An important aspect of VHDL is its ability to capture the circuit performance specifications in a form commonly referred to as a *test bench.* The test benches are VHDL descriptions of circuit stimulus and corresponding expected outputs for the verification of circuit behavior.

A top-down design methodology takes the HDL model of hardware, which is written at level of behavioral abstraction (system or algorithmic) down through the intermediate levels, such as the register transfer level (RTL), to a low (gate or transistor) level. The Fig. 4.16 pyramid represents the top-down design methodology and its various components.[10] The term *behavioral* implies the behavior (or functioning mode) of the intended hardware, independent of the abstraction level at which it is modeled. The major benefit of modeling complex hardware at higher levels of behavioral abstraction is to avoid describing a large number of unnecessary details and to reduce design complexity.

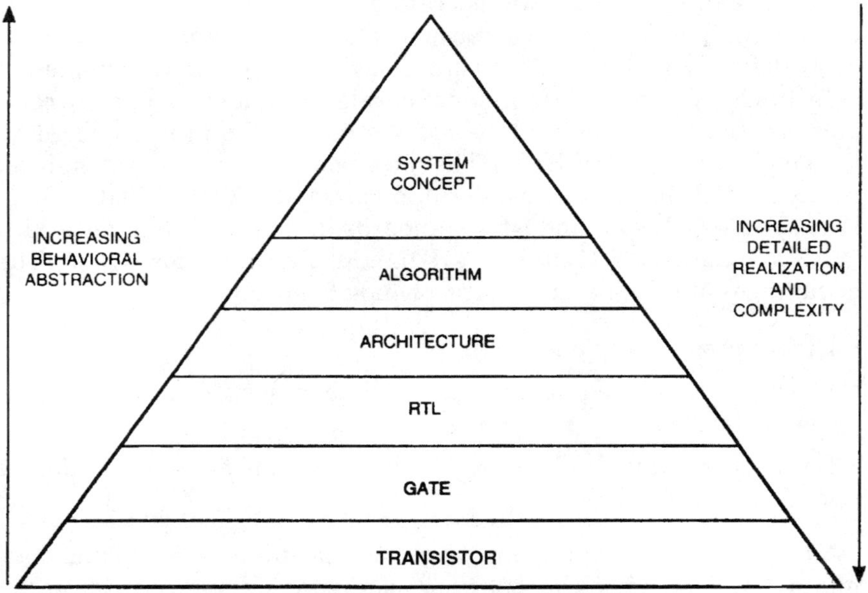

Figure 4.16 Top-down design methodology showing various levels of abstraction *(from Ref. 10).*

There are two basic aspects of a hardware design modeled by an HDL: true abstract behavioral modeling and hardware structural modeling. The HDL provides declarative statements to facilitate the abstract description of hardware behavior that meets the specification. During the design process also, the hardware behavior can be modeled and represented at various levels of abstraction. The following are some of the major attributes of high level behavioral design languages:

- *Modeling capability.* The hardware behavioral modeling can be performed in either VHDL or Verilog. To model abstract hardware in Verilog as efficiently as in VHDL, the programming language interface (PLI) is used. The modeling constructs of VHDL and Verilog cover slightly different spectras (or areas) of abstraction levels that are shown in Fig. 4.16. The choice of which behavioral description language to use, VHDL or Verilog, depends on the designer's preference, EDA tool availability, and other business and marketing issues.
- *Compilation.* Compilation is a way of speeding simulation without changing the original nature of the language. Therefore, one must be careful in writing both the compilation order of code for a single file and the compilation order for multiple files. Simulation results can change by simply changing the order of compilation. In VHDL, one

can separately compile multiple design units (entity-architecture pairs), which are in the same system file. In contrast, Verilog is still rooted in its native interpretive mode.

- *Data types.* In VHDL, one can use many language- or user-defined data types, which may require dedicated conversion functions to convert objects from one type to another. Therefore, one has to carefully consider the choice of which data type to use, especially with the enumerated (abstract) data types. In comparison, Verilog data types are simple, easy to use, and geared toward modeling hardware structure, as opposed to the abstract hardware modeling. Unlike the VHDL, the Verilog language (and not the user) defines all data types used in a Verilog model. For example, there are net-data types, such as *wire,* and a register data type called *reg*. Some designers prefer Verilog because of the simplicity of its data types.
- *Design reusability.* In VHDL, the procedures and functions can be placed in a package so that they are available for general use by any other design unit. In Verilog, which does not support package concept, one must define the functions and procedures that a model uses within the module statement itself. To make the functions and procedures generally accessible from different module statements in Verilog, they must be placed in a separate system file and included with an *'include* (pronounced *tick include*) compiler directive.
- *Structural replication.* The generate statement in VHDL replicates a number of instances of the same design unit or the same subelements of a design and connects them appropriately. In Verilog, there is no equivalent *generate* statement.
- *Low-level and high-level constructs.* The VHDL has simple two-input logical operators built into the language, such as NOT, AND, NOR, NAND, NOR, XOR, and XNOR. The timing must be specified separately using the *after* clause. One needs to use separate constructs defined in IEEE 1076.4 VHDL initiative toward ASIC libraries (VITAL) standard to define the cell primitives of ASIC and FPGA libraries. Verilog, which was originally developed for gate-level modeling implementation, also has good low-level constructs for modeling at this level for ASIC and FPGA library cell primitives.

  The VHDL has more constructs and features for high-level modeling as compared to Verilog. In VHDL, one can use the abstract data types along with following statements: model reuse, configuration statements for configuring design structure, generate statements for replicating structure, and generic statements for generic models that can be individually characterized. These language statements are useful in synthesizable models.

- *Language extensions.* The use of language extensions makes the model nonstandard and may affect the portability to other design tools. However, the extensions are sometimes necessary and in VHDL there is an attribute called *'foreign* that allows modeling architectures and subprograms in another language. In Verilog, the PLI serves as an interface mechanism between Verilog models and either models in another language or Verilog software tools.
- *Libraries.* In VHDL, a library is a storage area in the host environment for compiled entities, architectures, packages, and configurations. Verilog does not have the concept of library, because of its origin as an interpretive language.
- *Managing large designs.* The VHDL has *configuration*, *generate,* and *package* statements, along with the generic clause to help manage large design structures. There are no such statements in Verilog.
- *Procedures and tasks.* The VHDL allows concurrent *procedure* calls, whereas the Verilog does not allow concurrent *task* calls.
- *General attributes for VHDL versus Verilog.* VHDL is a detailed and verbose language with roots based in ADA programming language. It may seem less intuitive at first, because of following two reasons:

  1. It is very strongly typed language, which requires precisely coded models using defined and matching data types. This is an attribute that makes it robust and powerful for an advanced user—after a relatively longer learning curve.
  2. There are many ways to model the same circuit, especially those with larger hierarchical structures.

  In comparison, Verilog is more like C programming language, with constructs based approximately 50 percent on C and 50 percent on ADA. If one has no knowledge of either language, Verilog is probably easier to grasp and understand. This statement assumes an exclusion of the Verilog compiler directive language for simulation and use of the PLI.

4.4.2 Logic synthesis

In general, logic synthesis is the process of generating an optimized gate-level netlist for a logic design from some higher-level structural description. It is typically considered as a two-step process. The initial step is to translate the logic representation from a high level such as register transfer level (RTL) to a lower level, e.g., generic gates or cell primitives. The second step then optimizes the generic logic to take advantage of the specific logic resources in a target FPGA. This automated translation and optimization process provides several benefits to the designer.

The primary benefit of logic synthesis is to improve a designer's productivity by speeding up the translation of high-level design description into design implementation, verification, and time to market.[11] This allows the designer to assess various high-level design trade-offs and evaluate their impact at the gate level with much faster turnaround time than the traditional, schematic entry based approach. The productivity advantages of logic synthesis become more attractive as the designs get larger and more complex. The combination of top-down design methodologies and logic synthesis provides an efficient means of migrating designs between technologies, i.e., technology transparent designs. However, for a top-down design methodology to be useful, the synthesis tools must be effective in producing a gate-level design optimized for the target technology. The optimization algorithms for most FPGA architectures are very different from algebraic-based algorithms used for gate arrays. In addition, logic synthesis for FPGAs is complicated by variety of architectures available, since the synthesis tools must be capable of compiling efficiently the high-level circuit description into optimized circuits to fit the target FPGAs logic modules.

The synthesis is most effective when the designer has achieved some expertise with HDLs and logic synthesis tools. The synthesis-based design is further complicated by a wide choice of synthesis-related software tools, each with its own varying levels of interoperability, library support, and features. The various FPGA architectures often require different coding styles, synthesis, and optimization algorithms. For the top-down methodology to be successful, strong links are required between the logic synthesis, simulation, and physical layout tools. There are mixed-level simulators available that can support functional and timing simulation from both the HDL and gate-level descriptions. Synthesis tool efficiency also depends on the design description; i.e., a very abstract description may yield poor results, whereas a detailed description can overly constrain the synthesis tool.

In general, a synthesis compiler can map logic functions into an FPGA's resources (a) through inference or (b) through instantiation. It is highly desirable that the synthesis compiler infer the use of available resources as much as possible; i.e., the compiler should be capable of automatically mapping logic and I/O functions into the FPGA through its synthesis and optimization algorithms, beginning with generic HDL code. An example of available resources is the utilization of built-in clock enable functions, since most FPGA architectures include D-type flip-flops in their logic cells that have dedicated clock enable functions. The latest synthesis tools take this feature into account and select this resource whenever a multiplexer precedes the data pin of a register.

In comparison to the inference, sometimes specific FPGA resources can be accessed only through instantiation; i.e., the user explicitly specifies the resources that are to be used by embedding required directives within the HDL code. As an example, I/O characteristics of some FPGA architectures (such as I/O registers, internal pull-ups, and slew rate controls) must be specified explicitly in the HDL code. To do that, the designer must be familiar with target FPGA's architecture, and the resulting code is architecture specific and not easily ported to other device families.

In general, the current synthesis algorithms tend to perform well in the design implementation of logic elements such as look-up tables, state machines, and decoders. The FPGA synthesis tools often include the ability to implement state machines in various encoding styles. The CPLDs are best suited for heavily encoded state machines, whereas the register-rich FPGAs produce quite efficient implementation of small- to medium-sized state machines using One Hot Encoding (see Sec. 4.6.1) scheme, i.e., creating state machines with one flip-flop per state.

The synthesis of arithmetic and data path structures is difficult, especially if the target FPGA includes dedicated architectural resources for these functions, e.g., dedicated arithmetic carry logic. A new approach emerging is the use of the Library of Parameterized Modules (LPMs) standard to help alleviate this problem by allowing the synthesis compiler to pass high-level module description (as opposed to the gate level) to the target device vendor's tools for optimization. A high-level function description such as "16-bit binary up-down counter" allows the FPGA's vendor's device-specific LPM compiler to generate the optimized implementation (see Sec. 4.5).

The synthesis process can be effective only when the synthesis compiler has the knowledge of algorithms used to map the design logic into logic cells and accurate models of the interconnect delays, as well as the use of target FPGA's available resources. There have been improvements in FPGA synthesis compilers that allow the designer to perform area versus speed trade-offs. In synthesis of high-performance designs, some critical portions of circuits may have to be "manually tuned," e.g., through the instantiation of a particular FPGA resource available. Many FPGA development environments support mixed-mode design entry, making it possible to use different design methods for different portions of the design. This approach allows the designer to reuse previously designed schematic or HDL-based modules.

In synthesis-based design process, it is very important to use careful HDL coding style for high silicon utilization and improved performance. When multiplexing wide data paths, simple case-like, *if-then* coding style is preferable over deep *if-then-else-if-then* coding style. An example given below illustrates (a) original coding style of a VHDL based design and (b) improved coding style.

```
(a)   if (A = '1' and B = '0') then
      Q <= data1;
      elseif (A = '0' and B = '0') then
      Q <= data2;
      else
      Q <= data3;
      end if;

(b)   if (A = '1' and B = '0') then
      Q <= data1;
      else
      Q <= data3;
      end if;
      if (A = '0' and B = '0') then
      Q <= data2;
      end if;
```

These processes are logically same, but when synthesis was applied to a wide data path application with arithmetic operations in series, the improved coding style listed in (b) for above example yielded a 40 percent higher silicon utilization and a 15 percent increase in performance, relative to implementation using the original coding style of (a). Also, in HDL based designs, the coding styles may vary between that written for simulation versus the one for synthesis.

It is not difficult to create synthesizable code. However, there is no guarantee that the code will be more efficient than that of a schematic or an alternative design approach. In HDL based design, significant performance improvements are possible if, for example, efficient coding and synthesis result in conversion from two levels to one level of logic between the flip-flops, which can roughly double the clock speed.

Logic synthesis levels and tasks. The synthesis process and its practical applications include discrete tasks associated with the behavioral, register-transfer level (RTL), and logic synthesis levels. Each of these levels has specific tasks associated, although some tasks may overlap. The behavioral descriptions are at the most abstract level of synthesis and describe the module generic function, but not the details. The next level below are the RTL descriptions, also known as *data-flow descrip-*

tions, which define the system in terms of registers, switches (or multiplexers), and operations. The RTL description is different from the behavioral description, since it is based on an architecture and a clocking scheme. However, both RTL and behavioral descriptions are technology independent. Logic descriptions are the lowest level of abstraction, at which a Boolean network or netlist describes a design implementation. The various logic synthesis tasks are briefly described below:[12]

- *Partitioning.* The partitioning is a first synthesis task and divides the circuit design into smaller chunks (subcircuits) to be implemented as separate modules or boards. The partitioning may accommodate hard constraints such as the die size or package pinout limitations, or the implementation can be based on reducing circuit cost or minimizing signal delays.
- *Pipelining.* The pipelining partitions an algorithm's implementation flow into a number of sequential stages that can execute simultaneous operations, enabling a circuit to process data at a higher rate by working in parallel on different portions of the algorithm. A pipelined instruction may take longer to execute than a non-pipelined instruction. However, the overall program speed increases, since several pipelined instructions can execute simultaneously.
- *Scheduling.* The scheduling task assigns the operations in a behavioral description to a sequence of control states or clock cycles. Scheduling may not necessarily maintain the order of operations of the original design description, but it does preserve data and overall behavioral integrity.
- *Register allocation.* Register allocation (also called *register assignment*), although within the realm of behavioral synthesis, also straddles the RTL synthesis level. The register allocation task selects registers for storing values that get generated in one clock cycle and accessed in the later cycles. To minimize circuit area, each user variable does not get a dedicated register; i.e., the different values of a variable may be stored in different registers.
- *Resource allocation.* The resource allocation task involves selection of components (e.g., counters, adders, multipliers) to implement the operations of a behavioral description such as addition, subtraction, multiplication, and so on. Resource assignment selection is made from a set of available library components.
- *Register inferencing.* The register inferencing task instantiates a register (or a latch) to store each value and then connects the appro-

priate clocking, asynchronous-reset, and/or load-enable pins. Register inferencing determines which values must be preserved across cycle boundaries and under what conditions those values must be preserved.

- *State machine synthesis.* State machine synthesis translates a state table or graph into the binary encodings of the symbolic states of a finite state machine (FSM). Therefore, these encodings determine the logic functions and number of registers in the resulting implementation.
- *Multilevel logic optimization.* Multilevel logic optimization takes a netlist of the gate-level combinatorial logic and creates a new description that optimizes area or performance (faster circuit), or both. Typically, these improvements are made through a series of transformations called *restructuring* and *simplification.* The restructuring process finds common logic that multiple equations can share to reduce the logic network. The simplification finds simpler logic equations with the same behavior as the original input description.
- *Two-level logic optimization.* A two-level logic, also known as a PLA or AND-OR representation, is a specialized form of multilevel logic. For the designs that can be represented by two-level logic, special algorithms and tools, such as "Espresso" from University of Berkeley, are used for logic optimization.
- *Redundancy removal.* Redundancies waste circuit area, may affect performance, and can make test pattern generation more difficult, since redundant portions of the circuit are untestable. Redundancy removal task is the process of identifying and removing these redundancies.
- *Technology mapping.* Technology mapping transforms a technology-independent Boolean network into a netlist that is specific to a particular FPGA architecture. The purpose of technology mapping is to find the combination of logic elements that best meet the designer's goals for circuit performance, area, or power consumption. The technology mapping process may actually involve two steps: library mapping and technology mapping. The library mapping translates a generic logic description in terms of a specific library of components. For example, a truth table description may be mapped into a netlist of AND/OR gates. The following step of technology mapping transforms the netlist into specific logic structures existing within the target FPGA. As an example, if Xilinx FPGAs are the target devices, the netlist would be mapped into specific configurable logic blocks (CLBs) of the logic cell arrays.

- *Technology translation.* Technology translation, which is a specific application of technology mapping, converts a design from one technology specific implementation to another.

Logic optimization and technology mapping. The most commonly used approach to synthesis divides the process into two phases consisting of a technology-independent phase followed by technology-mapping phase.[13] The technology-independent phase attempts to generate an optimized abstract representation of the circuit, whereas the technology-mapping phase selects the set of logic elements from the target device library while optimizing area, delay, or some trade-off between the two goals. The operations performed in the technology independent phase can be divided into two classes as follows: (a) network restructuring operations and (b) node minimization.

The network restructuring includes operations that modify structure of the Boolean network by eliminating some nodes, introducing new nodes, and by adding and removing arcs, which simplifies equations associated with the nodes. The network restructuring operations include decomposition, extraction, factoring, resubstitution, and collapsing. The node minimization attempts to reduce the complexity of a given network by using Boolean minimization techniques on its nodes, e.g., tools such as Espresso for two-level minimization. After the network optimization, the technology mapping takes this optimized network and maps it into the available library and logic resources of the target FPGA, using algorithmic approaches or rule-based techniques.

Synthesis for LUT-based FPGAs. The LUT-based logic blocks, such as Xilinx Configurable Logic Blocks (CLBs), can implement any logic function of no more than a fixed number of variables. Since a traditional cell-based technology mapper requires an explicit library definition, it is not feasible to enumerate all functions that a CLB can perform. There are number of LUT-based technology mappers that have been proposed recently. The first group of mappers such as MIS-pga,[14] Chortle,[15] Chortle-crf,[16] and Xmap[17] focus on area minimization. These technology mappers try to minimize the number of CLBs in the mapped network. However, since an FPGA contains a fixed number of CLBs, reducing the number of CLBs is not a big payoff as long as the design can fit into the chip. It should be noted that minimizing area, which is the most common goal of many synthesis tools, may actually result in slower design implementation.

The second group of technology mappers, such as Chortle-d and FlowMap, focus on delay minimization or performance optimization. There are three basic approaches to synthesis for performance optimization:

1. Delay optimization is equated to minimizing the depth in the network.[18]
2. The circuit delay is approximated by a combination of the block delay and interconnect delay. Also, interconnect delay is estimated as a function of number of levels, nodes, and edges in the network.[19]
3. Critical path analysis is performed on the placed circuit using a simplified delay model.[19] In this approach, logic optimization is performed in conjunction with the actual layout.

The first approach is faster, since the number of levels in a circuit correlates well with the circuit performance, especially when the delay is mostly due to the block delays. The third approach is more accurate but may be computationally inefficient due to the complexity of mixed-layout synthesis algorithms used. The second approach represents a compromise between the need for better accuracy and compute-time requirements.

Synthesis for multiplexer-based architectures. The multiplexer-based architecture uses logic blocks that are combinations of a number of multiplexers and possibly few additional AND/OR gates. An example of this architecture is Actel FPGAs ACT 1, ACT 2, and ACT 3 families (see Sec. 3.3.1). As in the case of LUT-based architectures, the number of blocks, the logic function implemented by them, and the wiring resources are the main constraints. Therefore, the architecture-specific mapping also starts with a network that has been optimized using technology-independent operations. Two basic methodologies used are library-based technology mapping or direct approaches.

In library-based technology mapping, a library is created that has gates that represent all the functions obtained from the multiplexer-based logic block, either by tying the inputs to constants or by bridging some of them. Technology-mapping algorithms based on dynamic programming are quite effective for libraries with few hundred gates but are considered too slow for significantly larger libraries. Therefore, library-reduction techniques are applied to reduce the number of gates. An advantage of library-based mapping is that it is completely insensitive to changes in the logic block architecture. The only change that needs to be made is the creation of a new library. In addition, the same tool could be used for other target technologies.

In the direct approaches, no library is generated. The mapping is performed directly onto the logic blocks. There are various direct approaches proposed for MUX-based architectures, which are quite similar and not much different from the standard library-based approaches.[20–22]

Logic synthesis tool selection. A major benefit of the HDL synthesis is that it raises the level of design abstraction away from the device architecture details, which helps increase the designer's productivity. However, field programmable logic devices exhibit more complex and diverse architectures than the ASICs and therefore present significantly more difficult problem to solve for the synthesis tools. HDL synthesis tool suppliers such as Synopsys, OrCAD, and others have worked closely with FPGA/CPLD suppliers and developed specialized architecture-specific optimization algorithms for each programmable logic device supported. These are some of the factors that should be taken into consideration during the selection of a synthesis tool:[23]

1. *Automatic, complex I/O mapping.* The programmable logic device architectures use highly integrated I/O functions. The input and output registers are directly connected to I/O pins to optimize speed-critical performance parameters such as clock-to-output (T_{co}) and input setup time (T_{su}). If the synthesis tool does not have an automated process to infer the best I/O mapping, the designer must either manually insert the complex structure or select simple buffers for inputs and outputs. Therefore, the synthesis tool must provide automatic mapping of the best I/O structures for each design input and output to facilitate analysis of the impact of different device architectures on the design, and to ensure high performance.
2. *Automatic resource sharing.* Complex functions as performed by multipliers and adders commonly require many of the internal logic elements available in a programmable device. Designers can waste valuable logic resources if they instantiate multiple instances of a function that could otherwise be shared. Therefore, the synthesis tool must provide automatic resource sharing and create separate functional module only when necessary. In synthesis tools, resource sharing allows automatic analysis of the entire design to look for functions that can be shared, even those functions that may not have been obvious to the designers while describing their designs in HDL.
3. *Automatic module generation.* Programmable logic vendors enhance capabilities of their devices by providing dedicated silicon resources to common functions such as fast-carry logic and wide decoding. Schematic capture systems typically provide designers with a range of precaptured options for commonly used functions such as multipliers, adders, subtractors, and comparators. The synthesis tools that are capable of automatic module generation rely on a built-in knowledge base that guide the optimal realization of a complex structure for a specific architecture by taking advantage of any dedicated silicon resources. For example, the designer may

simply specify the operation such as A = B + C, and the synthesis tool calls from its knowledge base to provide an implementation that best fits the design requirements.

Automatic module generators built into synthesis tools also speed up the overall design flow by eliminating external module generation. In the absence of built-in module generators, functional modules are left as *black boxes* in the netlist. Then, an external module generator is run for each undefined black box in the design, which adds time and complication to the design flow.

4. *Constraint-driven synthesis and analysis.* Many synthesis tools offer a switch to control optimization for speed or area at the global level, but for not smaller parts of the design. This approach may not be best suited for designs that have some speed-critical functions, but others are not. In these designs, choosing speed optimization may result in an implementation that uses more logic elements than necessary for the functions that are not speed critical. The selection of area optimization may result in logic implementation that does not meet the performance requirements.

   Therefore, constraint-driven synthesis enables designers to identify external and internal performance requirements. To properly guide the synthesis tools, separate constraint settings for multiple system clocks, on-chip and off-chip delays, and internal path delay specifications are necessary. Then, using timing estimation models, the synthesis tools can evaluate whether design requirements are met during optimization. This approach delivers speed where it is needed and area efficiency where speed is not critical factor.

4.5 Technology-Independent Logic Design

Designers often have data paths and/or memory blocks in their designs that need to be optimized for the target FPGA architecture to meet the high-performance requirements of a design. Therefore, to fully utilize the data path and memory capabilities of the FPGA architecture, designers need parameterized modules that are optimized for the target architecture and for user's application.

Many vendors provide tools (e.g., SCUBA, from Lucent Technologies) that have synthesis algorithms that allow the designers to fully utilize the arithmetic, data path, and memory capabilities of the FPGA architecture. The designer can select a design implementation that minimizes the total gate count (and hence area), one that maximizes speed of operation, or a compromise between the two. The parameterized module generators have a thorough knowledge of the target architecture and are targeted specifically toward providing the best possible solution for the particular architecture rather than one-for-all solution.

A consortium of programmable logic and CAE software vendors have established a technical standard for logic design that provides a standard interface between the technology-independent and technology-dependent parts of the FPGA design process. This interface, called the *Library of Parameterized Modules (LPMs)*, is intended to allow designers access to various FPGA architectures without requiring detailed knowledge of each individual FPGA vendor's unique silicon architecture. A preliminary version of the LPM standard was presented to Electronic Design Interoperability Format (EDIF) Committee in March 1991. The EDIF Committee subsequently adopted it as an extension to the EDIF 2.0.0 standard (EIA-548-A). The result was a generic set of 25 parameterized templates for some of the most common data path functions including basic gates and multiplexer; arithmetic processing components such as adder/subtractor, counter, multiplier, comparator; and storage elements including latch, D flip-flop, T flip-flop, RAM, and ROM.

Figure 4.17 shows the LPM standard intermediate format represented as a high-level synthesis path as well as an interface between the technology-independent optimization algorithms and vendor-specific technology-mapping algorithms in the FPGA design process. The

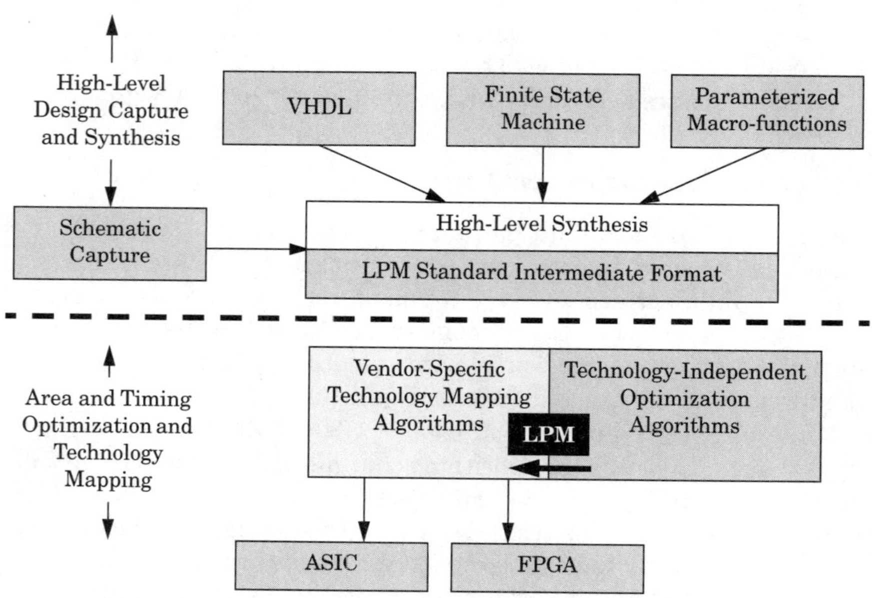

Figure 4.17 LPM standard providing an interface between the technology-independent and technology-dependent parts of the FPGA design process.

LPM comprises a set of modules that are used to describe the logical operation of a digital circuit. The LPM elements currently available range from basic gates and multiplexers to complex state machines and represent a fairly large number of logic functions, since each element can be modified or parameterized by the user. In addition to simplifying the design-entry process, LPM may also provide device technology independence.

Most of the large EDA vendors, including Cadence, Intergraph Electronics, Mentor Graphics, and Viewlogic currently support the LPM approach. Some EDA companies are teaming up with the silicon vendors to map LPM output into the programmable logic device's architecture. The silicon vendors also support the concept of LPMs that can be imported into a design and fit directly into the target device. This portable code concept implies that one can switch silicon vendors without having to modify the design. Some PLD vendors, such as Xilinx and Actel, have their own library of block elements that let the designer instantiate circuitry into their devices from the library. Mentor Graphics has worked with Altera to produce the LPM Design Kit, which uses the EDIF standard netlist readers and writers from both companies. The Design Kit, available from Altera, takes LPM input from the Mentor environment and maps it into an Altera device. Altera has also placed into the public domain VHDL versions of its descriptions of LPM blocks.

Although the LPM appears to be an enabler of synthesis, according to the vendors such as Xilinx, it is not as efficient as their own X-BLOX modules and does not map as well into their devices. One of the barriers to LPM's wider acceptance is some synthesis tool vendors' contention that the current version of the library does not synthesize well. EDA vendors have invested in their own high-level libraries, and they want to protect that investment. Also, PLD vendors have to make sure that the LPM provides quality results before they fully accept it. There are even supporters of the LPM approach who agree that some of the more complex functions are better left out of the synthesis process, since synthesis does not always map them efficiently to silicon. Therefore, the LPM approach and its implementation is in the evolutionary stages, and its future appears unclear.

4.6 FPGA/CPLD Special Design Topics

4.6.1 Finite state machine theory and design

A state machine is a digital device that traverses through a predetermined sequence of states in an orderly manner. State machines are widely used in control applications for functions such as control signal generation, arbitration, event monitoring, multiple condition testing, and timing delays. A simple state machine may consist of PAL-device based combinatorial logic, output registers, and buried (state) registers. The state in such a sequence is determined by the values stored in the buried and/or output registers. Figure 4.18*a* shows the general model of a finite state machine which, in addition to the device inputs and outputs, consists of two essential elements: combinatorial logic and memory (registers).[24] The combinatorial logic can be considered to be made up of two distinct functional blocks: the *next-state decoder* (also referred to as *transition function*) and the *output decoder* (or *function*). The next-state decoder determines the next state of the state machine, while the output decoder generates the actual outputs.

The basic operation of a state machine consists of following two steps:

1. It traverses through a sequence of states such that the next state is determined by the next-state decoder, based on the present state and the input conditions.
2. It provides sequences of output signals based on state transitions. The outputs are generated by the output decoder based on its present state and input conditions.

The use of input signals for deciding the next state is also known as *branching*. In addition to branching, complex sequencers provide the capability of repeating sequences (looping) and subroutines. The transition from one state to another is called *control sequencing*. The use of input signals in the decision-making process for output generation determines the type of state machine. There are two widely known types of state machines: Moore and Mealy. Figure 4.18*b* shows the Moore type of state machine, in which the machine outputs are functions of the present state only. Figure 4.18*c* shows the more general Mealy-type state machine in which the outputs are functions of both the state and the input signals. There are Mealy-Moore type state machines also that have both Mealy- and Moore-type outputs. Most practical state machines are synchronous circuits that rely on the clock signals to trigger the state transitions. A single clock is connected to all of the state and output edge-triggered flip-flops, which allows a state change to occur on the rising edge of the clock.

There are three methods of state machine representation: state diagrams, state tables and flowcharts. These are briefly described below.

1. *State diagram.* The behavior of an FSM may be specified in a graphical format called *state diagram* or *state transition diagram,* as shown in Fig. 4.19. Each bubble represents a state, and each ar-

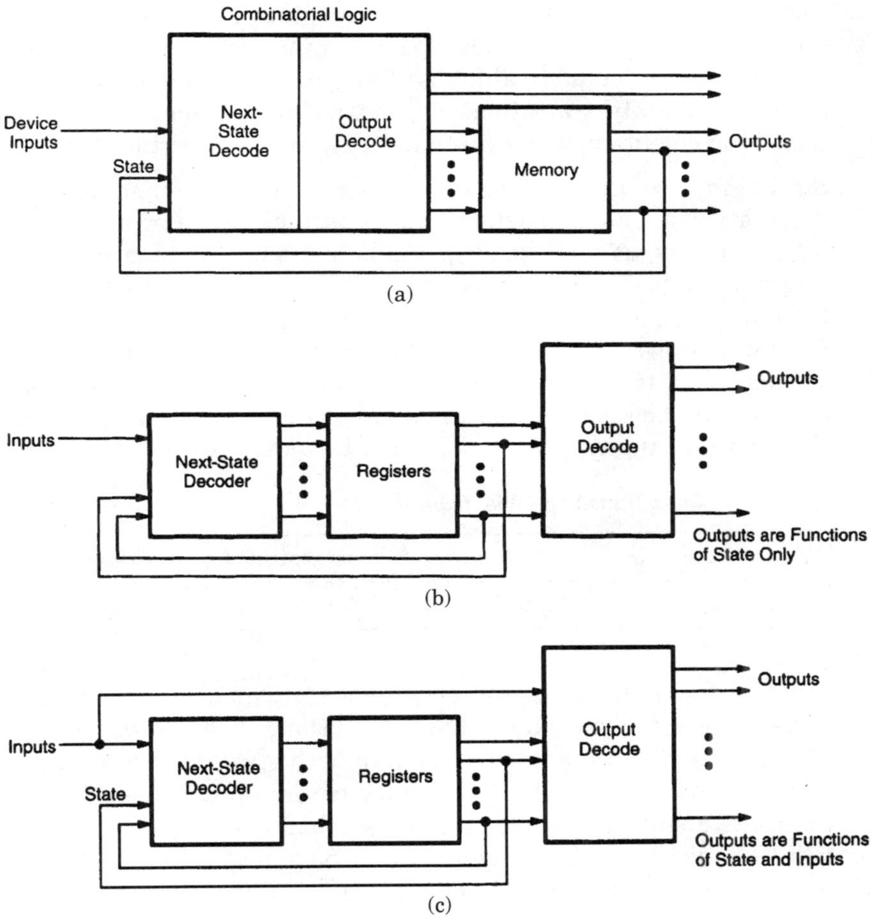

Figure 4.18 (a) General FSM model, (b) Moore state machine, and (c) Mealy state machine *(from Ref. 24)*.

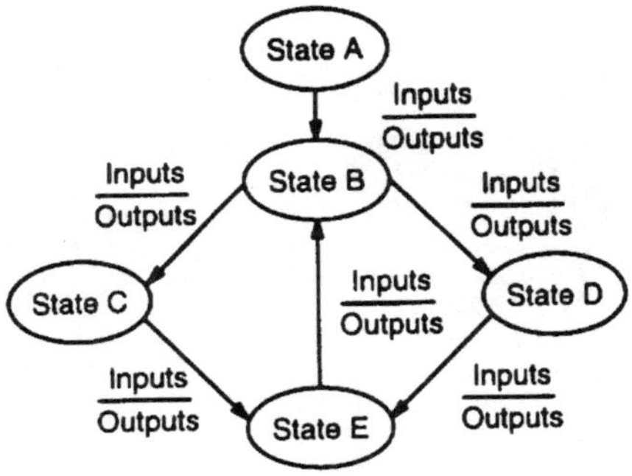

Figure 4.19 State machine representation *(from Ref. 24)*.

row represents a transition between the states. Inputs that cause the transitions are shown next to each transition arrow. State diagrams are quite popular, although they get more complex for problems where state transitions depend on many inputs, since the transition conditions are written directly on the transition arrows.

2. *State transition table.* The second method for state machine representation is known as state transition table, as shown in Table 4.14.[24] It lists, at the top, all possible input bit combinations and internal states. Each row gives the next state and the next output. Therefore, the table specifies both the transition and the output functions. However, this table is not suitable for specifying practical machines that have large number of inputs, since each input combination defines a row of the table. For example, in a state machine with 10 inputs, 1024 rows would be required.

TABLE 4.14 State Transition Table *(from Ref. 24)*

Present state	Inputs	Next state	Outputs generated
S0 – Sn	I0 – Im	S0 – Sn	O0 – Op

3. *Flowcharts.* Flowcharts are convenient state machine representations for designs that do not have more than about 10 states, and where up to 2 or 3 inputs are tested in each state. In flowchart notation shown in Fig. 4.20, rectangular boxes represent states, and alternative state transitions are determined by the strings of diamond-shaped boxes. The elements may have multiple entry points but, in general, have only one exit. The state name is written as the

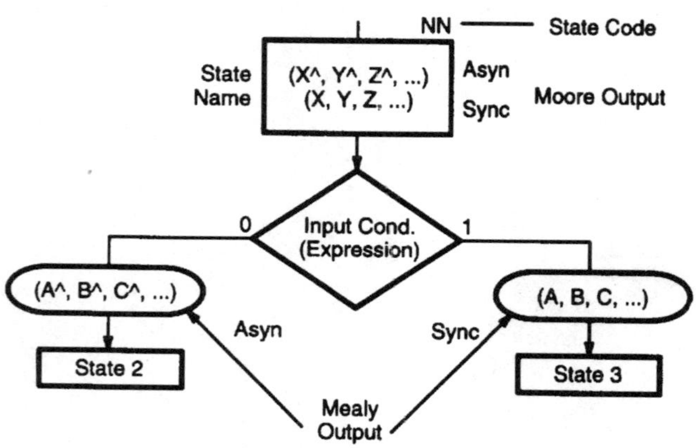

Figure 4.20 Flowchart notation *(from Ref. 24).*

first entry in the rectangular state box. Any Moore outputs present are written next in the state box, with a caret (^) following those that are unregistered. The state code assignment (if known) is written next to the upper right corner of the state box. The decision boxes are diamond or hexagonal shaped boxes containing either an input signal or logic expression. Two exits labeled "0" and "1" lead to either another decision box, a state box, or a Mealy output. The rounded oval is used for Mealy machine outputs. A caret follows those outputs that are unregistered. Each path, through the decision boxes from one state to another, defines a particular combination or set of combination of the input variables.

A designer's first objective in state machine design is writing a complete model that accounts for all possible stimuli and all directed responses. The second objective in state machine design is to generate an implementation that satisfies the required performance criteria and does so repeatedly. State machine performance is a function of both the next-state and output functions. Once the state machine is defined, software packages are used to implement the design in target device. Some of these software packages accept all three state machine representations directly as the design inputs. However, the more common methodology is to convert these state machine design representations into a textual format.

There are four key steps in state machine description entry and verification process that would include logic level reduction prior to synthesis and simulation, as follows:[25]

1. Enter state flow description using bubble diagrams.
2. Specify next-state transitions using equations.
3. Analyze next-state equations using Karnaugh maps.
4. Optimize next-state logic.

After complete state flow description and specifying all the next-state transitions, one can make a complete analysis of the next-state equations by simple application of Karnaugh maps. Although Karnaugh maps are most manageable for five or fewer variables, simple reduction and factoring techniques can make them effective in complex equations of six or more variables. The primary goal of the next-state optimization is the logic-level reduction and the first step in that direction consists of gathering up all the next-state terms, which is equivalent to accounting for all the transitions into a state bubble. Then, the next-state optimization can be performed manually or using sophisticated optimization tools. Neither Mealy nor Moore outputs are optimal for high-performance synchronous FPGA designs. Instead, the registered outputs derived from the next-state inputs usually yield higher performance.

The synchronous state machines can also be designed in VHDL. This approach provides the flexibility, since the behavioral model can be written without a priori knowledge of the type of the state machine to be used—Mealy, Moore, or Mealy-Moore. This is possible because VHDL processes that define the specific behaviors run concurrently, separating the output-conditioning logic into its own process such that the type of output (Mealy or Moore) is determined automatically. Therefore, the VHDL coding methodology is independent of the type of machine being developed. There are four basic steps in the VHDL-based design of a state machine with both Mealy and Moore outputs, as described below.[26]

1. The first step is to capture the design concept (its algorithm) using a state transition diagram.
2. The second step is to define the state machine's inputs and outputs. In VHDL, the state machine is treated as a *black box,* i.e., an entity-architecture pair, where the entity defines the interface of the design element to the external system, while the architecture defines the internal architecture and behavior of the entity.
3. The third step requires defining the states of the machine—the behavior (i.e., the body or architecture) of the FSM entity described. In VHDL, the states can be defined as enumerated types.
4. The fourth and final step is to write the body of the state machine. This allows the division of a state machine VHDL description into following three processes: (a) write a combinatorial process that performs the next-state conditioning logic, (b) write a synchronous process that creates the current-state variables, and (c) add the process that creates the output conditioning logic.

A state machine has a well defined sequence of states through which it traverses during normal operation. This implies the existence of a "first" state. One obvious problem is that some flip-flops—especially the older varieties—do not power up in a predictable state every time. Therefore, the flip-flops that power up into a random state must be initialized explicitly. However, lately, the flip-flops have become available that have a *power-up reset* feature that allows them to power up into a predictable state every time.

Another concern in the state machine design is the existence of *illegal states.* A properly designed state machine should not enter into an illegal state. However, there are situations when this may happen due to noise or other unpredictable occurrences. When it happens, there is an immediate need to return to a normal sequence of states. There are basically three ways to get out of an illegal state, as follows:[24]

- Reinitialize.
- Make sure that clocking can continue until the machine recovers.
- Design the machine such that the start-up state is reached from any illegal state in one clock cycle, independent of any conditional inputs.

In addition to the concerns listed above, the basic PLD design guidelines for high performance (see Sec. 4.7) are applicable to the state machine design as well and include the following:

- All state machine outputs should always be registered.
- State machine inputs, including resets, shall be synchronous.
- Clock skews must be minimized.
- Dead states must be avoided.
- All decode logic must be crafted carefully.

One-hot encoding state machine design. One-hot encoding (also called *bit-per-state* encoding) has become popular with the introduction of register-rich, fine-grain FPGA architectures. It is a technique that uses n flip-flops to represent a machine with n states. One-hot encoding is effective because it takes advantage of abundance of registers. In this approach, only one bit in the state vector is active, or *hot*, for any given state. This greatly simplifies state decoding, since only a single bit has to be examined to determine the state of the machine. An example of sample state machine design for the control section of a four-channel DMA controller, which is a super-macro for Actel FPGAs, will be discussed.[27]

Figure 4.21 shows the state diagram of the control section of this four-channel DMA controller. The state machine contains six states, seven inputs, and five outputs. Each circle represents a different state, and each arrow represents a transition between the states. Inputs that cause transitions are listed adjacent to the state transition arrows. Control outputs are labeled along with the states inside the circles. For example, in state S2, the state machine asserts the CNTD and CMREQ outputs. If the MACK input is low, the state machine remains at state S2. If the MACK input is high, the state machine goes to state S3 and, in the process, asserts the CE output.

The next step is determining the logic to generate the state sequence. For one-hot-encoding implementations, each state is assigned

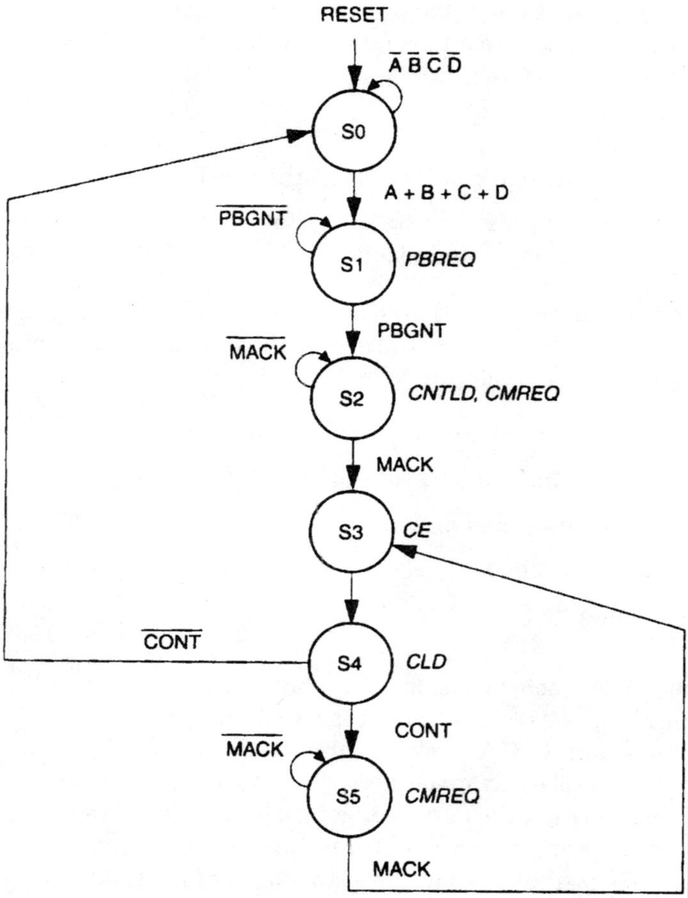

Figure 4.21 State diagram for control section of a four-channel DMA controller *(from Ref. 27 with permission of Actel Corp.)*.

to a separate register, and a state transition equation is written for each register. As shown in Fig. 4.21, the state S0 can be realized when state S4 is asserted and the input CONT is low, or it remains at state S0 if all four inputs (A, B, C, and D) are low. Therefore, the transfer equations of states S0–S5 can be written as:

$$\begin{aligned}
S0 &:= \overline{A} * \overline{B} * \overline{C} * \overline{D} * S0 + \overline{\text{CONT}} * S4 \\
S1 &:= (A + B + C + D) * S0 + \overline{\text{PBGNT}} * S1 \\
S2 &:= \text{PBGNT} * S1 + /\text{MACK} * S2 \\
S3 &:= /\text{MACK} * S2 + \text{MACK} * S3 \\
S4 &:= S3 \\
S5 &:= \text{CONT} * S4 + /\text{MACK} * S5
\end{aligned}$$

It should be noted that state S3 will go to state S4 unconditionally; therefore, the transition equation is written as S4 := S3. Once the state transition equations are determined, the output equations can be written by encoding the states. In some cases, the output is active only in one state and therefore is simply the function of that state.

For example, CE is active only in state S3, so the output equation for CE is simply CE = S3. For other outputs that are active in more than one state, the output equations of the state machine can be written simply as function of those states, as shown below.

```
PBREQ = S1
  CLD = S4
CNTLD = S2
CMREQ = S2 + S5
   CE = S3
```

The state machine can be captured using schematic entry, or it can be automatically mapped using synthesis tools such as Actel ACTmap VHDL Synthesis, for this example. Figure 4.22 shows the complete state machine VHDL file, which includes a reset signal (RST) to initialize the circuit.[27] Figure 4.23 shows the schematic of the state machine implementation using the one-hot encoding approach in the ACT 3 family.[27] It requires 18 logic modules after combining timing specifications for target device, Actel A1425A-2, as shown.

The one-hot encoding state machine design that uses one-bit-per-state methodology can be summarized as follows:

1. Draw a state diagram.
2. Assign each state to a separate register.
3. Write a state transition equation for each register.
4. Derive output equations based on active states.

This technique can be used for implementation of larger state machines by distributing control to several smaller state machines and using a single master machine to coordinate activities among the various state machines.

4.6.2 Digital signal processing (DSP) design

A general approach for DSP functions design is to use the programmable DSPs for low data rate applications, and fixed-function DSPs or function-and-algorithm-specific ICs (FASICs) for high data rate applications.[28] However, advances in PLD technologies and architectures pro-

```
--library ieee;
--use ieee.std_logic_1164.all;
entity DMASM is
        port (A,B,C,D:  in bit;
              PBGNT, MACK, CONT :  in bit;
              RST, CLK :  in bit;
              PBREQ, CMREQ, CE, CNTLD, CLD :  out bit);
        end DMASM;
architecture BEHAVE of DMASM is
        type STATE is (S0, S1, S2, S3, S4, S5);
        signal CURRENT_STATE, NEXT_STATE: STATE;
begin
SEQ: process (RST, CLK)
begin
        if (RST = '0') then
                CURRENT_STATE <= S0;
        elsif (CLK' event and CLK = '1' ) then
                CURRENT_STATE <= NEXT_STATE;
        end if;
end process;
COMB: process (CURRENT_STATE, A, B, C, D, PBGNT, MACK, CONT)
begin
  PBREQ <= '0';
  CMREQ <= '0';
  CE    <= '0';
  CNTLD <= '0';
  CLD   <= '0';
case CURRENT_STATE is
    when S0 =>
        if (A = '1' or B = '1' or C = '1' or D = '1') then
        NEXT_STATE <= S1;
        else
        NEXT_STATE <= S0;
        end if;
    when S1 => PBREQ <= '1';
        if (PBGNT = '1') then
        NEXT_STATE <= S2;
        else
        NEXT_STATE <= S1;
        end if;
    when S2 => CNTLD <= '1'; CMREQ <= '1';
        if (MACK = '1') then
        NEXT_STATE <= S3;
        else
        NEXT_STATE <= S2;
        end if;
    when S3 => CE <= '1';
        NEXT_STATE <= S4;
    when S4 => CLD <= '1';
        if (CONT = '1') then
        NEXT_STATE <= S5;
        else
        NEXT_STATE <= S0;
        end if;
    when S5 => CMREQ <= '1';
        if (MACK = '1') then
        NEXT_STATE <= S3;
        else
        NEXT_STATE <= S5;
        end if;
    end case;
  end process;
end BEHAVE;
```

Figure 4.22 Complete state machine VHDL file for control section of a four-channel DMA controller *(from Ref. 27 with permission of Actel Corp.)*

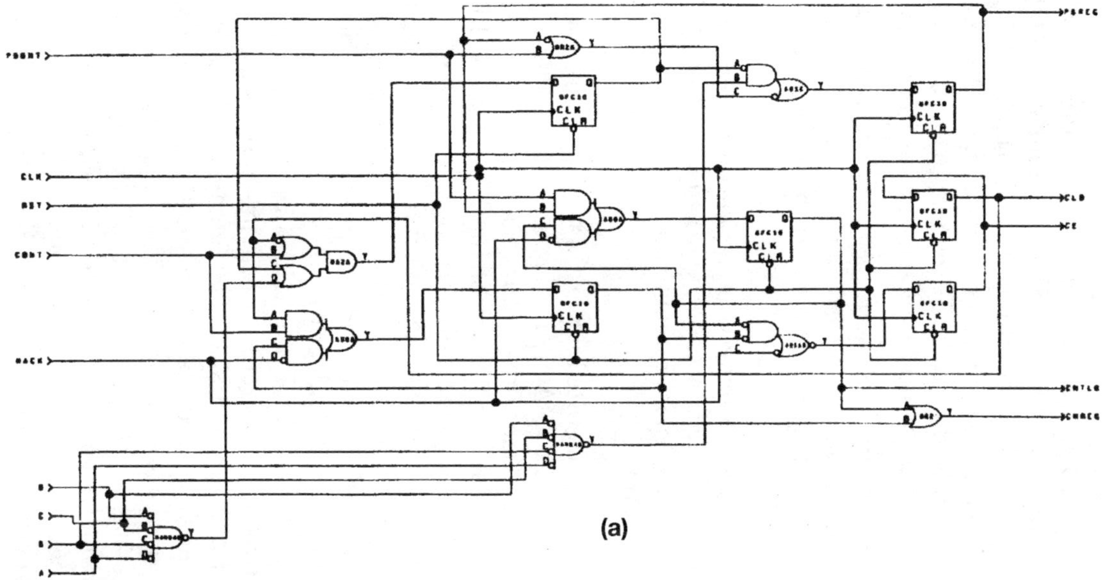

(a)

	DF1	NAND4D	AOI

Worst-Case Path	= tCO + tRD2 + tPD + tRD1 + tPD + tRD1 + tSUD
(from A, B, C, D	= 2.3 + 1.4 + 2.3 + 1.0 + 2.3 + 1.0 + 0.6
Registered Inputs)	= 10.9 ns

(b)

Figure 4.23 State machine design for a four-channel DMA controller: (a) schematic in ACT3 family and (b) timing using Actel A1425A-2 *(from Ref. 27 with permission of Actel Corp.)*

vide new options for the DSP design engineers to take advantage of parallel structures in FPGAs for implementation of the DSP algorithms, with performance exceeding that of the general-purpose DSPs. The following are some of the major advancements in PLD technology that make them attractive candidates for DSP design applications:

- *Improvements in density and performance, which are critical to implementing real-time system-on-a-chip interfaces.* Currently, 100,000-gate PLDs are available in production volumes and are capable of implementing designs with system speed faster than 75 MHz. Also, process improvements have enabled many PLD vendors to dramatically increase their yields and lower the prices for some product families.
- *Architectural enhancements that have made available, on-chip, large embedded blocks of RAM.* These are critical elements for DSP function implementation, such as discrete cosine transforms and fast Fourier transforms (FFTs).
- *Availability of libraries of megafunctions, predesigned and pretested building blocks that are optimized for performance and density.* Such architecture-specific optimization is critical to the implementation of high-performance DSP systems.

An SRAM-based FPGA is well suited for arithmetic DSP functions such as multiply and accumulate (MAC) functions, FFTs, convolution, and other filtering algorithms. One of the primary benefits is the flexibility offered by reprogrammability feature of the SRAM-based FPGAs. These FPGAs can also be dynamically reconfigured on the fly to perform different DSP tasks or other system-level functions in building of the adaptive hardware (see Chap. 6). However, a thorough understanding of the target FPGA architecture is needed for efficient implementation of FPGA-based DSP algorithms.

Some vendors offer high-capacity PLDs (HCPLDs) that are EEPROM-based and nonvolatile, yet offer in-circuit reconfiguration (ICR), such as Altera's FLEX10K and MAX9000 families.[29] These HCPLDs are used in designs that need the performance of an ASIC and the flexibility of a microprocessor in one chip. The advantage of HCPLDs in these applications is the enhanced flexibility that allows change of algorithm or implementation of several algorithms without increase in the component count. Long and complex algorithms can pose hardware implementation problems and, therefore, benefits can be obtained by breaking them into base functions such as adders and multipliers.

Some of the DSP functions that may benefit from HCPLD-based implementation, as opposed to the fixed-function DSPs, include FIR and IIR filters, Viterbi decoders, modulators/demodulators, correlators, and so on. In many cases, design may require elements common to both the HCPLD and microprocessor implementation. In these applications, it may make sense to utilize a hybrid approach in which the HCPLD acts as a *hardwired subroutine*. Accurate speed and density estimates are important factors in successful HCPLD- based DSP designs. This means that the designer should count the number of I/O pins required to move data and control signals for his design and then focus on the major components of his algorithm, i.e., adders, dividers, and multipliers. The HCPLDs required for these components can be computed by analyzing design requirement versus the architecture of target HCPLD. It should be noted that, for parallel processing, the computing resources need to be duplicated, whereas serial processing will most likely result in the reuse of logic blocks. Therefore, time sharing of the logic blocks means keeping track of the overhead.

A key consideration is the computation of logic layers (or resources) in the target HCPLD that the design implementation requires, and then accurately estimating the speed and timing delay. The longest delay from one layer to the next, plus the setup time of the registers, indicates the minimum clock cycle, and inverse of this clock cycle is the design's maximum clock frequency. The total number of layers between the input and output will determine the latency or pipelining delays. It is important to select HCPLD with predictable delays that can accurately predict speed through logic and interconnects.

The following are some of the important design considerations in HCPLD-based DSP implementations:

- Synchronous design practices should be followed. In some "hybrid system" designs, where the interface to the processor cannot be made synchronously, it is important to use synchronizing registers to reduce the possibility of an asynchronous event causing a metastable condition.
- Pipelining should be used as a method of maintaining the clock speeds, since most DSPs use pipelining internally without any performance sacrifice.
- The global clock offered by the HCPLDs should always be used in DSP design, since the clock skews in a complex algorithm can cause error propagation throughout the signal flow, which may be difficult to debug later on.

A Viterbi decoder used in a variety of communication applications for providing forward error correction is a good example of an FPGA design implementation that can accelerate a DSP function. A Viterbi decoder requires a noise cancellation function, and the computation intensive portion of this function can be implemented using an FPGA.

In this case, the FPGA works alongside a DSP that implements the remainder of the Viterbi-decoder algorithm. The FPGA implementation of the function requires a simple combination of adders, multiplexers, and registers.

FPGA-based DSP function implementation is becoming popular. Therefore, many FPGA vendors claim to offer DSP support. However, the level of support varies considerably from vendor to vendor. The FPGA vendor tools provide some of the fundamental building blocks for developing DSP applications. More advanced top-down design methods—such as high-level graphical capture, language-based design, and multilevel simulation—play an important role in achieving the rapid design cycle for developing complete systems. DSP function design level support provided by some major FPGA/CPLD vendors is discussed below.[30]

Actel does not provide direct support for high-level DSP implementations, although third-party DSP functions are available that work well with Actel's FPGAs. Actel provides a variety of data path building blocks with its function generator tool, ActGen Macro Builder. ActGen allows the use of procedural descriptions to create parameterizable functions that include the adders, multipliers, FIFOs, and register files.

These functions can be specified in a variety of bit widths and several control options, such as carry-in/carry-out polarity, add/subtract, and FIFO full/empty flags. After creating the functional macro, a netlist can be generated to combine with Actel's HDL code. ActGen supports LPM and offers a tool called ActMap that allows use of VHDL code for the design entry.

Altera, using the library of parameterized modules (LPM) standards, offers low-level DSP-related functions that include adders, multipliers, and comparators. The company also offers a DSP design kit that includes a macrofunction generator. This generator supports a FIR filter program called Firgen that converts filter coefficients to create a fixed-coefficient FIR filter in Altera hardware description language (AHDL). The AHDL is specific to company's Flex FPGA architecture. Altera's design and development system Max+Plus II can be used to compile the AHDL. Firgen provides either parallel or serial FIR implementations and can accommodate 8-, 16-, 24-, and 32-tap functions with user-settable parameters for data coefficient widths, symmetry, and pipelining.

Atmel's automatic component generator (ACG) creates many DSP related functions that include absolute value, adder, ALU, barrel shifter, comparator, increment/decrement by n, Boolean datapath, multiplier (both array and serial-parallel), and subtractor. The ACG allows parameterization of these functions using a menu-driven screen with options for pipelining, bus width, and other function specific operations. ACG generates hard macros that are placed, routed, and characterized on creation time, as opposed to the generator tools that develop only relative placements. ACG also outputs schematic, symbol, and netlist files for a variety of Unix- and PC-based tools. For higher-level DSP functions, the ACG generates hard macros for 8-bit digital filters using a bit-serial implementation. The macros include 8- and 16-tap asymmetrical FIR filters; 8-, 16-, 32-, or 64- tap symmetrical FIR filters; and second-order IIR-stage filters.

Lattice also has a workstation- and PC-based design tool, ispDS+, that enables design with Boolean text macros and HDLs. Lattice ispDS+ HDL Synthesis-Optimized Logic Filter is a universal compiler engine designed for use with third-party CAE vendor synthesis, capture, and design verification tools.

Lucent's macrocell generation software tool, SCUBA, supports the company's ORCA FPGAs. It can be used to create VHDL instantiations for basic DSP functions such as adders and multipliers. For higher-level DSP functions, Lucent provides customer solution core (CSC) that implements DSP functions in VHDL and Verilog. Filter parameters options in DSP CSC include clock parity, input and output datapath width, coefficient datapath width, multiplication-product datapath width, variable pipelining control, symmetric/antisymmetric filter option, number of filter taps (1 to 256), various data types (unsigned, two's complement, offset binary, and signed magnitude). On entering the appropriate parameters, the CSC automatically creates the desired filter type by calling on SCUBA to supply the required arithmetic and multiplication elements. CSC also calls on Synopsys' EDA tool to create an ORCA-specific netlist. This netlist is then sent to ORCA Foundry (Lucent's place-and-route tool), which provides functions such as mapping, placement, routing, and static timing analysis.

For DSP-related applications, QuickLogic provides a library of macrofunctions that includes various types of adders, accumulators, and multipliers. Although these functions are fixed macros, one can also use the VHDL or Verilog for design description, which allows scaling operations by changing a few lines of codes. To create higher-level DSP functions with QuickLogic FPGAs, one must use VHDL or Verilog and synthesize the functions into QuickLogic devices using a third-party synthesizer tool such as Synplify-Lite from Synplicity Corp.

Xilinx's Xblox is the company's version of LPM and can be used to synthesize the datapath logic for low-level, DSP-related functions including adders, subtractors, multipliers, MACs, and barrel shifters. The Xblox created functions can be scaled in one-bit increments to generate functions in a variety of sizes. Xilinx also provides an extensive support of high-level DSP blocks through tools such as the Logiblox for developing blocks that include FIR, IIR, FFT, decimating, interpolating, video-space conversion, convolution, 2-D FIR, inverse FIR, correlator, convolution, comb filter, gradient filter, and noise cancellation. Xilinx offers schematic- and HDL-based design files for these blocks. Another design approach is to parameterize the DSP blocks at a higher level of abstraction using third-party EDA tools that call the Logiblox during compile time.

4.6.3 PCI bus interface design

The PCI bus that was originally developed to interconnect high-speed peripherals to the host CPU is currently serving variety of other applications from industrial control systems to network bridgers and routers. A broad range of applications served by PCI or PCI-like interface allow designers to create custom versions of the PCI interfaces. However, PCI is not a simple interface to implement. For full PCI specification compliance, the interface must operate at 33 MHz with no-wait states. Also, the signal interface must meet the 11-ns clock-to-output time, the 7-ns clock-to-input (setup) time, and the 0-ns hold-to-system-clock time required by the PCI specification. In addition, the signal interface must provide the current drive and low leakage current to meet the load requirement. If it cannot meet the timing or drive specifications, the interface is considered PCI-like but not fully PCI compliant.

Therefore, to implement PCI bus interface designs, the target FPGA devices must have the required logic density, routability, and the ability to meet the tight timing specifications. For the PC market, PCI cards are typically implemented in one of following two main variations: target only or master/target.[31] The target-only interface is the simpler of two PCI options, requiring typically 2000 to 3000 gates for implementation. The more complex master/target interface, which can initiate operations, perform DMA transfers, and handle other more complex data management functions, requires 6000 to 8000 gates. In both cases, larger FPGAs may be required to accommodate other system-specific logic beside the PCI interface.[32]

The PCI target interface is more widely available block for FPGA implementation and requires the creation of read and write buffers, configuration registers, address control, and general control logic for the PCI bus signals. This megafunction is available from several FPGA suppliers, although some implementations may not offer full compliance with the PCI specification. An example of PCI target interface is megafunction developed by Eureka technology through Altera megafunction partner program (AMPP) alliance set up by Altera for its FLEX8000 FPGA family (see Sec. 8.1.2). This megafunction is fully compliant with the PCI specification (rev. 2.1) and has three user-bus options to support peripheral devices with different data-transfer characteristics.

In comparison, the master/target interface is more complex and typically requires FIFO buffers, a DMA engine, and a full set of configuration registers. For master/target applications, Altera has introduced a megacore for its FLEX 10K embedded PLD family. The initial version of this core will be compliant with the PCI specification (rev. 2.1), delivering 33 MHz operation.

Lucent Technologies' ORCA OR2CxxA series of RAM-based FPGAs are supported by several PCI megafunctions both target and master/target in VHDL form, licensed from Logic Innovations. The megafunction design kits provide the description in either VHDL or Verilog HDL format, which makes it easy to modify functionality. The master/target includes parity generation and checking, configuration registers, a 128-byte SRAM for data buffering, and bus control logic to achieve 33-MHz data transfer rate over the PC interface. This more complex master/target block requires 6000 to 8000 logic gates and can only be implemented in high density ORCA 2CA devices or the newly released ORCA OR3C family. Figure 4.24 shows the block diagram of master/target PCI interface offered by Lucent Technologies.

Xilinx RAM-based FPGAs also offer PCI building blocks that can be implemented in XC4000 and XC6000 families. The first planned offer-

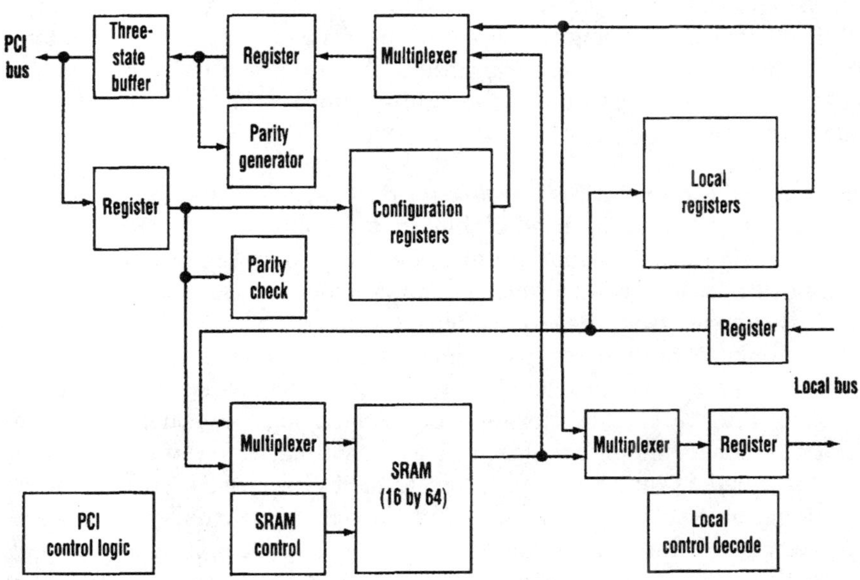

Figure 4.24 Block diagram of master/target PCI interface offered by Lucent Technologies for its 2CxxA and 3C families of ORCA FPGAs *(from Ref. 32)*.

ing is master/target megafunction in the Logic-Core family that provides a fully compliant PCI interface block, capable of 33 MHz operation.

In the antifuse based FPGA families, Actel is offering PCI support for its high-density ACT 3 family devices such as A1460, with 6000 usable gates, and A14100, with 10,000 usable gates. Actel provides following PCI bus interface designs: three target variations (one with memory and I/O, the second with memory only, and third with DMA master); two master versions; and a PCI-to-PCI bridge. The PCI support is also being extended to Integrator series 3200SX family. In the new ES FPGA system array family, Actel is considering a prediffused PCI implementation in the base array that will eventually be capable of supporting the next-generation 66-MHz version of PCI.

In addition to Lucent Technologies, Xilinx, and Actel providing PCI design support, a number of third-party IP suppliers, such as Technical Data Freeway and Toucon Technologies, are also marketing PCI functions.

4.7 PLD Design Guidelines for High Performance

In the PLD design, one of the most important issues is to consider design as a set of optimizations and then list the variables that are highest on the list for optimization. These variables may include high-performance design, cost of the part, ease of manufacture, time to market, software support, and tool requirements. Some of the rules and basic guidelines for high-performance designs are discussed below.[33]

1. *Use synchronous design techniques.* A synchronous circuit operation is proper and predictable in all modes of operations, from static dc up to the maximum clock rate. Since the circuits have to operate over a wide range of temperature, voltage, and manufacturing processes, these variables can affect asynchronous designs. A major benefit of proper synchronous design is that testability techniques, such as scan chains, can be incorporated to ease debugging and testing of the device. Modern logic synthesis requires truly synchronous designs and will assume that the circuit being described is synchronous. Therefore, if the circuit includes additional products to cover race conditions and hazards, synthesis programs will find the redundant logic and eliminate it. Analysis tools such as static timing analyzers require synchronous circuits for obtaining complete timing information. As the PLD designs migrate toward high-level descriptions and synthesis, asynchronous tech-

niques become more hazardous. In general, the following are some of the rules for synchronous designs:

- The design should have a minimal number of clocks, with the optimal number being one.
- Ensure that all inputs are registered, and metastability has been taken into consideration.
- No combinatorial logic may drive a flip-flop's set, reset, or clock input pin.
- Flip-flops must not be allowed to asynchronously set/reset themselves.
- Clocks cannot be delayed using buffers or delay elements.
- Logic should not be used to delay signals.
- Logic delays are not assumed to be longer than routing delays.

2. *Minimize clock skews.* A major concern in the PLD design is the clock skew, i.e., the amount of time the clock is different at different points in the design. This problem is magnified when multiple clocks are used, since each clock is skew sensitive. Many FPGA architectures use special clock lines and buffers to minimize the clock skew. However, the use of multiple clocks can exceed these special resources. The flip-flops can be root source of many problems, and the basic rules for minimizing timing skew are (a) use the same clock signal for every flip-flop and (2) do not drive the asynchronous set, reset, or clock line with the combinatorial circuits.
3. *Take metastability into consideration.* A common concern for designers is the effect of metastability on the flip-flops. If a flip-flop's setup or hold time is violated, the output of the flip-flop can be in the unknown state for a finite period of time. Therefore, if an input is being used by the system to make decisions, the signal must be synchronized to the internal clock. If the signal is not synchronized, a timing violation can occur, causing improper operation. Therefore, the metastability should be considered as another path delay element. If the combined propagation delay and the metastability time together are less than a clock period, only a single synchronizing flip-flop is required. For a faster clock rate, a second flip-flop is added after the I/O flip-flop. Now, if the metastability time plus the setup time is less than the clock period, the system is synchronous.
4. *Do not preassign pins.* It is easy to get into trouble if arbitrary package pin assignments are made before taking device architecture into consideration.[34] A well planned pin assignment can make the difference between successful or unsuccessful routing, low- or

high-performance margins, and the number of future FPGA modifications that can be absorbed without PCB redesigns. Therefore, temporary assignments of I/Os while verifying the routability of the entire design ensures the best opportunity for implementation success.

5. *Perform route budget analysis.* Route budget analysis provides a method for predicting which circuit implementations will need special attention, either in redesign, mapping, or placement constraints. During this analysis, the logic element delays are subtracted from the target clock period to calculate the allocatable time remaining for all intermediate delays. If the average allocatable time is found to be reasonable and roughly equal to a typical logic cell delay, then that implementation can be routed without further consideration.

6. *Use pipelining where possible.* The pipelining is basically the introduction of a sequential element into a combinatorial path. It is an easy way to improve performance by trading resources and latency for speed and routability. This is especially true for predecoding. Pipelining of state machine next-state functions, counter decodes, and arithmetic intermediate carries can produce significant performance improvements.

7. *Use linear-feedback shift registers.* For those functions that require a terminal count and not a binary output (e.g., divide-by-n) counters, linear-feedback shift registers (LFSRs) are more efficient from a space saving and routing point of view. Since LFSRs have minimal next-state logic and no carry logic, they are even more logic resource efficient than the binary counters. They produce 2^{n-1} states for n bits.

8. *Use prescale binary counters.* If a binary count sequence is required, consider implementing a prescaled counter in which the prescalar terminal count (PTC) can be used as the count-enable (CE) to the remaining counter stages. One advantage of this technique is that the remaining stages can then be implemented with a daisy-chain-style carry that will operate at a fraction of the clock frequency, which is a function of the prescaling.

9. *Use duplication for performance enhancements.* A duplication of critical pins, gates, flip-flops, and macros is another simple performance-enhancement technique. When a critical input signal needs to reach several distant destinations, consider assigning multiple input pins. If the clocked version of a synchronous signal is used by several functions, then duplicate flip-flops should be designed in for placement near their destinations.

10. *Customize macros.* It helps to use customized macros, which can be tailored copies of the library versions to facilitate custom mapping, placement, and naming. They cost less in compile time to expand and have fewer deletions to verify. Customized macros are easy to create by using cut-and-paste or auto-array commands.

4.8 PLD Design Conversion

Field programmable logic devices including CPLDs and FPGAs have enjoyed tremendous growth and popularity over the last few years, primarily because of their quick turn-around time and cost savings. The designs can be implemented in hours and easily modified, allowing for system performance evaluation in matter of days. Therefore, programmable logic devices provide an ideal solution for low- to moderate-production volumes and for fast prototyping of more complex logic designs. However, as the volumes increase, programmable devices may become prohibitively expensive, and gate arrays may provide more efficient implementation. They offer superior performance, higher density, and lower cost per gate in production volumes, as compared to the PLDs.

The design tools that support gate arrays (and ASICs) are typically more comprehensive and expensive than the PLD design tools. However, many ASIC design platforms now support CPLD/FPGA designs also. There are several conversion options available: full-custom ASICs, traditional gate arrays, and mask programmable gate arrays (MPGAs). A variation on the MPGAs is a new technology called *laser programmable gate arrays (LPGAs),* introduced by a company called Chip Express.

Therefore, in a high-volume production environment, substantial cost savings may be obtained by using programmable logic devices for logic verification and prototyping and then converting to a gate array design. The gate arrays have lower standby and operating currents, plus they offer higher performance than the PLDs. The designer also has a greater selection of buffer types, drive currents, and higher-order logic and memory (SRAM) functions. Also, converting several PLDs by consolidating their designs into a single gate array reduces the component count, consumes less power, and improves overall reliability of the system.

The gate array/ASIC conversion process typically requires a nonrecurring expense (NRE) for design implementation, and design revisions may require additional time and expense. Gate array/ASIC prototypes can be delivered in days or weeks, which is a much longer period than the design and verification cycle for a programmable logic device. Therefore, when considering PLD design conversion, one must take into consideration the design trade-offs and certain factors in gate array/ASIC vendor selection. A critical factor is the engineering time required for development or conversion. In a custom ASIC solution, the customer pays a substantial up-front NRE charge.

In a traditional gate array approach, the customer is usually required to purchase the supplier's cell library, perform the design work, and often pay a substantial NRE charges. In the MPGA/LPGAs design conversion, one needs only to provide the PLD netlist to the vendor and critical path information for the asynchronous designs. The NREs for MPGA approach are usually smaller than for the custom ASIC or traditional gate array approach.

Another key factor in design conversion is the design maturity. If a design is being altered or updated on a regular basis, the best choice is to keep it as an FPGA. If the design is stable, a conversion process may present the best choice. Once the decision has been made to convert a PLD design, an appropriate vendor selection requires consideration of several issues, such as:[35]

- What is the fabrication process technology and packaging being offered? How smoothly will the design conversion to available resources be achieved?
- What are the up-front lead times for prototypes and volume production?
- Does the supplier have its own fabrication facility, or will the design be outsourced to another vendor?
- What additional services are available, such as addition of scan chain insertion to make the design more testable, and is there any automated test pattern generation (ATPG) capability?
- What is the level of engineering support available as a follow-up before, during, and after the conversion process?

In design conversions from FPGA to ASIC, a considerable amount of time and energy can be saved if FPGAs are initially designed with ASIC migration as a potential long-term goal. This means that the FPGAs should be designed for ASIC testability. The designing for testability means making the circuit nodes more controllable and observable from all major I/Os. The dedicated pins in FPGAs used for programming the chip are of special concern when migrating to an MPGA or ASIC design. It is usually desirable to have the migrated device to be pin-for-pin compatible, drop-in replacement. To accomplish this, the FPGA dedicated pins must be handled carefully during the design conversion process. For example, there may be an unspecified timing parameter that allows the FPGA to boot up properly in the system and may need to be taken into consideration. To avoid this and other problems, the designer must understand the configuration-pin-related timing issues all the way from device power-up through configuration.

4.8.1 FPGA/CPLD design conversion process (example)

A summary of design conversion process is provided as an example by outlining the conversion methodology offered by Atmel Corporation. Figure 4.25*a* shows the typical FPGA conversion process flow.[36] The conversion of programmable logic device's netlist into an Atmel cell library is accomplished via Synopsys™ tools that can read a variety of formats, including Atmel's preferred EDIF™ and Verilog™. Figure 4.25*b* outlines the process by which Synopsys' tools convert the FPGA netlist into Verilog-XL™ format. The FPGA design database is supplied as an input to a proprietary Atmel mapping file for translation into Atmel cells. When the design is mapped in its entirety, a Verilog netlist in Atmel cells is produced.

Design conversion process requires the designer to supply a database that is unique to the specific choice of FPGA, as well as information that is independent of the FPGA choice. Some FPGA architectures are similar to the gate array in both design and physical implementation, and use only a few logic blocks, which are typically found in most gate array cell libraries. This allows the FPGA netlist to be mapped directly into gate array primitive cells, simulated, validated, and routed in the standard gate array design flow. Any design optimization that is required to meet the performance objectives may also be implemented. A general list of database requirements for Atmel conversion process includes:

- *Netlist.* The preferred netlist format is EDIF, but Verilog is also acceptable. However, other formats, such as those generated by Viewlogic™ or Mentor™ platforms have to be converted to the EDIF for use with Synopsys tools.
- *Test vectors.* All vectors must be in the same five groups (input, output, tristate, bidirectional, and enable) and have a stated purpose. The outputs are sampled once per clock cycle at 75 percent cycle point. Test vectors must include a 1 MHz set for wafer probe and an *at-speed* set for the final test. This *at-speed* may be a 1 MHz set with certain critical paths identified for testing at speed. The test vectors must pass Atmel's Test Vector Checker (tvc), a tool provided with Atmel libraries to verify the format of the vector set.

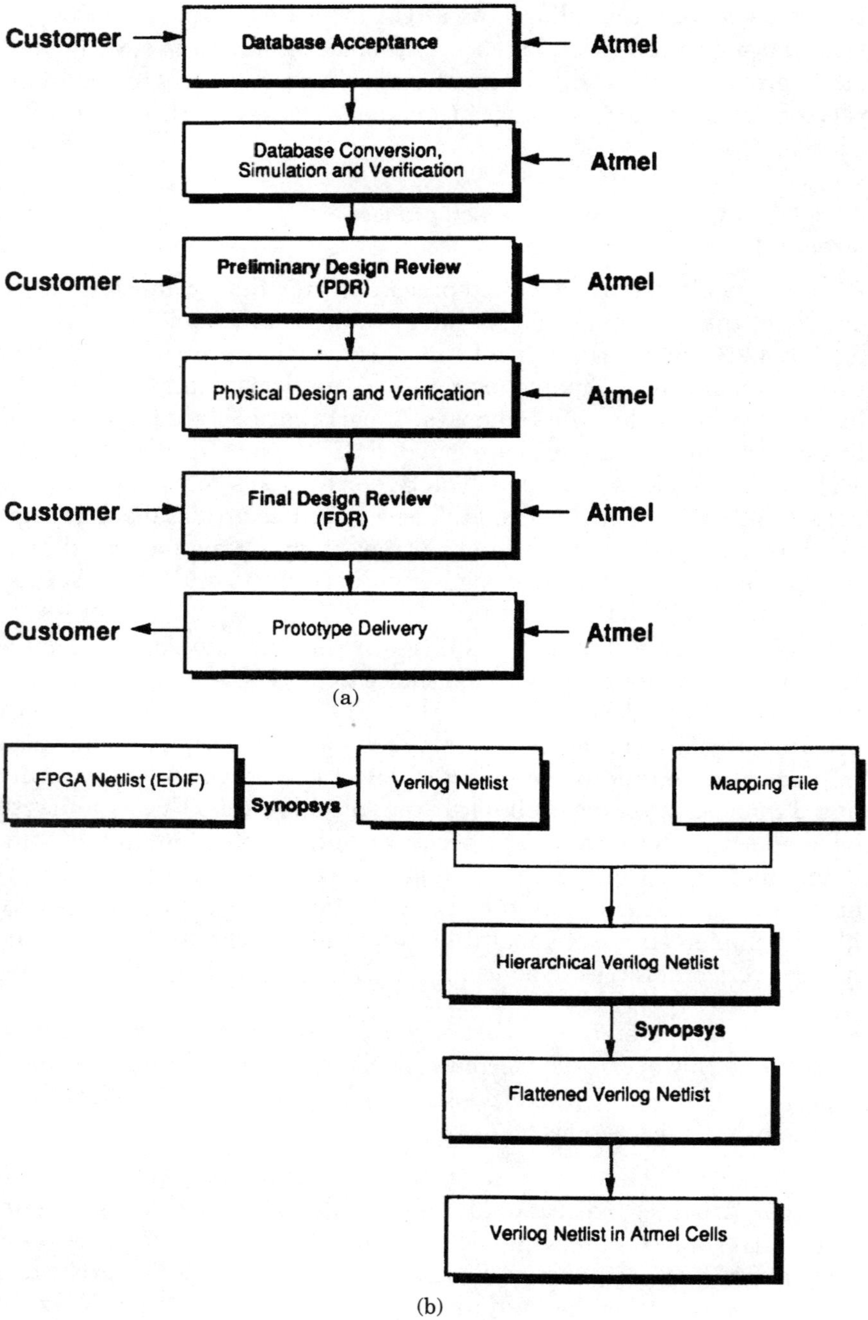

Figure 4.25 (a) Typical FPGA conversion process flow offered by Atmel Corp. and (b) FPGA netlist (EDIF) conversion into Verilog netlist using Synopsys tools *(from Ref. 36, with permission of Atmel Corp.)*

For converting simple or complex PLDs to gate arrays, two different methods can be used: deterministic conversion and timing match conversion. Each of these methods, discussed below, involves trading gate array utilization for matching the timing of the original design.

1. *Deterministic conversion.* Since simple PLDs have a uniform (deterministic) architecture, every signal within a PLD traverses a constant-length path and avoids race conditions. For the converted design to be a drop-in replacement, it must meet the specifications of the original design, including minimum and maximum signal arrival times, setup times, and hold times. In the deterministic approach, the gate array logic is implemented using blocks similar in structure to the product and sum blocks used in the PLD design. This approach is moderately efficient in terms of gate utilization and timing matching.
2. *Timing matching conversion.* For completely synchronous PLD designs, timing optimization techniques can be used to find the minimum propagation delay for each path. This information can then be used for delay adjustment to paths until the maximum signal arrival time of the path, as implemented in the gate array, matches that of the PLD. This approach offers the best timing match between the gate array and PLD, at the expense of additional buffers to adjust specific path timing.

After the FPGA/CPLD database acceptance, the design is converted into an equivalent netlist of primitive cells from Atmel's gate array library. The vectors from the original FPGA or CPLD design are also converted and used as functional simulation vectors to validate the gate array netlist. On completion of the validation process, any additional circuitry, such as memory blocks, testability improvement elements, or higher-order logic functions, can be incorporated into the netlist. At this stage, any further optimization necessary to match the timing or to improve performance can be performed. Also, boundary and internal scan can be added for ATPG vector generation. A preliminary design review (PDR) is held with the customer to review and approve the results of the design conversion.

The PDR is followed by the gate array implementation process in which the cells are placed and routed, and a post–routing simulation is performed. Design rule check (DRC) and electrical rule check (ERC) are performed to confirm the logic versus schematic (LVS) verification. A final design review is held with the customer to review and approve the post-routing data and to authorize mask making and prototype fabrication.

References

1. *AMD PAL Devices Data Book and Design Guide,* 1996 Application Notes: Combinatorial Logic Design and Registered Logic Design.
2. *Actel FPGA Data Book and Design Guide 1996,* Sec. 3: Macro Libraries.
3. Bradley Fawcett, "Tools to speed FPGA Development," *IEEE Spectrum,* Nov. 1994.
4. M.A. Breuer, "Min-cut placement," *J. Design Automation and Fault Tolerant Computing,* Oct. 1977.
5. S. Kirkpatrick, et al., "Optimization by simulated annealing," *Sci.,* 13, May, 1983.
6. S. Goto, "An efficient algorithm for two-dimension placement problem in electrical circuit layout," *IEEE Trans. Circuits Syst.,* Jan. 1981.
7. Larry Saunders, "Effective design verification," *Integrated System Design,* April 1997, pp. 19–28.
8. Doug Conner, "Boost programmable logic performance with timing driven tools," *EDN,* Feb. 4, 1993, pp. 47–55.
9. *Altera Data Book,* 1995.
10. Douglas J. Smith, "HDL basic training: top-down chip design using Verilog and VHDL," *EDN,* Oct. 24, 1996, pp. 103–112.
11. Bradley Fawcett, "Synthesis for FPGAs: An overview," *WESCON '94 Proc.,* pp. 576–580.
12. Steve Carlson and Emil Girczyc, "Understanding synthesis begins with knowing the terminology," *EDN,* Sept. 3, 1992, pp. 125–131.
13. A. Sangiovani-Vincentelli, et al., "Synthesis methods for field programmable gate arrays," *Proc. IEEE,* vol. 81, no.7, July 1993, pp. 1057–1083.
14. R.K. Brayton, et al., "MIS: A multiple-level logic optimization system," *IEEE Trans. Computer-Aided Design,* 1987, pp. 1062–1081.
15. R.J. Francis, et al., "Chortle: A technology mapping algorithm for look-up table based field programmable gate arrays," *Proc. Design Automation Conf.,* 1990, pp. 613–619.
16. J. Rose, et al., "Chortle-crf: Fast technology mapping for look-up table based FPGAs," *Proc. Design Automation Conf, 1991,* pp. 227–233.
17. K. Karplus, "Xmap: A technology mapper for table look-up field programmable gate arrays," *Proc. DAC '91,* June 1991, pp. 240–243.
18. Z. Vranesic, et al., "Technology mapping of look-up table based FPGAs for performance," *Proc. Int. Conf., Computer-Aided Design,* 1991, pp. 568–571.
19. R. Murgai, et al., "Performance directed synthesis for table look up programmable gate arrays," *Proc. Int., Conf., Computer-Aided Design,* 1991, pp. 572–577.
20. S. Ercolani and G. DeMicheli, "Technology mapping electrically programmable gate arrays," *Proc. Design Automation Conf.,* 1991, pp. 234–239.
21. K. Karplus, "Amap: A technology mapper for selector-based field-programmable gate arrays," *Proc. Design Automation Conf.,* 1991, pp. 244–247.
22. R. Murgai, et al., "An improved synthesis algorithm for multiplexer-based FPGAs," *Proc. 1st Int. ACM/SIGDA Workshop on FPGAs,* 1992.
23. Kirk Owyang, "High level design for high density programmable logic," paper received at PLD Conference,, 1996.
24. *AMD Databook,* Application Note: State Machine Design.
25. Stephen L. Wasson, "High speed state machine design," *Integrated System Design,* July 1993, pp. 19–28.
26. Sundar Rajan, "Practical state machine design using VHDL," *Integrated System Design,* Feb. 1995, pp. 60–70.
27. *Actel FPGA Data Book and Design Guide 1996,* Application Note: Designing State Machines for FPGAs.
28. Gregory R. Goslin, "Implement DSP functions in FPGAs to reduce cost and boost performance, *EDN,* Oct. 10, 1996, pp. 155–164.
29. Leo Petropoulos, "Replace digital signal processors with HPCLDs," *Electronic Design,* Sept. 5, 1995, pp. 99–104.
30. Markus Levy, "DSP design tools target FPGAs," *EDN,* June 20, 1996, pp. 75–82.

31. Barry Britton, et al., "Design an FPGA-based PCI bus interface," *Electronic Design,* March 6, 1995.
32. Dave Bursky, "FPGAs take on the PCI performance challenge," *Electronic Design,* April 14, 1997, pp. 111–120.
33. Tom Leach and Bob Hackett, "Modern synchronous design," *ASIC & EDA,* Jan. 1993, pp. 44–52.
34. Stephen L. Wasson, "Design tips for high-performance FPGA designs," *ASIC & EDA,* Oct. 1994, pp. 41–52.
35. Brian Liddiard, "Reducing time to revenue," *Computer Design,* Feb. 1997, pp. 20–21.
36. *Atmel Configurable Logic Databook,* Application Note: Converting FPGAs and PLDs to Atmel Gate Arrays.

Chapter

5

PLD Performance, Testing, and Reliability Issues

5.1 PLD Performance and Benchmarking

PLD selection basic issues were discussed in Chap. 1. A major consideration in PLD selection and performance evaluation is to realize that each vendor's device may exhibit fundamental architectural differences in individual logic block structure, as well as routing resources. In many cases, the maximum performance specification of a device provided by the manufacturer may be applicable only to the simplest portions of the design. For example, an FPGA manufacturer's data sheets may have some critical parameters specified differently from those of the other vendors, making it hard to make a general capability and performance comparison. Therefore, predicting design performance based on the manufacturer's data sheet specifications and published performance results for a given application requires careful consideration.

For a given design, the selection of a PLD based on density requirements depends on determining the true capacity of a device, i.e., the number of usable gates versus the number of available gates. For vendors of both CPLDs and FPGAs, data sheet specifications quote the number of gates in numerous ways, some of which are misleading. Therefore, the gate count is a fairly useless specification except when comparing specific devices from the same manufacturer. In addition, the final design performance depends not only on the target FPGA resource utilization, but also on the quality of place-and-route software.

In 1991, eight PLD vendors, in association with the programmable logic users, formed the Programmable Electronics Performance Corp. (PREP) to provide a comparison standard for the relative capacity of

CPLDs and FPGAs.[1] The current members of the PREP participating companies include Actel Corp., Atmel Corp., Cypress Semiconductor Corp., Lattice Semiconductor Corp., Lucent Technologies, QuickLogic Corp., and Xilinx Inc. The PREP organization is responsible for the development of benchmark circuits, measurement methodology, and reporting criteria. The PREP benchmarks provide the designers with a means to evaluate different architectures, their relative logic capacity, and performance information for appropriate device selection for design implementation. The organization is also responsible for verification and certification of the data derived from the benchmarks and the test results that are publicly reported.

The benchmark circuits have been selected to be representative of typical applications so that the logic designers can relate them to actual designs. PREP has developed a standard reporting form, which all vendors must use when reporting results. The logic capacity of a device is reported by specifying the number of instances (repetitions) of the benchmarks and filler circuits along with the percentage of total logic available on the target device. In addition, the reported information includes speed performance (both internal and external), whether the benchmark circuit has been optimized for logic capacity or performance, and the tools used.

PREP has established following rules and procedures for reporting results:

- The entire suite of PREP benchmarks must be performed on a device before the results can be reported.
- The PREP reporting form specifies that, for all benchmark circuits, both maximum logic capacity and worst-case internal and external speed performance must be reported.
- All data reported must represent worst-case process conditions as well as the worst-case commercial-grade, temperature and power supply voltages.
- Each benchmark report must contain results based entirely on automatic place and route, although critical path assignments are permitted. The results may also be reported for benchmark circuits that are implemented manually.
- The reported device data sheet timing parameters and simulation models must be available and guaranteed by the manufacturers.
- All tools used to obtain results must be reported and available from the manufacturers.
- The PLD manufacturers can report on all available devices, but the largest device in the family must always be represented.

- All benchmark results must be reproducible by an independent third party. The PREP member companies have agreed to cooperate in a process of verifying each other's benchmark results before they are published.

5.1.1 PREP Benchmarks Description

The nine PREP benchmarks are briefly described below:[2]

- *Benchmark #1: data path.* This circuit consists of combinatorial logic functions between the registers. The first function is a four-to-one multiplexer whose output drives a register, and the register output drives a shift register.
- *Benchmark #2: timer/counter.* This circuit is an eight-bit timer/counter with a loadable compare (time) value and a multiplexer that allows the counter to be preloaded from either a latched value or a direct bus.
- *Benchmark #3: state machine.* This benchmark is a typical finite state machine with seven states, consisting of eight inputs and eight registered outputs. All output and state transition occur off a single, positive clock edge. The state machine and all its outputs can be reset by a reset signal. The reset signal is global to all instances of the state machines and can be of any polarity. Any encoding methodology may be used for encoding the state information.
- *Benchmark #4: large state machine.* This benchmark is a 16-state, 40-transistor state machine with 10 input signals and 8 outputs. The state encoding is not specified. The output decoding is unregistered.
- *Benchmark #5: arithmetic circuit.* This circuit accumulates the output of a combinatorial multiplier in an eight-bit accumulator with reset. The combinatorial circuit multiplies two four-bit data inputs as positive integers, generating an eight-bit positive integer product, which is accumulated in the eight-bit accumulator.
- *Benchmark #6: 16-bit accumulator.* This benchmark circuit adds 16-bit binary positive integer numbers to the content of a 16-bit register and clocks the sum into the same register. It is concatenated by connecting the 16-bit accumulator output to the 16-bit data input of the next circuit. The clock and reset signals are common to all circuits.
- *Benchmark #7: 16-bit counter.* This is a 16-bit binary up-counter with synchronous load, synchronous count enable, and asynchronous clear.

- *Benchmark #8: 16-bit prescaled counter.* This benchmark circuit is a 16-bit, prescaled counter that enables the division of a frequency by up to 65,536.
- *Benchmark #9: memory map.* This benchmark circuit implements a memory mapped I/O scheme of different-size memory spaces common to microprocessor systems.

Each of the benchmark circuits listed above has its own certified reporting form. There are two PREP certified databases, one that complies with the rules and methodologies of Benchmark Suite #1, version 1.2, and the newer Benchmark Suite #1, version 1.3. There were several changes made in the updated Benchmark Suite #1, version 1.3, including the removal of a "filler circuit" and its requirements for measuring capacity, and a new alternate, pin-constrained version of Benchmark #1.

5.1.2 PREP test bench method

The following are the definitions used in describing PREP's Test Bench Method:[2]

- *Test bench*—test description that uses a simulator to control and observe a behavioral design, or back-annotated netlist with timing information
- *Behavioral design*—functional description of hardware for design verification and test bench development
- *Optimized design*—hardware description that is optimized for a target synthesis tool and target PLD
- *APR*—place, route, timing, and netlist process
- *Netlist*—structural design resulting from APR. Timing may be included in a separate file, e.g., SDF format.

Figure 5.1 shows the test bench method flow chart.[2] These are the various steps in reporting a test bench:

1. Exercise test bench with simulation tool and behavioral design.
2. Develop optimized design for best results with target synthesis tool and a target PLD.
3. Run synthesis, APR tools to produce a back-annotated netlist with timing information.
4. Exercise test bench with simulation tool on back-annotated netlist with timing information.
5. Report timing margin and device utilization.

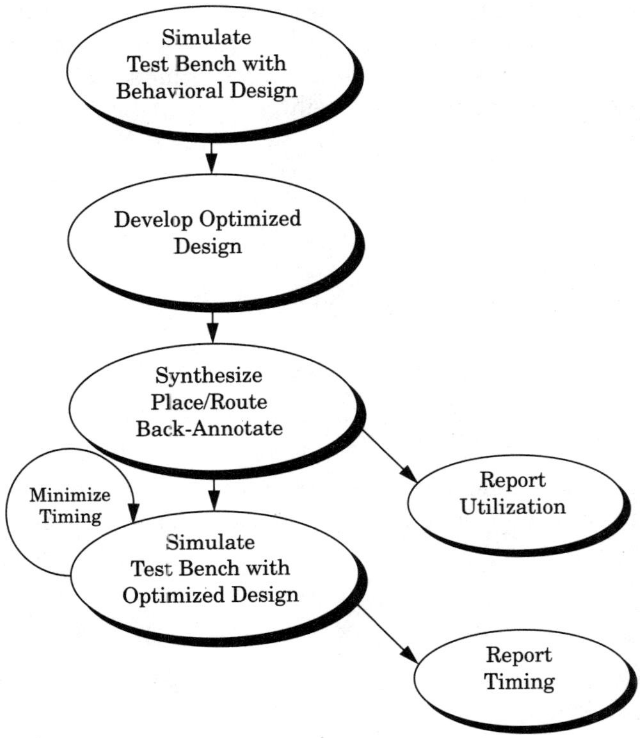

Figure 5.1 PREP test bench method flowchart *(from Ref. 2).*

The benchmarks and benchmark data developed by PREP Corp. have been controversial, not only in the industry, but also among the PREP member companies. The issues under debate include questioning of the PREP data relevance to the "real-world applications" of programmable logic users community, and the meaningfulness of "step and repeat" methodology in filling devices with as many benchmark instances as possible. According to the arguments in favor, PREP benchmarks are relevant to the real-world applications, since they represent collection of low level functions popularly used in programmable logic design applications. The effectiveness of PREP benchmark data in predicting performance and capacity in real designs depends primarily on the expertise of the designer in evaluating similarity of the benchmark functions to application design, and estimating the effect of differences.

5.1.3 Benchmarking Analytical Approach

A quantitative approach to the measurement of circuit performance and efficiency was first developed by E.F. Rent, of IBM, by preparing log plots of numerous circuit pin requirements for a given partitioning of the circuit. This relationship of pins to the logic partitioning is known as *Rent's rule.*[3] Several studies have been performed that have shown similar relationships and results that support Rent's rule.

According to Rent's rule, a relationship exists between P, the average number of pins per module in a given partitioning of the circuit, and B, the average number of blocks per module in that same partition, as follows:

$$P = KB^r$$

where K is the average number of pins per block and $0.57 \le r \le 0.75$.

Rent's rule quantifies the relationship between the pins to partition density and provides two different measures as follow:

1. *Pin measure.* The rule yields optimal number of pins required, P (either input, output or bidirectional), for a selected logic partition, given by

$$P = KB^{2/3}$$

where K is the average fan-in to fan-out ratio, and B is raised to the power 2/3.

As an example, a simple PLD with 22V10 architecture has 22 inputs to 10 outputs, which yields a fan-in to fan-out ratio of 2.2. The number of modules per partition block, B, is 10. Thus, the value of P is the product 2.2 times the number of modules raised to power 2/3, which is 10.

2. *Capacity measure.* An efficiency of partition can be measured in terms of the capacity if Rent's rule is solved for B. This measure is based on a unit of modules, where the module may be considered as a basic logic unit of the given architecture, such as a macrocell, flip-flop, AND/OR gates, and so on.

$$B = \sqrt{(P/K)^3}$$

In a PLD, since both the values for P and K are known, we can measure capacity B in terms of the modules. For example, in a simple 22V10 PLD, the value of K as 2.2 is known. The device has 12 input pins and 10 output pins, thus P is 22. Therefore, solving the above equation yields a value of 32 for B.

The work performed by Rose and Brown[4] and Sturges[5] provides a set of metrics for measuring relationship between the pins (signals, either input, output, or bidirectional) and the nets (channels or tracks that provide interconnection resources for the partitions). The overall network capacity (B_f) is the product of all bus networks of the interconnect. It is measured with base units in nets and reflects the total routing resource pool available within a PLD.

$$B_f = \frac{(F_s n)(F_c m)}{W(vh)}$$

where $(F_s n)$, and $(F_c m)$ are the fan-in and fan-out amounts and counts, respectively, and W is the average width of nets with v and h reflecting the counts for vertical and horizontal channels of nets.

The metric B_f for routing capacity in an architecture can also provide information about the effective number of routable signals and the interconnect metric, which represents the number of signals per net. The signal-to-net ratio is given by the interconnect matrix (I_c), as follows:

$$I_c = \frac{S}{B_f}$$

where S is the total amount of resources fan-in or fan-out (signals, either inputs, outputs, bidirectional) that can be routed to form a circuit.

5.2 Reliability Issues and Screens

5.2.1 General reliability issues

A semiconductor device manufacturing yield may vary from one manufacturer to another, although it is commonly expressed as the ratio of the passing devices during an initial test (such as wafer sort or wafer-level inspection testing) versus the total number manufactured. Quality is measured by the proportion of die (or packaged parts) that fall out during a screening test flow (e.g., screening per MIL-STD-883, Method 5004). Reliability is measured by the device failures that occur during the qualification and quality conformance inspection (QCI) testing. *Field failures* refers to devices that fail in the actual operating environment subsequent to all processing and assembly steps.

The reliability of a programmable logic device is the probability that the device will perform satisfactorily for a given time at a desired confidence level under the specified operating and environmental conditions. The mathematical theory for reliability includes fundamental concepts such as the reliability function, the probability density function, the hazard rate, the conditional reliability function, and *mean time to failure* (MTTF).[6] A device's reliability is often expressed in a measurement called *failure in times* (FITs), where 1 FIT equals one failure per billion (10^9) device operating hours. The FITs are usually expressed at an elevated temperature, e.g., 70° C or higher. The generally reported failure rate is a 60 percent confidence level of the observed FITs. The 60 percent confidence level is a statistical analysis of the actual failure rate, which accounts for the observed failure rate and sample size. For larger sample sizes, the 60 percent confidence level approaches the observed failure rate. The failure rate at this confidence level is calculated using the Poisson distribution, which is valid for a low failure occurrence in a large sample size.

A semiconductor device failure is typically grouped into one of three categories: (a) infant mortality, (b) useful life, and (c) wearout failures. The infant mortality failures for a device type occur during the early product life cycle and, by definition, the rate of failure decreases with age. The major causes for early failures include process and design related defects, poor manufacturing assembly and workmanship, improper process monitoring and lack of quality controls, and inadequate screening (e.g., burn-in). The second type of failures occurs during the useful life of the device, for which the rate of failure is assumed to be constant in the widely used Arrhenius model. During this stage, the causes of failures include inadequate derating, improper start-ups and voltage transients, and higher than expected thermal and electrical stresses. The third type includes wearout failures, which occur late in the device operating cycle due to causes such as the aging phenomenon, degradation and material fatigue. By definition, the rate of failure during this cycle increases with age.

According to this failure model, if the failure rate is plotted against the operating life test data taken from a statistically large sample size placed in operation at $t = 0$, then the resulting curve for a medium- to low-level stress takes the shape of a "bathtub curve." For this bathtub curve, the high initial rate of failure represents infant mortality failures, the flat central portion of the curve indicates constant failure rate corresponding to the useful life of the product, and the increasing failure rate reflects wearout mechanisms. Figure 5.2 shows a series of curves corresponding to three different levels of stresses: high, medium, and low.

A semiconductor device failure is a function of the circuit design techniques and materials and processes used in fabrication, beginning from wafer-level probing to assembly, packaging, and testing. In general, the failure mechanisms and reliability issues for bipolar and MOS technologies include (a) dielectric-related failures from the gate-oxide breakdown, time dependent dielectric breakdown (TDDB), and ESD failures, (b) dielectric interface failures such as those caused by

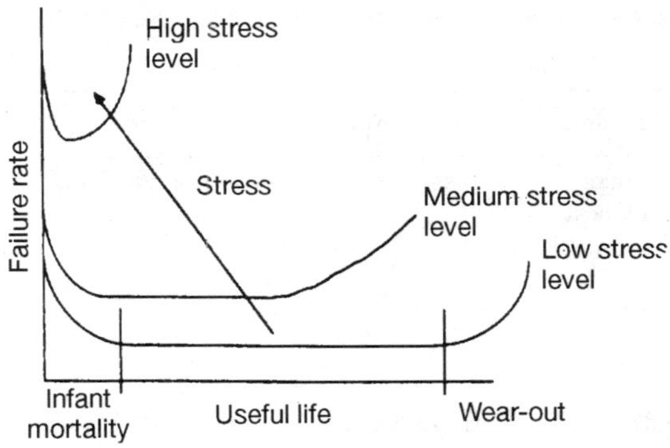

Figure 5.2 A series of curves corresponding to three levels of stresses: low (bathtub curve), medium, and high.

ionic contamination and hot-carrier effects, (c) conductor and metallization failures, e.g., electromigration and corrosion effects; and (d) assembly- and packaging-related failures. Table 5.1 shows general failure-related causes, mechanisms, and modes for microelectronic devices such as programmable logic devices.[7]

In addition to the general failure modes and mechanisms listed in Table 5.1, there are some PLD-specific failure modes related to the programmable elements used, such as antifuse, SRAM, or EPROM cell. The reliability assurance testing for these specific failure modes depends on the programming technology used and varies from a manufacturer to manufacturer. For example, Actel performs special stress tests for integrity and reliability assurance of its ONO based PLICE fuses. Quicklogic, for its pASIC FPGAs that use metal-to-metal ViaLink antifuses, performs high-V_{cc} static life tests on unprogrammed antifuses and low-temperature operating life test on the programmed antifuses. Altera, with its EPROM, EEPROM, or flash cell based devices, performs voltage margin testing for monitoring the ability of a device to be programmed or erased and for evaluating the long-term stability of a device after programming or erasure. These special tests are discussed in Sec. 5.2.3.

5.2.2 Screening, Quality Conformance Inspection, and Qualification Testing

MIL-STD-883, Method 5004, establishes screening procedures for microelectronic devices such as PLDs. Most PLD manufacturers have screening flows that are either in direct compliance with this procedure or have in-house reliability screens based on these test procedure requirements. The programmable logic devices used for high-reliabil-

TABLE 5.1 General Failure Modes, Mechanisms, and Causes for Bipolar and CMOS Device Technologies[9]

Failure modes	Failure mechanisms	Failure-related causes	
Withstanding voltage reduction, shorts, leakage current increases, hFE degradation, threshold voltage variation, noise	Pinholes, cracks, uneven thickness, contamination, surface inversion, hot carrier injection	Passivation	Surface oxide film, insulating film between wires
Opens, shorts, resistance increase	Flaws, voids, mechanical damage, breaks due to uneven surface, non-ohmic contact, insufficient adhesion strength, improper thickness, electromigration, corrosion	Metallization	Interconnection, contact
Opens, shorts, resistance increase	Bonding runout, compounds between metals, bonding position mismatch, bonding damage	Connection	Wire bonding, ball bonding
Opens, shorts	Disconnection, sagging, shorts	Wire lead	Internal connection
Withstanding voltage reduction, shorts	Crystal defects, crystallized impurity, photoresist mismatch	Diffusion, junction	Junction diffusion, isolation
Opens, shorts, unstable operation, thermal resistance increase	Peeling, chips, cracks	Die bonding	Connection between die and package
Shorts, leakage current increase, opens, corrosion disconnection, solder failure	Integrity, moisture ingress, impurity gas, high temperature, surface contamination, lead rust, lead bend or break	Package sealing	Packaging, hermetic seal, lead plating, hermetic and plastic package, filler gas
Shorts, leakage current increase	Dirt, conducting foreign matter	Foreign matter	Foreign matter in package
Short, open, fusing	Electron destroyed	I/O pin	Electrostatics, excessive voltage surge
Soft error	Electron hole pair generation	Disturbance	Alpha particle
leakage current increase	Surface inversion	Disturbance	High electric field

ity military and space applications usually have been procured to MIL-M-38510, which specifies two quality levels for microcircuits. These are *class S* and *class B*, each with its own manufacturing and quality control inspection flows. MIL-M-38510 references MIL-STD-883, Method 5004, as a standard screening procedure for both class S and B level screening.

These screening flows include wafer lot acceptance requirements, nondestructive bond pull, internal and external visual inspection, temperature cycling, constant acceleration, particle impact noise detection (PIND), burn-in test with pre- and post-electrical measurements, seal leak, radiographic, and radiation latch-up screening tests. Class S screening flow is more stringent (and costly) than class B. Table 5.2 shows MIL-STD-883, Method 5004, screening flow for class S and B levels, each with its own percent defect allowable (PDA) requirements.[8] This table does not duplicate test paragraph references and notes contained in Method 5004. For test method details, including applicable test conditions and requirements, refer to MIL-STD-883, Method 5004.

MIL-STD-883, Method 5005, establishes qualification and (QCI procedures for microcircuits to assure that the lot quality conforms with the specification requirements. The full qualification and QCI requirements of MIL-STD-883, Method 5005, include Groups A, B, C, D, and E testing. Group A consists of electrical testing (dc and ac parametrics, functional, and dynamic tests) over a specified operating temperature range; groups B and C combine various types of environmental tests including 1000 hour life test, typically at 125° C or higher for hermetic packages; group D testing consists of package related tests; and group E tests are required for radiation hardness assurance levels, as specified. For detailed test conditions and other applicable requirements, refer to MIL-STD-883, Method 5005.

MIL-M-38510 is gradually being replaced by a new military specification, MIL-PRF-38535, *General Specification for Integrated Circuits (Microcircuits) Manufacturing.* This new specification, also referred to as the *qualified manufacturer's listing* (QML) specification, emphasizes the use of in-line process monitors and statistical process control (SPC), self-audits, and continuous process improvements as quality management tools rather than end-of-the-line qualification/QCI testing. The QML approach detailed in MIL-PRF-38535 emulates many of the *best commercial practices* (BCPs) employed by high-volume microcircuit manufacturers producing plastic encapsulated modules (PEMs). In general, the BCP has the following characteristics:[9]

- It includes a design, process, assembly, and test methodology that assures consistent achievement of user's requirements.

TABLE 5.2 MIL-STD-883, Method 5004 Screening Flow[8]

	Class S		Class B	
Screen	Method	Req.	Method	Req.
Wafer lot acceptance	5007	All lots		
Nondestructive bond pull	2023	100%		
Internal visual	2010, test cond. A	100%	2010, test cond. B	100%
Temperature cycling	1010, test cond. C	100%	1010, test cond. C	100%
Constant acceleration	2001, test cond. E (min) Y_1	100%	2001, test cond. E	100%
Visual inspection		100%		100%
Particle impact noise detection (PIND)	2020, test cond. A	100%		
Pre–burn-in electrical measurements	Per device spec.	100%	Per device spec.	100%
Burn-in test	1015 240 hr @ 125° C min	100%	1015 160 hr @ 125° C min	100%
Interim (post–burn-in) electrical measurements	Per device spec.	100%		
Reverse-bias burn-in*	1015, test cond. A or C, 72 hr @ 150° C min	100%		
Interim (post–burn-in) electrical measurements	Per device spec.	100%	Per device spec.	100%
Percent defective allowable (PDA) calculation	5% (3% on functionals) @ 25° C	All lots	5%	All lots
Final electrical measurements @ 25° C, min and max rated operating temperatures	Per device spec. including static, functional, and switching tests	100%	Per device spec. including static, functional, and switching tests	100%
Seal tests	1014	100%	1014	100%
Fine and gross leak				
Radiographic	2012, two views			

*The reverse-bias burn-in is required only when specified in the applicable device specification and is recommended only for certain MOS linear or other microcircuits where surface sensitivity may be of concern.

- It includes continuous process improvements with in-line controls, SPC monitors, self-audits, and a goal of eliminating the end-of-line testing and screens by reducing the product variability.
- In the microelectronic industry, the BCP is driven by the high volume markets, and the QML system has no significant impact on the BCP.

An example of this BCP approach is the reliability testing performed by Xilinx Inc. on its various PLD families, nonhermetic and hermetic product types. Table 5.3*a* shows the reliability testing sequence for nonhermetic devices, where tests 1–2 are for die qualification, and tests 3–10 are package integrity and assembly qualification tests.[10] Table 5.3*b* lists reliability test sequence for hermetic package integrity and assembly qualification testing, consisting of subgroups D1–D8, performed in accordance with test methods and procedure of MIL-STD-883. For each of the test specified, Table 5.3 also lists the lot tolerance percent defective (LTPD) or minimum sample size (*s*)/maximum acceptable failures (*c*). Some of the important tests are briefly described in the following section. For details of these test procedure and conditions, see MIL-STD-883.[8]

5.2.3 Qualification and QCI Test Descriptions

This section briefly describes some of the important tests referenced in Table 5.3. For further details, refer to MIL-STD-883.[8]

- *High-temperature storage.* High-temperature test is an unbiased bake at 150° C for 1000 hours. This test is performed to accelerate failures due to mobile ionic contamination and bond ball intermetallic formation.
- *High-temperature life.* This test is performed to evaluate the long-term reliability and life characteristics of the die and is a listed QCI requirement as part of MIL-STD-883, Method 5005, testing. This can be a static or dynamic operating life test for 1000 hours or more, at maximum recommended supply voltage (e.g., 5.5 V) and 125° C or higher ambient temperature. The purpose of performing the test at high temperature is to reduce the test time and extrapolate the failure rate to typical operating conditions. The Arrhenius equation shown below is used for this extrapolation:

$$R = R_0 \exp\left[\frac{E_a}{kT}\right]$$

where R is the failure rate, R_0 is constant for a particular process, T is the absolute temperature in degrees Kelvin, k is Boltzmann's con-

TABLE 5.3a Reliability Test Sequence for Nonhermetic Devices, Including Die Qualification, Package Integrity, and Assembly Qualification[10]

Name of test	Test conditions	Lot tolerance, % defective Min. sample size/ Max. acceptable failures
	Die qualification	
1. High-temperature life	1000 hr min equivalent at 125° C Life test circuit equivalent to MIL-STD-883 test method	LTPD = 5
2. Biased moisture life	1000 hr min exposure at 85° C, 85% RH Max rated operating voltage Biased moisture life circuit equivalent to MIL-STD-883 test method	LTPD = 5
	Nonhermetic package integrity and assembly qualification	
3. Unbiased pressure pot	96 hr min exposure, T = 121° C, P = 2 atm H_2O saturated	LTP = 5
4. Thermal shock	MIL-STD-883, Method 1011, Cond. C, –65 to +150° C, 200 cycles	LTP = 5
5. Temperature cycling	MIL-STD-883, Method 1010, Cond. C, –65 to +150° C, 500 cycles	LTP = 5
6. Salt atmosphere	MIL-STD-883, Method 1009, Cond. A, 24 hr	s = 25, c = 0
7. Resistance to solvents	MIL-STD-833, Method 2015	s = 4, c = 0
8. Solderability	MIL-STD-883, Method 2003	s = 3, c = 0
9. Lead integrity	MIL-STD-883, Method 2004	s = 2, c = 0
10. Physical dimensions	MIL-STD-883, Method 2016	s = 15, c = 0

stant (8.62×10^{-5} eV/°K), and E_a is the activation energy of the process in electron volts.

To determine the acceleration factor for a given failure mode at temperature T_2 as compared with temperature T_1, the Arrhenius equation can be written as

$$A(T_1, T_2) = \exp\left(\frac{E_a}{k}\right)\left(\frac{1}{T_1} - \frac{1}{T_2}\right)$$

TABLE 5.3b Reliability Test Sequence for Hermetic Package and Assembly Qualification[10]

Name of test	Test conditions	Lot tolerance, % defective Min. sample size/ max. acceptable failures
1. Subgroup D1 Physical dimensions	MIL-STD-883, Method 2028	LTPD = 15
2. Subgroup D2		LTPD = 15
a. Lead integrity	MIL-STD-883, Method 2028	
b. Seal (fine and gross leak)	MIL-STD-883, Method 1014 (not required for PGAs)	
3. Subgroup D3		LTPD = 15
a. Thermal shock, 15 cycles	MIL-STD-883, Method 1011, Cond. B	
b. Temp. cycling, 100 cycles	MIL-STD-883, Method 1010, Cond. C	
c. Moisture resistance	MIL-STD-883, Method 1004	
d. Seal (fine and gross leak)	MIL-STD-883, Method 1014	
e. Visual examination	MIL-STD-883, Method 1004 and 1010	
f. End-point electricals	Group A, subgroup 1	
4. Subgroup D4		LTPD = 15
a. Mechanical shock	MIL-STD-883, Method 2002, Cond. B	
b. Vibration, variable freq.	MIL-STD-883, Method 2007, Cond. A	
c. Constant acceleration	MIL-STD-883, Method 2001, Cond. E min, Y1 only (Cond. D for large PGAs)	
d. Seal (fine and gross leak)	MIL-STD-883, Method 1014	
e. Visual examination	MIL-STD-883, Method 1010	
f. End-point electricals	Group A, subgroup 1	
5. Subgroup D5		LTPD = 15
a. Salt atmosphere	MIL-STD-883, Method 1009, Cond. A	
b. Seal (fine and gross leak)	MIL-STD-883, Method 1014	
c. Visual examination	MIL-STD-883, Method 1009	
6. Subgroup D6 Internal water vapor content	MIL-STD-883, Method 1018, 5000 ppm water @ 100° C	s = 3, c = 0 or s = 5, c = 1
7. Subgroup D7 Lead finish adhesion	MIL-STD-883, Method 2025	LTPD = 15, s = 25 leads (3 device min), c = 0
8. Subgroup D8 Lid torque	MIL-STD-883, Method 2024 (for ceramic quad flat pack only)	LTPD = 5, s = 5, c = 0

Therefore, acceleration factor $A(T_1,T_2)$ can be calculated for a failure mode, given the activation energy of that failure mode.

- *Biased moisture life (or temperature, humidity and bias) test.* This test is performed to evaluate the reliability of the die (as well as package) under the conditions of long-term exposure to severe, high-moisture environments that may cause corrosion. The device is operated at maximum recommended operating supply voltage (e.g., 5.5 V) and is exposed to temperatures of 85° C and a relative humidity of 85 percent throughout the test. This test is also called *temperature, humidity, and bias test* and is performed in accordance with JEDEC STD 22-B. An alternative temperature, humidity, and bias test is the highly accelerated stress test (HAST). This test is similar to the 85° C/85 percent RH test, except that the test is performed at 130° C and 85 percent RH. The vapor pressure at this condition is 33.5 psia.

- *Unbiased pressure pot test.* This test is performed at a temperature of 121° C and a pressure of 2 atm of saturated steam in an unbiased state. This test forces moisture into the plastic package and tests for corrosion of the bonding pads and wires, which are not protected by the passivation layer. A corrosion can also occur in the passivated areas, where there are microcracks or poor step coverage.
- *Thermal shock.* This test is performed to evaluate the resistance of the package to cracking and resistance of the bonding wires and lead frame to separation or damage. The devices are cycled over temperature extremes (e.g., –65 to +150° C for Cond. C) in accordance with MIL-STD-883, Method 1011, with almost zero transition time allowed from cold to hot transfer or vice-versa.
- *Temperature cycling.* This test is performed to evaluate any package problems induced from alternating exposures to temperature extremes. The package (ceramic or plastic), lead frame, silicon die and attach material all may expand and contract at different rates. This mismatch can lead to cracking, peeling, or delamination of the layers and/or package. The range of temperatures is –65 to +150° C. The transition time is longer than that for the thermal shock test, but the test is conducted for many more cycles (e.g., 500 to 1000).
- *Salt atmosphere.* This test was originally designed by the U.S. Navy to evaluate military-grade shipboard electronics' susceptibility to corrosion from the seawater. It is used more generally for nonhermetic industrial and commercial products as a test of corrosion resistance of the package marking and finish. This test is performed in accordance with MIL-STD-883, Method 1009.
- *Resistance to solvents.* This test is performed to evaluate the integrity of the package marking during exposure to a variety of solvents, such as those used during the board-level assembly and automated cleaning before system assembly. This test is performed in accordance with MIL-STD-883, Method 2015.
- *Solderability.* This test is performed to evaluate the solderability of leads under conditions of low soldering temperature, following expo-

sure to the aging effects of the water vapor. The test is performed in accordance with MIL-STD-883, Method 2003.

- *Lead integrity.* This method provides various test procedures and conditions for determining the integrity of microelectronic device leads (terminals), welds, and seals. This test is performed in accordance with MIL-STD-883, Method 2004.
- *Internal water vapor content.* The internal water vapor content test is intended to detect the amount of water vapor present inside the package. It is performed in accordance with MIL-STD-883, Method 1018.

In addition to the tests listed above, certain process monitors and tests are performed to assure the integrity of the nonvolatile programming elements such as EPROM, EEPROM, and flash memory cell based PLDs. There have been various studies performed to evaluate the charge loss mechanisms for an EPROM cell that uses ONO stack as an interpoly dielectric.[11,12] Figure 5.3 shows the schematic cross-section of an EPROM cell with an ONO stack as the interpoly dielectric. In general, there are four main charge loss mechanisms for the EPROMs, which can also be used for device qualification and quality screens. These are briefly described, as follows:[13]

- *DC erase (or word-line stress):* In this case, when a high voltage is applied to the polysilicon word line (poly 2), it causes electrons to be emitted through the poly 1 floating gate through the interpoly dielectric to the poly 2, resulting in a loss of charge and reduction in the programmed threshold voltage of the cell. This may occur during

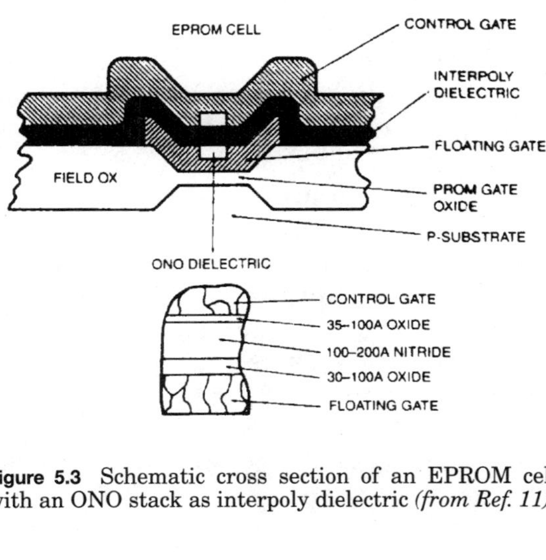

Figure 5.3 Schematic cross section of an EPROM cell with an ONO stack as interpoly dielectric *(from Ref. 11).*

the device programming or under extended read cycling. This stress can be used to examine the quality of the interpoly dielectric.

- *Program disturb (or bit-line stressing):* This may occur when a high voltage is applied to the bit line (n^+ diffusion) of the cell and charge can be transferred from the floating gate to the drain, causing a reduction in the programmed threshold voltage. This stress examines the quality of the interfaces between the gate and the drain.
- *DC programming:* In this charge loss mechanism, an erased cell may have its threshold voltage raised by an inadvertent charge from the transistor channel on to the poly 1, which makes an unprogrammed cell become programmed. This stress provides information on the quality of the gate oxide between the cell substrate and the floating gate.
- *Charge loss due to bake stressing:* EPROMs tend to lose floating-gate charge over a period of time, and an increase in temperature tends to accelerate the rate of charge loss. This high-temperature bake stress can be used to evaluate the overall quality of all oxides surrounding the polysilicon floating gate.

These four stresses are often used by the vendors as in-line process monitors for an overall quality assurance plan that includes the statistical process control (SPC) and quality conformance inspection (QCI) requirements. The objective of the reliability testing for the program disturb in EPROM cells is to determine the voltage and temperature acceleration factors that would make it possible to predict the defect levels and to develop screens to identify the potential failures (marginal devices). The use of this testing can accelerate program disturb defects by high-voltage stress on the drain and grounding the gate through internal device modes.

The three principal memory disturbs for the flash memory cell based devices that may occur during the programming are the same as for EPROM cells, i.e., dc erase, dc program, and program disturb.

An example of reliability assurance for nonvolatile programming element is the voltage margin testing performed by Altera on its PLD families. This voltage margin testing is performed to measure the threshold voltage of EPROM, EEPROM, or flash memory cells. It is considered essential for monitoring the ability of the device to be programmed or erased and for evaluating the long-term stability of a device after programming or erasure. Altera devices incorporate special test mode circuitry that allows measurement of the threshold voltage of each EPROM, EEPROM, or flash cell on a device. This circuitry is used to determine the programmability of all devices and for reliability screens that detect charge loss from the programmed devices. During the screening process, any degradation of threshold voltage is measured for each cell as a function of time, temperature, and voltage.

Altera performs this voltage margin testing during the wafer sort operations. At wafer sort 1, the devices are tested for combinatorial and registered logic functionality and for cell programming to ensure that all EPROM, EEPROM, and flash cells are fully programmable. Each cell is "voltage margined" to ensure that its threshold voltage exceeds 5.5 V, the recommended operating supply voltage. This voltage margin test is performed immediately after programming and before the DC erase stress. After the DC erase stress is applied, each cell's voltage margin is tested again. If any cell shows a significant reduction in threshold voltage, the device is rejected from the manufacturing flow. After wafer sort 1, to detect material defects, every device is baked at a high temperature with all cells in the charged state. This bake is performed at 300° C for 12 hr or 245° C for 96 hr, depending on the device type. After this bake operation, the threshold for each cell is again measured during wafer sort 2 and compared to the prebake values. The devices that exhibit a reduced threshold voltage are rejected from the manufacturing flow. Figure 5.4 shows the wafer sort, assembly, and final test flow for Altera devices.[14]

5.3 Antifuse Reliability Characterization and Testing

Antifuse programming technology was discussed in Sec. 3.2.1, including oxide-nitride-oxide (ONO) based antifuses such as PLICE used in FPGAs by Actel Corp., and amorphous silicon (or metal-to-metal) antifuse structures such as the ViaLink used by QuickLogic Corp. in their pASIC family FPGAs. The major reliability considerations for antifuses are leakage currents and their dependence on the temperature, antifuse resistance, breakdown voltage, time-dependent dielectric breakdown (TDDB), and resistance disturb during the read of a programmed antifuse. A study was performed to compare the following antifuse structures: n^+ diffusion/ONO/polysilicon, n^+ diffusion/NO/polysilicon, metal/NON/metal, metal/amorphous-silicon/metal, and metal/N/metal. The results of this study are briefly summarized below:[15]

- A comparison of the leakage currents and breakdown voltages of several different types of antifuses showed that, in general, the metal-to-metal antifuses have higher leakage currents, while n^+ diffusion to polysilicon antifuses have lower leakage currents. The leakage currents can be reduced by increasing the antifuse film thickness, but then the thicker antifuse increases the breakdown voltage, which limits the scaling. Metal/dielectric/metal antifuse

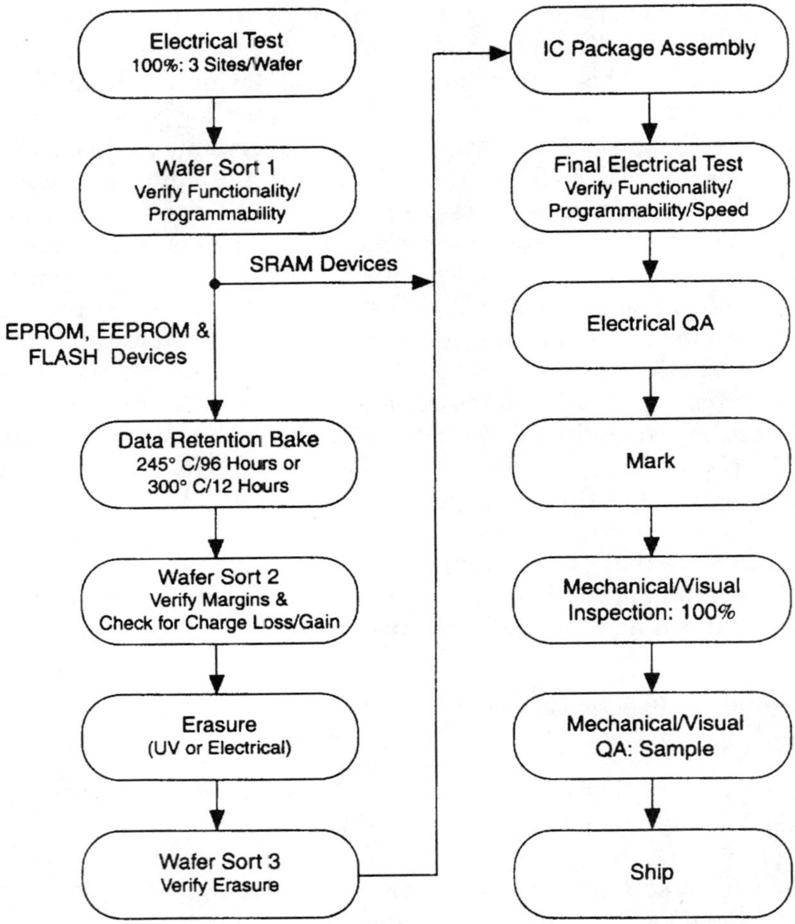

Figure 5.4 Wafer sort, assembly, and final test flow for Altera devices. *(Reprinted from Ref. 14 with permission of Altera Corp.)*

structure can have lower leakage current and lower breakdown voltage than the amorphous silicon antifuse. The choice of electrodes, antifuse materials, cell structures, and process conditions allow further leakage current reduction.

- Programmed antifuse resistance as well as the parasitic resistance associated with the electrode are lower for metal-to-metal antifuses. While the lower resistances obtained for the metal-to-metal antifuse with lower programming current are attractive, the read disturb phenomenon requires antifuses to be programmed with higher currents.
- The TDDB data showed intrinsic antifuse lifetimes at 5.5 V as exceeding 10 years for most antifuse types except for metal/N/metal,

which had a very low breakdown voltage. This low breakdown voltage can be improved by increasing the nitride thickness.

- The programmed metal-to-metal antifuses were found to exhibit a disturb phenomenon in which the antifuse can be turned off when the DC read current is comparable to, or larger than, the programming current in a ramp voltage test. This turn-off behavior in both the forward and reverse read disturb direction is similar to electromigration. This read disturb phenomenon was not observed in n^+ diffusion/ONO/polysilicon antifuses in which the link is made of single crystal silicon or polycrystalline silicon, and not metals.
- Antifuse thickness also appears to affect metal-to-metal antifuse characteristics. Increasing the antifuse thickness reduces the capacitance and leakage but increases the programming voltage and increases the antifuse lifetime. Also, reducing the antifuse thickness will reduce the antifuse on-resistance and the possibility of a read disturb.

The following section discusses the reliability characterization issues associated with two dominant technologies used: dielectric-based ONO and amorphous silicon antifuses.

5.3.1 Oxide-nitride-oxide (ONO) antifuses

An ONO antifuse basically consists of a thin layer of oxide, which is thermally grown on top of the N^+ surface, followed by LPCVD nitride and the reoxidized top oxide. The antifuse is sandwiched between N^+ diffusion and N^+ polysilicon gate to form an array with density limited by the metal pitches. An example of the ONO antifuse is Actel PLICE structure described in Sec. 3.2. A reliability characterization study was performed on these fuses to address the following three major concerns:[16]

1. The unprogrammed antifuse should survive a 5.5 V, 40-year operating condition.
2. During programming, all unprogrammed antifuses are subject to a momentary stress of half the programming voltage ($V_{pp}/2$).
3. The programmed antifuses should have a very low resistance that will not increase over the device lifetime.

One of the major reliability concerns with sub-10 nm ONO antifuses is the time-dependent dielectric-breakdown (TDDB) characteristics over 40 years. In this study, performed on 0.04 mm^2 capacitors, typical electrical field and temperature accelerated tests were performed to extrapolate the dielectric lifetime under the normal operating conditions. The results showed that ONO antifuses have a lower and tighter resistance distribution than that for the oxide antifuses. The ONO-based antifuse intrinsic lifetime failure data was observed to have a better fit with 1/E model, rather than the E model. This exp(1/E) field dependence is explained mainly with the Fowler-Nordheim tunneling mechanism.[16]

The temperature acceleration effect on the 0.04 mm^2 ONO area capacitors for 5.5 V lifetime estimate showed the activation energy as close to 0.9 eV and function of the electric field (1/E model). Based on an extensive reliability characterization testing of the ONO antifuse structures, Actel has developed a voltage test to screen out the programming yield loss due to the breakdown of defective unselected antifuses. During programming, all antifuse electrodes are precharged at a given voltage V_{pre}, which is usually set to half the programming voltage (V_{pp}/2). To program the antifuse, its polysilicon electrode is raised to V_{pp} while its N^+ diffusion is grounded. The unselected antifuses are subjected to a stress of either V_{pre} to ground, or V_{pp} to V_{pre} for an average of 100 times the single antifuse programming time. If, due to this stress, the unselected antifuses fail (i.e., become programmed), they will show up as programming failures. These defective antifuses can be screened out at a wafer sort by a 1 s stress at 10 V (10V/1 s). This screen is performed twice during the wafer sort, first (FS-1) to screen out the defective dielectric distribution, and second (FS-2) to simulate the percentage yield loss during programming.

In the study of programmed ONO antifuses, it was found that the lifetime is limited by the contact electromigration, not by the antifuses. In addition, the resistance remained low throughout the test, indicating that the resistance does not increase over the lifetime. Actel has developed the following tests for its antifuse-based FPGA products, based on the reliability studies of ONO antifuse structures:

- *Antifuse shorts test (or blank test).* This test is performed to ensure that all antifuses are in unprogrammed state prior to shipping. In this test, the array is precharged, and then the vertical tracks are driven to ground while the horizontal tracks are read to confirm they are still high. A programmed or leaky antifuse would drive a horizontal track low. This test is repeated by driving horizontal tracks low and then reading the vertical tracks.
- *Antifuse stress test.* During this test, the voltage applied to the V_{pp} pin can be applied across all antifuses on the chip, while the other side is grounded. The voltage applied is the precharge voltage plus a significant guard band. After the voltage is applied, the antifuse shorts test is again used to ensure that no antifuses have been programmed. This test is effective in detecting the antifuse defects.

Since the reliability of an antifuse is much more voltage- than temperature-dependent, this test can also be used as an effective antifuse infant mortality screen.

5.3.2 Amorphous silicon (or metal-to-metal) antifuses

The metal-to-metal antifuses use an amorphous silicon layer (instead of a dielectric) sandwiched between the two metal layers as the programming material. An example of a metal electrode amorphous silicon antifuse is the ViaLink used in pASIC family of FPGAs developed by Quicklogic Corp. and discussed in Sec. 3.2.1. Figure 5.5*a* shows the cross-sectional drawing of amorphous silicon antifuse structure, which

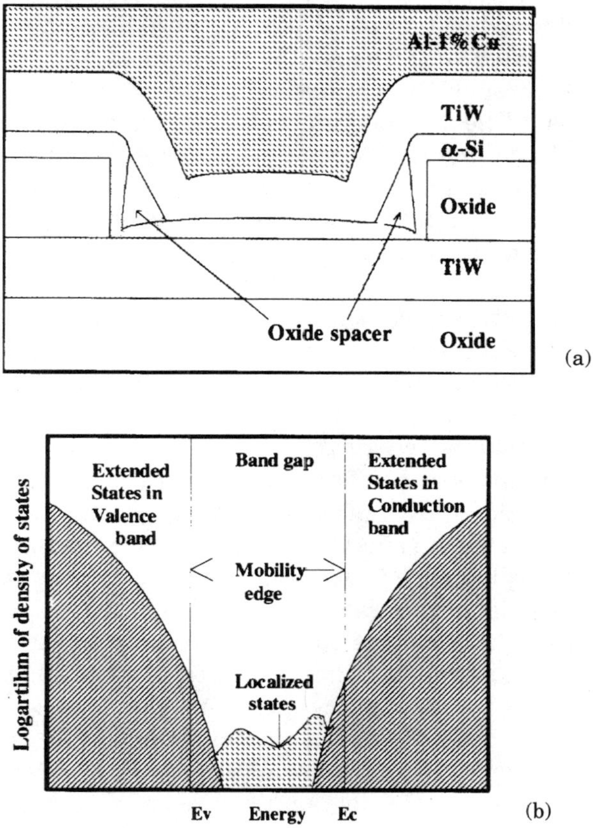

Figure 5.5 (a) Cross section of a metal-to-metal amorphous silicon antifuse *(from Ref. 17)* and (b) density of states for amorphous silicon as a function of energy *(from Ref. 18)*.

is added to a standard 1.0 μm double-level metal CMOS process after the first metal.[17]

The lower TiW electrode is deposited and patterned on top of the intermetal dielectric.A 3000 Å oxide is deposited and etched back to create spacers on the sidewall. These spacers are used to isolate the thinnest and most variable area of the amorphous silicon from the top TiW electrode to improve control of the programming voltage, which is a function of the amorphous silicon layer thickness. A 1.2 μm via defines the antifuse area.

During programming, when a negative voltage is applied to the top electrode while the bottom electrode is grounded, the TiW metal moves from the top electrode into the amorphous silicon. When the programming pulse polarity is reversed, the filament is formed with the bottom TiW material, and it is composed of polycrystalline tungsten silicide. The filament formation is a result of both the electromigration and self-heating. A study performed on the electrical characteristics and reliability of the unprogrammed amorphous silicon antifuses showed that the leakage current through the antifuse will increase during a high electrical field stress.[18] The degradation is accelerated by electric field, but it is controllable and predictable with in an array and on the wafers from different fabrication runs. This degradation eventually saturates and does not lead to a catastrophic failure of the amorphous silicon antifuse. The leakage current can be returned to the initial value by annealing.

This amorphous silicon antifuse leakage current phenomenon has been explained with a reverse conduction mechanism model. The conduction mechanism in the amorphous silicon is different from that in the single crystalline silicon. Figure 5.5*b* shows the density of states for amorphous silicon as function of energy.[18]

In an amorphous silicon antifuse, the forbidden gap of the electron energy states is essentially closed by the localized states, as shown in Figure 5.5*b*. The extended states are similar to that for the crystalline silicon but do not have sharply defined regions at the conduction and valence band edges; the density of these states falls off rapidly. According to the conduction model, the leakage current in the antifuse creates localized states in the amorphous silicon by breaking the weak silicon bonds. As the localized state density increases, the dominant conduction process changes from the carrier conduction in the extended states to carrier hopping among the localized states. When most of the weak bonds are broken, the degradation process saturates. The localized states are annealed at high temperatures, and the conduction returns to the initial mechanism.

The reliability characterization testing for amorphous silicon antifuses, such as the ViaLink elements in the unprogrammed state, must determine their ability to withstand applied voltages over the operating conditions without showing significant increase in resistance or becoming programmed. When a ViaLink element is stressed by high electric fields, its resistance can decrease from the initial 1 GΩ value. In a reliability testing performed on the unprogrammed antifuses, time for the resistance to reach 50 MΩ at different stress field and data was modeled using the following equation:[18]

$$t_{50\mathrm{M}\Omega} = t_0 \exp(-BE)$$

where the time to 50 MΩ resistance decreases exponentially with the increasing applied electric field. The constant t_0 is 3×10^{15} s, and the field acceleration factor, B, is 20 cm/MV. This model is valid for electric field (E) below 1.5 MV/cm, whereas, at higher fields, the amorphous silicon has a different programming mechanism. This high field effect inherent to the amorphous silicon in ViaLink element is predictable and reversible.

An increase in resistance of the programmed ViaLink elements may not be catastrophic, but a higher resistance can affect the device operating speed because of an increased switching (RC) delay. In the study performed, programmed ViaLink elements were stressed under severe capacitive currents using ac stresses instead of dc to accelerate the failures. The test data showed the mean number of cycles to disturb, N_{50}, as being identical for 25° C and 250° C. This lack of an observable temperature dependence indicates that the activation energy is 0. The number of cycles to disturb can be modeled as

$$N_{50} = N_0 \exp(-PJ)$$

where $N_0 = 7 \times 10^{41}$ is a constant

P = current density acceleration factor, 1.2 cm^2/mA

J = peak ac current density through the link

A high ac current density can occur at low frequencies where there is sufficient time for the network to be fully charged or discharged. At frequencies above 50 MHz, ac current through ViaLink decreases due to incomplete charging and discharging cycles.

Accelerated stress test for QuickLogic pASIC devices uses low-temperature operating life with V_{cc} = 6.0 V at 15 MHz for 1000 hours at –55° C. The cold temperature is selected because it accelerates the stress by increasing the drive current and stresses the programmed ViaLink elements at 40×10^6 A/cm^2 for 5.4×10^{13} cycles. The acceleration factor calculated from the equation given above is 380. This test is equivalent to 2.0×10^{16} switching cycles, or continuous operation under the worst-case condition at 50 MHz for 12 years at the worst-case current density of 35×10^6 A/cm^2.

A new metal-to-metal antifuse has been developed for 3.3 V FPGA operation in which the dielectric breakdown reliability is improved by using amorphous-like WSi_x film as a bottom electrode having extremely smooth surface and 10 nm thick SiN_x film.[19,20] In scaling down the source voltage from 5 V to 3.3 V, it is desirable to reduce the programming voltage down to approximately 6 to 10 V. Thus, the thickness of dielectric film must be reduced down to 10 nm in case of SiN_x film. However, it is difficult to achieve sufficient OFF-state reliability of such an extremely thin dielectric film formed on the barrier metal electrode due to the electric field intensification caused by the roughness of metal surface.

It was shown that higher reliability of the dielectric breakdown (OFF-state reliability) and tighter distribution of the ON-state resistance (ON-state reliability) can be obtained by using sputter-deposited WSi_x as a bottom electrode. A lower ON-resistance can also be obtained by using constant-current pulse stress for programming.

Figure 5.6 shows the schematic cross section of a metal-to-metal antifuse.[20] Table 5.4 lists the various material combinations used for Metal 1 and Metal 2 electrodes.[20] Each film was deposited by sputtering method, and the surface of TiN film was pretreated with a solution (NH_4OH/H_2O_2) to reduce the surface microroughness. Antifuses were programmed with a constant voltage pulse stress with resistor and constant-current pulse stress limited up to 12 V.

The dielectric breakdown characteristics of the antifuses using WSi_x, TiN and Al-Cu film as a bottom electrode were investigated. The test data showed that the breakdown voltage of antifuses using TiN film as a bottom electrode decreases rapidly as the device area increases. However, the breakdown voltage for antifuses using WSi_x film

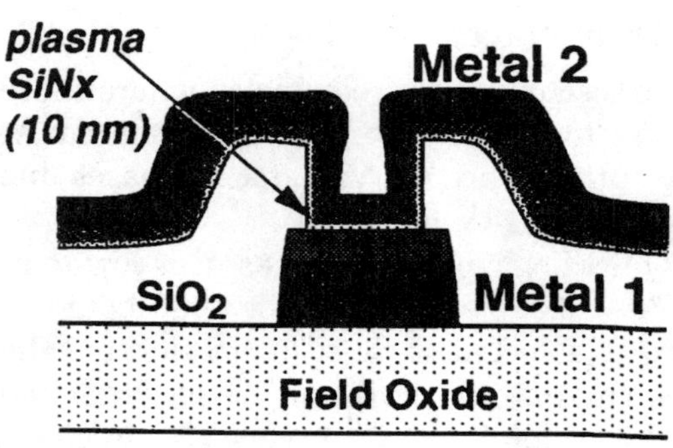

Figure 5.6 Schematic cross section of a metal-to-metal antifuse *(from Ref. 20).*

TABLE 5.4 Various Metal Combinations Used for Metal 1 and Metal 2 Electrodes[20]

Metal 1	Metal 2
WSi_x/Al-Cu	Al-Cu/TiN
TiN/Al-Cu	Al-Cu/Ti
Al-Cu	Al-Cu

as a bottom electrode was kept constant. This result suggests that the defect density of SiN_x film formed on WSi_x film is lot smaller than on TiN film. Also, the atomic force microscopy (AFM) image comparison of WSi_x and TiN films deposited on Al-Cu film showed the surface of sputter deposited WSi_x to be far smoother than that of the TiN film.

According to this study, improvements in the insulating properties and tighter distribution of breakdown voltage are strongly related to surface smoothness of the WSi_x film as a bottom electrode. By using amorphous-like WSi_x as a lower electrode, the lifetime to dielectric breakdown was found to be much longer and had a tighter distribution than that by using the Al-Cu or TiN. The lifetime to breakdown was extrapolated to be 1E13 years at 3.3 V. The 10 nm thick SiN film formed on the WSi_x film provided a low programming voltage, below 10 V. A low ON-resistance (8.2 Ω) having a tighter distribution was obtained for WSi_x/SiN_x/Al-Cu antifuses without switch-off failure.

5.4 FPGA Metastability

Metastability is a region of uncertainty exhibited by a flip-flop when the timing characteristics of setup and hold times have been violated. Metastability may arise even in synchronous systems because of some asynchronous conditions, such as a signal not being fully synchronized to the system clock. The metastable events are also associated with the data transitions occurring close to the active edge of the clock. Metastability causes the output transition delay to be unpredictable. For example, the output of a flip-flop driving two different parts of a circuit may be latched differently because of the metastable behavior. Figure 5.7*a* shows an example of the data changing within a timing window, resulting in a timing violation. When the flip-flop is metastable, it causes the output to hover at a voltage level between high and low, causing the output transition to be delayed beyond the specified clock-to-output delay (t_{CO}). The additional time beyond t_{CO} that a metastable output takes to resolve to a stable state is called the settling time (t_{MET}).

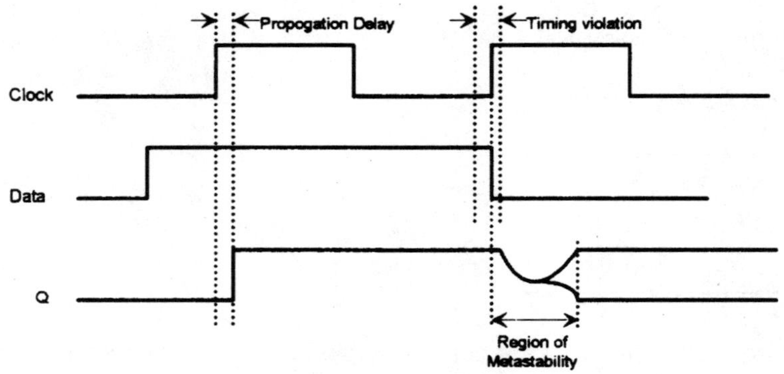

(a)

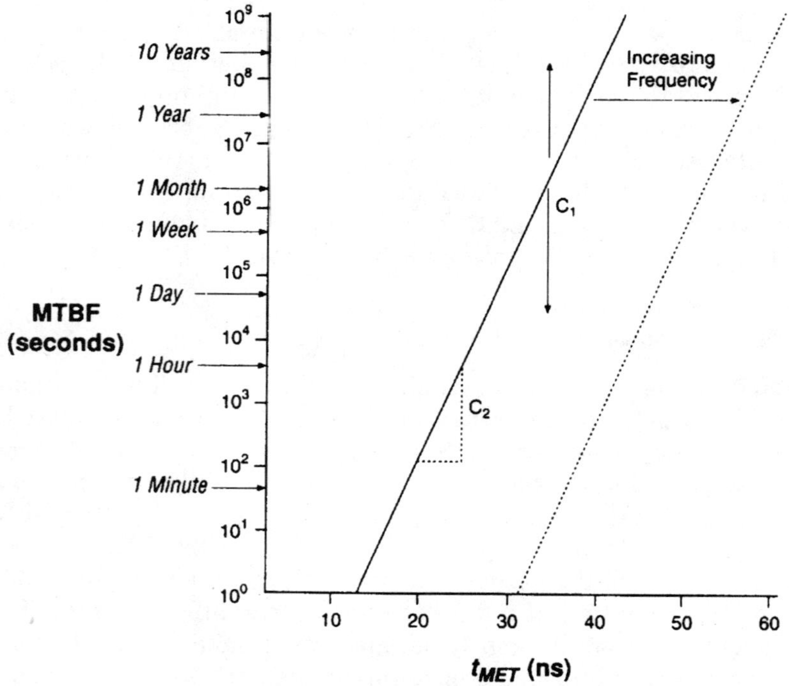

(b)

Figure 5.7 (a) Example of timing violation resulting in metastability condition and (b) relationship between MTBF and t_{MET} *(reprinted from Ref. 22 with permission of Altera Corp.).*

Metastability is typically described by the MTBF, which provides an estimate of the mean time between the probabilistic occurrence of two successive metastable events. The MTBF of a synchronizing flip-flop is typically given by following equation:[21]

$$\text{MTBF} = \frac{1}{f_{CLK} \times f_{DATA} \times C_1 \times \exp(-C_2 \times t_{MET})}$$

where t_{MET} is the additional time required by the flip-flop to settle down to a steady state.

Figure 5.7*b* shows the relationship between MTBF and t_{MET}. The constants C_1 and C_2 depend on the manufacturer's process technology used and are determined by plotting the natural log of MTBF versus t_{MET} and then performing a linear regression analysis on the data. The slope and the y-intercept of the resulting line determines values of the constants C_1 and C_2.

$$C_1 = \frac{\exp(-C_2 \times t_{MET})}{\text{MTBF} \times f_{CLK} \times f_{DATA}}$$

$$C_2 = \frac{\ln(\Delta\text{MTBF})}{\Delta t_{MET}}$$

The constant C_1 scales linearly with MTBF equation so that the larger the value of C_1, the higher the MTBF. The constant C_2 affects the slope of the MTBF/t_{MET} curve and determines how fast the flip-flop settles to a stable state. Once the constants C_1 and C_2 are determined for a particular device, the MTBF of the system can be calculated for a given settling time t_{MET}. Alternatively, the metastability equation can also be used to find the t_{MET} delay required for a given MTBF value, as follows:

$$t_{MET} = \frac{1}{C_2[\ln(\text{MTBF}) \times f_{CLK} \times f_{DATA} \times C_1]}$$

Figure 5.8*a* show the typical test circuit used for evaluating the metastability characteristics of the Altera devices.[22] The logic that generates the metastable event and the logic that detects it are both located in the same device under test (DUT). One flip-flop shown has asynchronous clock and data inputs. The output of the synchronizing flip-flop is fed directly to one of the resolving flip-flops and through an inverter to the other resolving flip-flop. The outputs of the two resolving flip-flops feed an XNOR gate, which is at a high logic level when the values of the outputs (both the signal and its complement) are the

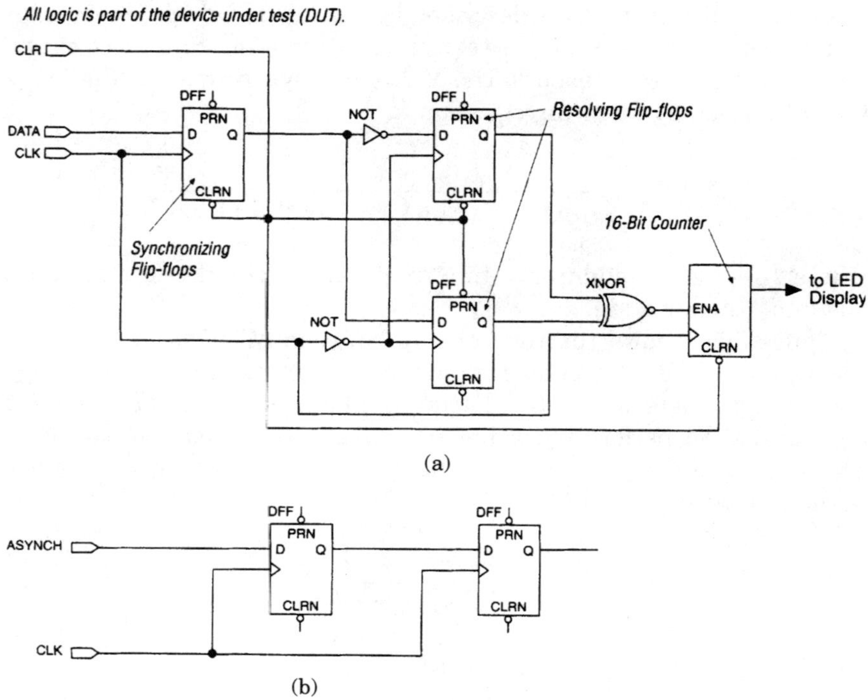

Figure 5.8 (a) Typical metastability test circuit used for Altera devices and (b) multiple-stage synchronizer for reducing metastability conditions (*reprinted from Ref. 22 with permission of Altera Corp.).*

same. If the resolving flip-flops detect the condition that the signal and its complement have the same logical value, a metastable event has occurred, and the counter is incremented.

Since the resolving flip-flops are controlled by the falling edge of the clock, the required settling time can be controlled by changing the clock high time (Δt). The t_{MET} is the minimum time allowed under normal circuit operation and can be determined by the equation given below.

$$t_{MET} = \Delta t - t_{ACNT}$$

where t_{ACNT} is the minimum delay from the clock edge to the output of the synchronizing flip-flop, plus the delay from the output of the synchronizing flip-flop to the input of the resolving flip-flops, plus the setup time of the resolving flip-flop.

Several techniques are used for reducing the metastability in a system design. The probability that a metastable event may occur greatly increases if an asynchronous signal is fed to several flip-flops. In this case, the metastability conditions can be avoided by using the output of the synchronizing flip-flop throughout the system, rather than the asynchronous signal itself. A commonly used method for minimizing the effects of metastability is to use a multiple-stage synchronizer in which two or more flip-flops are cascaded to form a synchronizing circuit, as shown in Fig. 5.8*b*. If the synchronizing flip-flop produces a metastable output, then the metastable condition may be resolved before the signal is clocked by the second flip-flop. Although this method does not guarantee that the second flip-flop will not clock an undefined value, it doubles the probability that the data will go to a valid state before propagating down to the rest of the circuit.

The metastability effects can also be reduced by adding the calculated settling time t_{MET} for a specific MTBF to the worst-case timing delay calculations to inhibit the transitions occurring during the failure windows. This allows the output of the synchronizing flip-flops enough time to settle down. Another approach is to use a *metastability hardened* device, whenever possible. These are fast switching devices with very small failure windows.

References

1. Dennis McCarty, "PREP benchmarks for programmable logic devices," *IEEE Custom Integrated Circuits Conf. Proc., 1993,* pp. 7.7.1–7.7.6.
2. Programmable Electronics Performance (PREP) Corp. WWW home page.
3. Jay Sturges, "A quantitative approach to benchmarking programmable logic architecture," *IEEE Custom Integrated Circuits Conf. Proc., 1993,* pp. 407–415.
4. J. Rose and S. Brown, "Flexibility of interconnection structures for field-programmable gate arrays," *IEEE J. Solid State Circuits,* vol. 26, no.3, March 1991, pp. 277–282.
5. J. Sturges, "An analytical approach to PLD/FPGA benchmarking," *ASIC & EDA,* June 1992, pp. 36–46.
6. Ashok K. Sharma, "*Semiconductor Memories: Technology, Testing and Reliability,*" Ch. 6. Piscataway, NJ: IEEE Press.
7. *Hitachi IC Memory Data Book,* 1992.
8. Military Standard 883D (MIL-STD-883D): Military Standard Test Methods and Procedures for Microelectronics. Washington, DC: U.S. Government Printing Office.
9. E. B. Hakim, et al., "Beyond the qualified manufacturer list (QML)," in *IEEE Reliability and Maintainability Symp. Proc.,* 1995, pp. 362–369.
10. *Xilinx Databook,* 1996.
11. K. Wu, et al., "A model for EPROM intrinsic charge loss through the oxide-nitride-oxide (ONO) interpoly dielectric," in *IEEE IRPS Proc., 1990,* pp. 145–149.
12. C. Pan, "Physical origin of long-term charge loss in floating gate EPROM with an interpoly oxide-nitride-oxide stacked dielectric," in *IEEE IRPS Proc.* 1991, pp. 51–53.
13. D. A. Baglee, et al., "Building reliability into EPROMs," in *IEEE IRPS Proc.,* 1991, pp.12–18.
14. *Altera Data Book,* 1996.
15. Steve Chiang, et al., "Antifuse structure comparison for field programmable gate arrays," *IEEE IEDM 92 Proc.,* pp. 24.6.1–24.6.4.
16. Steve Chiang, et al., "Oxide-nitride-oxide antifuse reliability," in *IEEE IRPS Proc.,* 1990, pp. 186–190.
17. *QuickLogic Data Book,* 1994, Section 6.

18. Richard J. Wong, et al., "Reliability mechanism of the unprogrammed amorphous silicon antifuse," in *IEEE IRPS Proc. 1994,* pp. 378–382.
19. Y. Tamura, et al., "Most promising metal-to-metal antifuse based 10 nm-thick p-SiNx film for high density and high speed FPGA application," *IEEE IEDM,* 1994, pp. 10.5.1–10.5.4.
20. Y. Tamura, et al., "ON-state programming and OFF-state reliability of metal-to-metal antifuse based 10 nm-thick SiNx film for 3.3 V operation," *IEEE IEDM,* 1995, pp. 36–41.
21. *QuickLogic Data Book,* 1996/1997: Metastability Report for FPGAs.
22. *Altera Data Book,* 1995, Application Note 42: Metastability in Altera Devices.

Chapter

6

PLDs and Reconfigurable Logic Applications

6.1 Reconfigurable Logic

The generic term *reconfigurable PLD* refers to a programmable logic device whose function can be customized to a particular system or application. The advent of SRAM-based FPGAs in the late 1980s and early 1990s has presented the programmable logic designers with advanced capabilities of dynamically reconfigurable logic, which means that the designs can be reconfigured *on the fly*. In the SRAM-based FPGAs, the state of the internal static memory cells determines the logic functions and interconnections within an FPGA device. Therefore, by downloading different configuration programs into the memory cells, the device functions can be changed as desired, while the device is still resident in the target system. These devices can be "reconfigured" an unlimited number of times. Therefore, systems with such reconfigurable FPGAs are also referred to as the *soft-hardware*, i.e., the hardware whose function can be changed easily during a normal system operation.[1] The ability to reconfigure FPGAs resident in the target system can significantly ease the debugging process, reducing the overall development time cycle and cost.

Reconfigurable logic differs from reprogrammable logic, which includes devices with limited reprogramming life, such as EPROM, EEPROM, or flash-technology based PLDs. The reprogrammable devices offer some of the features of reconfigurable logic, such as ease of development and field upgrades. However, they are not suitable for dynamic reconfiguration, which is finding extensive use in FPGA-based reconfigurable custom computing applications. In the past, the term *reconfigurable computing architecture* referred to multiprocessor systems with a fixed architecture implemented in each processor and a fixed topology for the interprocessor communication, but with reconfigurable connections between the processors and the network. The current developments in reconfigurable FPGA devices have enabled system-level designs of processors and coprocessors whose internal architectures and connections can be reconfigured to match the needs of a given application. These architectures and applications will be discussed in the following sections.

In general, configurable (or reconfigurable) hardware may be defined as a product whose function may be customized several (or many) times. FPGAs based on SRAM technology are available from several manufacturers, including Atmel, Lucent Technologies, Motorola, and Xilinx.[2] In these FPGAs, the logic functions and interconnections are determined by the configuration program stored in internal static memory cells. Since these devices are SRAM based, a configuration program must be loaded into each FPGA when the system is powered up. Thus, the configuration program must reside in an external system such as an EPROM or on a disk. There are usually several configuration program loading options available to accommodate various system requirements. The automatic modes use the on-chip control logic for direct loading of the FPGA from an external memory device, typically a serial EEPROM. In peripheral loading modes, the configuration programs are written to the FPGA under the control of an external processor DMA device.

The main limitation with the majority of older SRAM-based FPGAs was the necessity of loading up the whole device and interruption of the circuit operation while the devices were being reconfigured. Some of the recent-generation FPGAs support the dynamic reconfiguration of the selected portions of the internal logic and have additional features as follow:[3]

- No disruption to the FPGA's inputs and outputs
- No disruption to the contents of internal register during reconfiguration, even in the area being reconfigured
- Continuous operation of any portions of the device that are not undergoing reconfiguration
- No disruption to the system-level clocking

A simple application of reconfigured logic may be for a system with the built-in self-diagnostics that can be implemented by using FPGA configurations dedicated to testing functions. In the traditional approach, the additional logic needed to implement diagnostic operations can significantly increase the system's size, complexity, and cost.

The reconfigurable logic implemented through the SRAM-based FPGAs offers designers an economical means to implement diagnostic test functions without the additional hardware. The same FPGAs that implement the system's logic can be reconfigured to hold diagnostic logic to test the system. When the system is powered up, or placed in a test mode, the resident FPGAs can be configured with logic functions that are dedicated to testing other circuitry in the system. For example, a program configuration dedicated to implementing the boundary scan test can be loaded for board-level assembly testing.

Another common application for reconfigurable logic is the implementation of a single hardware design that can be adapted for various tasks or applications. This is referred to as *adaptable hardware.* The adaptive hardware concept is based on the development of a number of potential configuration programs that can be downloaded into the system's FPGAs to alter the logic for particular applications or operations, as needed. This adaptable hardware design has been particularly popular among test equipment manufacturers, since the reconfigurable logic provides the flexibility to use the same basic hardware design for various applications. This adaptive hardware concept has also benefited EDA manufacturers, such as Quickturn Design Systems, Zycad, and others in offering ASIC emulation systems based on Xilinx FPGAs. These systems use multiple FPGAs to emulate the ASIC logic designs consisting of tens of thousands of gates.

Another application area currently in evolutionary development is the design of *multipurpose* hardware systems in which reconfiguration of the programmable logic is normal part of the system's operation. These systems have some *mutually time-exclusive* functions that can be swapped into and out of the FPGAs as needed.

The reconfigurable logic is already shaping the future of high-performance computing. The term *reconfigurable computing* refers to the use of reconfigurable chips in a system, which allows the devices to be reprogrammed *on the fly* in real time to perform multiple functions that would normally require individual chip for each function. In some applications, these reconfigurable computing engines can provide performance exceeding that of the supercomputers.

These capabilities have also been described with terms such as *virtual hardware* and *cache logic.* The term *virtual hardware* implies that the hardware is not really there, even though its functions are being implemented through software. Similarly, the term *cache logic* (trademark of Atmel Corp.) is derived from its similarity to the concept of cache memory, since the active data refers to the data or instructions that a program is currently using, or which the operating system identifies the program will be using next. The virtual hardware concept is based on the fact that each large macrofunction in a device is usually implemented through a number of smaller microfunctions (or elements) such as counters, shift registers, and multiplexers. This implies that (a) there is a functional redundancy, so that an element such as a counter may be used several times in different places, and (b) there is a functional latency, since during any given clock cycle, only a portion of the elements may be active. Therefore, by tracking the occurrence and use of each element, the overall functionality of the device can be consolidated to eliminate redundancy. The virtual hardware allows implementation of far more complex tasks than it would appear from the amount of amount of available logic and macrofunctions.

Two examples of FPGAs that are dynamically reprogrammable and considered suitable for reconfigurable logic applications are:

- *The Atmel AT6000 FPGA family.* These devices promote the concept of cache logic and were briefly discussed in Sec. 3.3.2. These are partially reconfigurable so that one need not program the entire device when a change of function is required. This AT6000 architecture has been licensed by IBM Microelectronics to develop a family of devices that utilize a medium-grained, fixed library cell in a symmetrical array that supports dynamic reconfiguration.
- *The Xilinx XC6200 family.* This is specifically designed for coprocessing applications, since the members of the XC6200 family have a built-in microprocessor interface that can be programmed to accommodate 8-, 16-, or 32-bit data. In addition, a high-speed interface provides instant read/write access to all internal registers of the FPGA.

These two families of dynamically reconfigurable FPGAs are discussed in Sec. 6.2.

One of the considerations in selection of the reconfigurable logic for an application is whether the FPGA will be reconfigured once per function or multiple times per function.[4] The designs that configure logic only once per function can tolerate relatively long reprogramming times. However, the designs that need to be reprogrammed multiple times during a single function are substantially affected by the long programming times. For example, if an FPGA needs to be reconfigured several hundred times per second, the target FPGA selected should be capable of reconfiguration in few milliseconds or less.

There are practical limitations to efficient implementation of the reconfigurable logic, since commercially available FPGAs do not offer unlimited connectivity. FPGA designers have to make trade-offs between the available logic utilization and routing resources. More routing resources means fewer usable logic gates, which in turn, drives up the cost per gate. In applications that require multiple reconfigurable chips, packaging technology also limits the number of pins to few hundreds. One approach to multichip interconnect problem is to have a large, reconfigurable bus connecting the reconfigurable devices. Instead of having single bus architecture, one can use a reconfigurable bus that allows multiple bus types. A limitation in some applications is that the FPGAs clock rates are not as high as for the mask-programmed devices. Also, the interconnect problems on some designs can cause clock rates to further drop. Compensation for these clock rates and speed deficiencies is accomplished by the use of pipelined design techniques and parallel architectures.

Another limitation in reconfigurable logic implementation is that the current design methodologies and tools are predominantly focused on the static designs and are ill equipped to adequately manage dynamically reconfigurable designs. Therefore, for wider applications of virtual hardware implementation, the design methodologies will have to be improved, and more efficient tools will have to be created.

6.2 Dynamic Reconfiguration and Device Technology

In general, dynamic reconfigurability is the ability to modify part of an FPGA design, while the remainder (unchanged) portion carries out the system functions. There are various terms used to describe dynamic reconfiguration technology and associated hardware. A summary of these definitions is given below:[6]

- *Dynamic reconfiguration (DR).* This is also called *run-time reconfiguration (RTR),* or *on-the-fly reconfiguration*, and is a type of reconfiguration whereby a system can be reprogrammed during normal operation.
- *Dynamically reconfigurable FPGA (DR FPGA).* This is a field programmable gate array that can be reprogrammed on the fly during the system operation. The DR FPGA may support partial configuration, or it may be only reprogrammed in a full reconfiguration.
- *Dynamically reconfigurable logic (DRL).* These are digital logic circuits (e.g., DR FPGAs) that can be reprogrammed during system operation.
- *Dynamically reconfigurable hardware (DRHW).* This term denotes hardware capable of dynamic reconfiguration.
- *Dynamically programmable gate array (DPGA).* This is another name for *dynamically reconfigurable FPGA.*

- *Cache logic.* This is another term used to denote the DRL, based on similarities between the DRL operation and cache memory operation. Cache logic is trademark of Atmel Corp.
- *Partial reconfiguration.* This is also referred to as the *run-time configuration* and is a form of reconfiguration, which allows only a portion of the reconfigurable system to be reconfigured. It can be (a) nondisruptive (i.e., the portions of the system that are not being reconfigured remain fully operational during the entire reconfiguration cycle), or (b) disruptive (i.e., partial reconfiguration affects other portions of the system). The nondisruptive partial reconfiguration (often abbreviated to *partial reconfiguration*) is more practical and common in current-generation DR FPGAs.
- *Full reconfiguration.* This is also referred to as the *global run-time reconfiguration,* where the entire reconfigurable system is reconfigured.
- *In-circuit reconfiguration.* This is another term for dynamic reconfiguration, used to emphasize that the reconfiguration is performed without a need to physically remove the reconfigurable elements for programming.

The implementation of dynamic reconfigurability requires not only special FPGA architectures but also specific design tool capabilities, in both the front-end logic design and back-end design implementation. These are some of the important considerations in the front-end and back-end tool selection requirements:[5]

1. *Front-end tools.* The front-end tools must be able to isolate portions of the design, identify parts to be reconfigured, and allow designers to work independently on selected areas of the chip without disturbing the rest of the circuitry. The design can be "windowed" after physical design by graphically identifying a physical area of the chip, or it can be sectioned at any point in the design flow by selecting appropriate circuits in logical design representation. The following are some of the desired methodologies and features:
   - *Floorplanning* enables the logic designer to partition a design from the beginning, separating the logic functions for reconfiguration and assigning them into specific physical entities. This physical and logical linkage simplifies the reconfiguration, since it allows the designer to focus on the logical entities that require updating and isolates the corresponding physical areas, which helps the automatic place-and-route (APR) software.
   - Hierarchical design tools provide an additional benefit, since they allow the floorplanned sections of the design to be further divided

into subsections so that the designer can make updates on smaller portions of logical design. For complex designs in which there is significant data sharing between the constant and reconfigured portions of the chip, functional separation can be difficult. In these cases, it is critical that the reconfigured circuitry be functionally isolated. It means that the signals to and from the reconfigured logic must be clearly identified and their operation halted during reconfiguration.

- The ability to isolate clock and other control signals is critical in ensuring that their operation is not affected by the reconfiguration. One method to achieve signal isolation is by assigning separate control signals to the constant and reconfigured portions of the design. Another method is to buffer the control signals to minimize the impact of changes in the network loading that results from the wiring or circuit modification.

2. *Back-end tools.* The APR software must also be able to operate on isolated portions of the design, updating placement and routing according to the reconfigured logical description, without disturbing constant portions of the chip. Some other necessary features are:

   - *I/O assignment.* The pin-locking capability is critical, so that a portion or all of the chip I/O assignments remain fixed during configuration.
   - *Bitstream generation.* The bitstream generation tool, which provides the sequence of bits to program the FPGA, must be able to generate a partial bitstream for only the reconfigured portion of the design. Therefore, the bitstream tool and FPGA architecture must support addressing of specific configuration bits.
   - *Additional features.* In addition to the above features, more sophisticated APR capability is required to handle designs where reconfigured logic is mixed with portions of the design that remain constant. An iterative automatic place-and-route capability enables the tool to make progressive APR attempts based on the results of the previous runs. A timing-driven place-and-route feature is essential for many designs incorporating reconfiguration, since the complex timing requirements often make the place-and-route job difficult for a manual analysis. However, if the design is particularly complex, such that a small number of paths cannot be routed successfully by the APR tools, manual place-and-route capability may also be required.

     A comprehensive circuit library that includes complex functional macros can help make it easier for the designer to achieve APR implementation and timing closure for the reconfigured designs. Typically, these complex macros are more densely packed and efficient than the equivalent functions built from automatically placed simple logic elements.

6.2.1 Atmel AT6000 FPGAs for cache logic implementation

The cache logic implementation is conceptually similar to the cache memory functions. The active functions of an application are performed by an FPGA that can be reconfigured as it operates, while the inactive functions are stored in an inexpensive configuration memory (e.g., an EPROM). It is even possible to compile variations of a design in real time. As logic functions are required, they can be loaded into cache logic, replacing or complementing the logic already present.

The Atmel AT6000 series FPGA architecture was discussed in Sec. 3.3.2. It has a symmetrical array with identical cells, divided into 8×8 cell sectors, programmable interconnects, and surrounded by I/Os.[7] The ability to implement cache logic requires following:

1. FPGAs that are capable of being dynamically reconfigured in the system, either completely or partially, without disrupting the operation of the balance of logic in the device
2. FPGAs that should have architectural symmetry, which is necessary for easy modeling of the device characteristics and arbitrary placement of generic blocks in a location that is available at the time required

Two types of cache logic implementation schemes have been defined: predetermined and dynamic cache logic. The predetermined cache logic involves the use of predefined functions and macros that are stored in an external, nonvolatile memory (e.g., EPROM, disk, loading from a remote communication link).

In this scheme, the predefined functions should have already been placed and routed, and have previously generated bit streams. The implementation of these functions is controlled by a resident manager in the cache logic or in an external control such as a microcontroller or processor routine. New functions may be downloaded to the cache logic in the background without disrupting the operation of the cache (logic, I/O, and register data). Figure 6.1 shows the block diagram of a new task being downloaded from the configuration memory on to the dynamically configurable AT6000 FPGA.

The second type of cache logic implementation scheme is called *dynamic cache logic,* which is the basis for building adaptive hardware and virtual products. This scheme involves determination of logic, its placement and routing, bit stream generation, and programming of

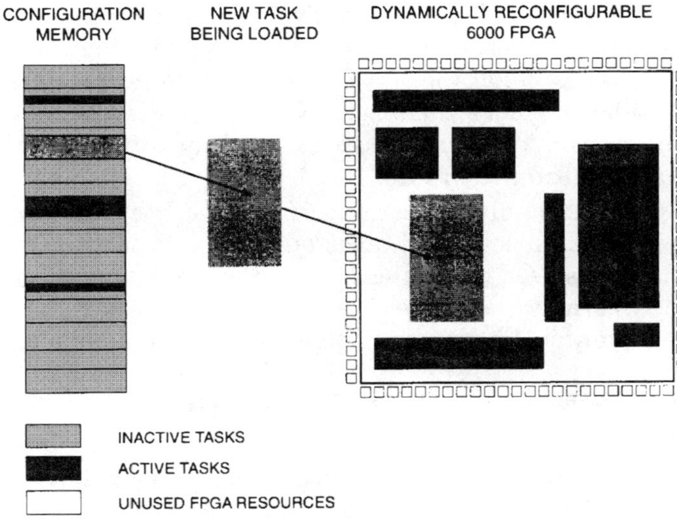

Figure 6.1 Block diagram of cache logic concept showing configuration memory and a new task being loaded into a dynamically configurable AT6000 FPGA *(from Ref. 7 with permission of Atmel Corp.).*

the logic cache in real time. The dynamic cache logic for implementation of the virtual products is currently in the evolutionary stage. The major issues to be addressed include the scheduling and allocation of functions, random logic collection, and collision handling and avoidance within the cache.

Atmel has introduced an FPGA design tool for in-system reconfiguration of its AT6000 FPGAs used in reconfigurable DSP coprocessor types of applications. This design tool, called QuickChange, allows designers to interactively specify multiple parameters for the implementation of FIR filters, convolvers, FFTs, and other DSP functions. After completion of a design with an initial set of parameters, the designer can invoke QuickChange from the design environment. QuickChange searches the design for DSP filter coefficients and other parameters, logically grouping and displaying them so that the designer can interactively specify as many replacement sets of parameters as desired. The QuickChange tool then generates the FPGA logic stream for each new specified parameter.

The AT6000 architecture has been licensed by IBM Microelectronics to develop a family of FPGAs to support dynamic reconfiguration applications.[8] The areas of focus for improvement were logic cell functionality, routing structure, I/O resources, and the global clock and reset structure. This IBM 10000 FPGA family utilizes a symmetrical array of medium-grained, fixed library cells in an architecture similar to the sea-of-gates approach. The array is divided hierarchically into sectors, each of which contains an array of 8 × 8 cells and a hierarchy of buses. There are repeaters at the sector boundaries to repower signals traversing the sectors and switching signals from one level of busing hierarchy to another. The array is surrounded by a flexible I/O busing network and multifunction I/Os.

The IBM 10000 FPGA family utilizes a logic cell that has eight configuration bits allocated for logic and takes advantage of nonuniform distribution of functions used by synthesis to implement various logic functions and their permutations. The array uses four types of routing buses, differing in quantity, length, connectivity and speed, as follows:

1. Direct buses are used for the fastest adjacent cell-to-cell connections.
2. Local buses are used for localized logic cluster connections that are not possible with direct bus connections. They span each sector in both the horizontal and vertical directions and connect to each cell along a cell row or column.
3. Express buses are used for longer connections; they span either one or two sectors and do not connect directly to the cell.
4. The super bus spans the entire array and connects to the express and local buses at selected repeaters.

6.2.2 Xilinx XC6200 FPGAs for reconfigurable logic implementation

Xilinx XC6200 series is a family of fine-grain, sea-of-gates architecture FPGAs designed to operate with a microprocessor or microcontroller. These devices can be used to implement functions such as providing interfaces to external hardware and peripherals, glue logic, and custom coprocessors. The XC6200 family consists of devices such as XC6216 with typical available gate range of 16,000 to 24,000 gates, 4096 as number of cells (or maximum number of registers), 256 input-output blocks (IOBs), and 65k as maximum number of RAM bits. Other devices in this family are planned for future introduction, with the highest density device in the available gate range of 64,000 to 100,000, with the number of cells as high as 16,384 and 512 IOBs, and 262k as the maximum number of RAM bits.[9]

The XC6200 device consists of a large array of simple, configurable cells. Each basic cell contains a computation unit capable of simultaneously implementing one of a set of logic level functions and a routing area through which intercell communication can take place. The XC6200 part is configured by the content of an integral, six-transistor

SRAM cell for control store, which allows the part to be reconfigured an unlimited number of times. The SRAM control store can be mapped into the address space of a host processor, and additional logic is provided to allow for rapid configuration for all or part of the device. In addition, the outputs of function units within the device can be read by a processor through the RAM interface. A host processor can read or write registers within the logic implemented on the device. The data transfer can be 8, 16, or 32 bits wide, even when the register bits are distributed over a column of cells. These capabilities allow XC6200 FPGAs to support dynamic reconfiguration applications in which sections of the device can be reconfigured without disturbing the circuit operation in other sections. Thus, an XC6200 FPGA in a coprocessor application can be time-shared by several processes running on the host computer. The following sections provide the functional description, routing resources and the general architecture for XC6200 family.

- *Logical and physical organization.* The XC6200 is a fine-grain, hierarchical cellular architecture in which neighbor-connected cells are grouped into blocks of 4 × 4 cells. Figure 6.2*a* shows the 4 × 4 cell block.[9] These blocks form a cellular array, communicating with the neighboring 4 × 4 cell blocks. A 4 × 4 array of these 4 × 4 blocks forms a 16 × 16 cell tile, as shown in Fig. 6.2*b*. In the XC6216 part, a 4 × 4 array of these 16 × 16 tiles forms the central 64 × 64 cell array, which is then surrounded by I/O pads, as shown in Fig. 6.2*c*.
- *Routing resources.* In XC6200 family architecture, each level of hierarchy (basic cells, 4 × 4 cell blocks, 16 × 16 cell tiles, 64 × 64 array) has its own associated routing resources. The basic cells can route to their nearest neighbors, or through the neighboring cell to its neighbor. The cells used for interconnect in this manner can be still used to provide a logic function. The wires of length 4 are provided to allow 4 × 4 cell blocks to route across themselves without using the basic cell resources. Similarly, 16 × 16 cell tiles provide additional wires of length 16 cells and the 64 × 64 array provides wires of length 64.

  The switches at the edge of the blocks and tiles provide for connections between the various levels of interconnect at the same position in the array (e.g., connecting length-4 wires to neighbor wires). The longer wires provided at each hierarchical level are called *fastlanes* because it is convenient to visualize the 3-D structure with routing at each hierarchical level being conceptually above that in lower hierarchical levels, with cellular array as the base layer. The length-4 fastlanes are driven by special routing multiplexers within the cells at 4 × 4 block boundaries. All routing wires are directional and labeled according to the signal travel direction. As an example, S4 is

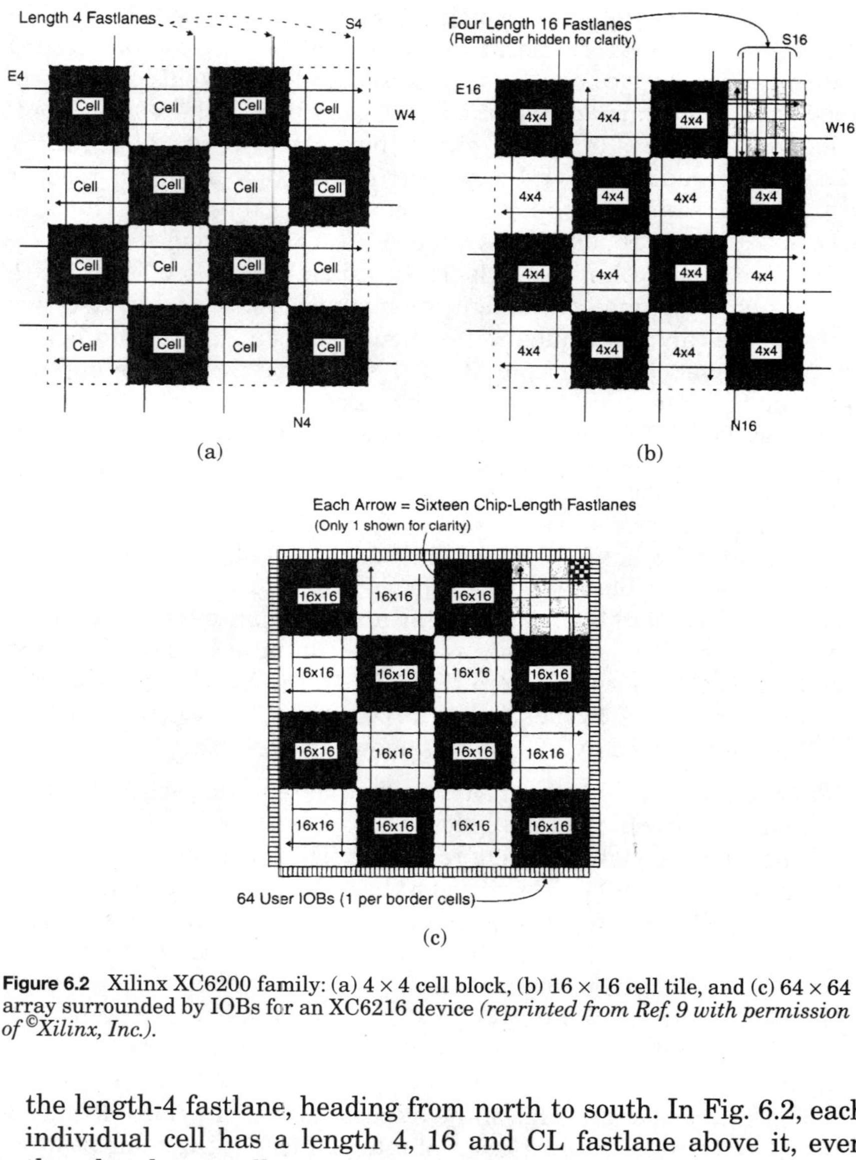

Figure 6.2 Xilinx XC6200 family: (a) 4×4 cell block, (b) 16×16 cell tile, and (c) 64×64 array surrounded by IOBs for an XC6216 device *(reprinted from Ref. 9 with permission of ©Xilinx, Inc.).*

the length-4 fastlane, heading from north to south. In Fig. 6.2, each individual cell has a length 4, 16 and CL fastlane above it, even though only a small number are shown.

The majority of interconnections are routed using the nearest-neighbor and fastlane wires. Also, each cell has an output (labeled *magic*) that provides additional routing resources. A cell's magic output is not always available for routing, and its availability depends on the logic function implemented inside the cell. Four global wires (G1, G2, GClk, and GClr) are provided for global clock and clear or other high fan-out signals and are distributed throughout the array in a low skew pattern. The global wires enter the part through the dedicated input pins and are distributed in a special low-skew "H" pattern.

- *Basic cell and functional unit.* Figure 6.3*a* shows the basic XC6200 cell details. The inputs from the neighboring cells are labeled N, S, E, and W, and those from length-4 wires as N4, S4, E4, and W4, in accordance with the signal direction. Additional inputs include clock and asynchronous clear for the function unit D-type register. The output from the cell function unit, which implements the gates and registers required by the user's design, is labeled F. The magic output is used for routing. The multiplexers within the cell are controlled by the bits within the configuration memory. The basic cells within the array have inputs from the length-4 wires associated with 4×4 cell blocks as well as their nearest neighbor cells.

  Figure 6.3*b* shows the implementation of the XC6200 function unit design so that any function of two Boolean variables can be computed by a 2:1 multiplexer, if suitable values are chosen from the input variables and their complements are placed on its inputs. The Y2 and Y3 multiplexers provide for this conditional inversion of the inputs. The CS multiplexer selects a combinatorial or sequential output. The RP multiplexer allows the content of the register to be "protected" so that, if the register protection is enabled, then only the programming interface can write to the register. It will not change when the X inputs to the function unit change, even if it is clocked or cleared. This feature is useful in designs containing the control registers, which are only to be written by an external microprocessor.

  Each cell can be configured as any two-input gate function, a 2:1 multiplexer, constant 0 or 1, single-input functions (buffer or inverter), or any of these, in addition to a D-type register. The gate names given to these functions correspond directly to the standard Xilinx library part names for these primitives.

- *Input/output blocks.* The user-configurable input/output blocks (IOBs) provide the interface between the device internal logic and external package pins. One IOB is provided for every cell position around the array border. IOBs are connected to fixed pad locations. However, there are some padless IOBs, since there are more IOBs than available pads. Every IOB has the capability of routing either an array signal or a control logic signal to/from the device pin. Every signal, including all the control signals (e.g., $\overline{\text{CS}}$, Address Bus, Data Bus, and so on.) passes through an IOB. This means that all the control signals can be routed into the logic array for use, if needed. Similarly, the user logic can control the XC6200 internal control circuitry.

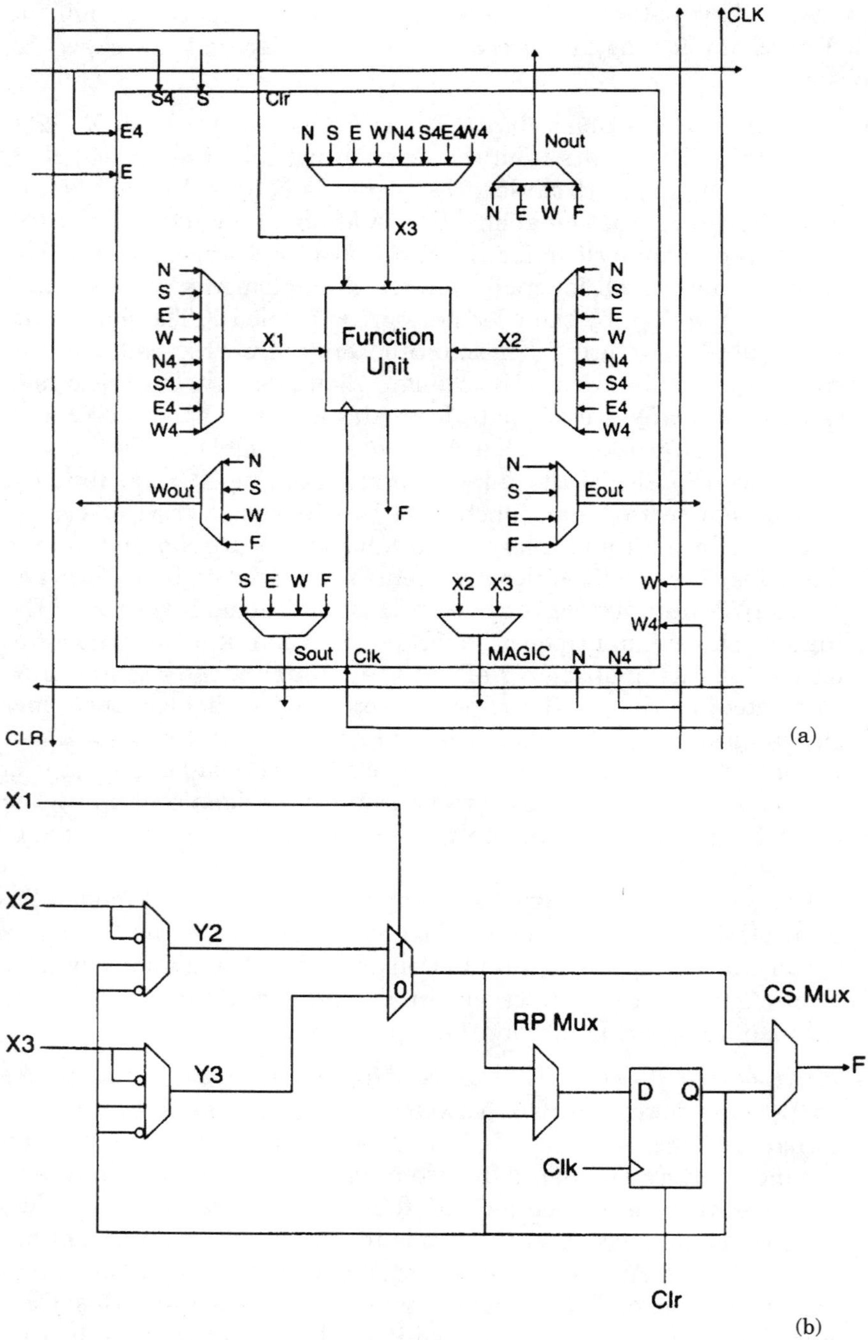

Figure 6.3 XC6200 (a) basic cell and (b) function unit *(reprinted from Ref. 9 with permission of ©Xilinx, Inc.).*

Each IOB has an array data input and a control data input, labeled ArrayDToPad and ControlDToPad, and associated with these inputs are two enable signals, ArrayEnable and ControlEnable. These signals control whether the pad associated with this IOB is the input or output mode. Each IOB also supplies ArrayData and ControlData when acting as an input. Three configuration RAM bits within each IOB control the programmable aspects of its I/O pad. These RAM bits have no effect for padless IOBs.

- *XC6200 design for reconfigurable logic applications.* An XC6200 part may be used for a board-level design as a microprocessor peripheral, as an ASIC type device, or both. In the first instance, the part will have conventional SRAM data, address, and control signals, whereas, in the other cases, it may require the user defined I/O signals of an ASIC. The design methodology can be high-level hardware description language such as VHDL used with logic synthesis targeted to XC6200 architecture. Alternatively, schematic capture with commonly used front-end design tools (e.g., Viewlogic PROcapture/Viewdraw) may be used in conjunction with Xilinx Unified Library. These tools produce an EDIF netlist, which is subsequently passed on to XC6200 place-and-route software. This software automatically maps the user's design to the XC6200 architecture and provides individual node delays that can be passed back to the high-level simulation tools such as Viewlogic PROsim/ViewSim for simulation. The simulation may be carried out prior to the placement for design logic verification using nominal delays. A binary configuration file that can be written to the XC6200 device via the programming interface is produced automatically.

  The binary data for configuring an XC6200 device, which is generated by the CAD software from the textual description of a user's design, must be downloaded into the part. This may be performed in several ways. However, the fastest and most efficient way is by writing directly to the control store, mapped into the address space of a host processor. If a microprocessor or other parallel loading data source is not available, a serial programming interface may be used.

  In dynamic reconfiguration applications, XC6200 design interacts with a design running on the host processor such that various registers within the XC6200 design appear as locations within the processor's memory map. Also, the configuration memory of the device appears within the memory map, and portions of the device can be reconfigured as required. The XC6200 architecture supports direct accesses from the processor to nodes within the user's circuit, so that the output of any cell's function unit can be read, and the flip-flop within any cell can be written. These accesses are carried out through the control store interface and involve no additional wiring within the user's design. The CPU interface signals involved in addressing the cell state can be routed into the configurable array so that the user circuits can detect that an access has been made and take appropriate action. In many applications, this access to internal nodes will be the main path through which the data is transferred to the processor, and in some coprocessor type applications, it may be the only external I/O method.

6.3 Reconfigurable Computing

The term *reconfigurable computing* (RC) denotes the use of reconfigurable hardware to speed up the execution of software. Therefore, reconfigurable computing systems are defined as those machines that use the reconfigurable aspects of FPGAs to implement an algorithm. The incorporation of reconfigurable logic into the general-purpose microprocessor allows applications to specialize the processing hardware to match the application requirements. "Soft" coprocessors and processor cores are capable of executing computationally intensive tasks with speeds outperforming the most powerful general computing machines at the fraction of the cost. In-system programmable SRAM-based FPGAs can be used to create processors and coprocessors whose internal architectures as well as interconnections can be reconfigured to match the needs of a given application. There are many computationally intensive algorithms that can benefit from being partially or fully implemented in hardware.

Figure 6.4 illustrates a very simple example of reconfigurable computing and how the computation of a very simple algorithm can be accelerated by seven times using an FPGA instead of a classic microprocessor.[10]

Most early attempts to implement reconfigurable computing focused on replacing the traditional DSP-based coprocessors used in applications such as 3-D graphics, real-time video image processing, telecommunications, and networking applications. However, recent research in the area of custom computing has demonstrated that the reconfigurable logic can be effectively employed to accelerate many computational tasks. Since FPGAs can be programmed as parallel processing computers with wide data paths and high-bandwidth I/Os, they offer tremendous performance advantages over the traditional CPU architectures. Thus, with an FPGA-based reconfigurable processor, the potential exists for the user to be able to design and precisely implement the special hardware required for a given task without having to construct new hardware for each application.

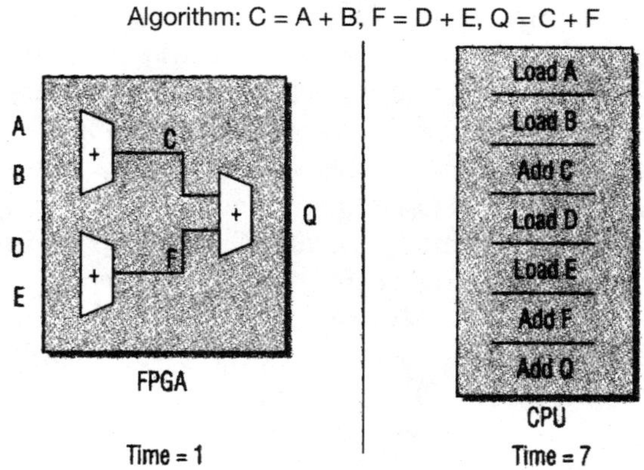

Figure 6.4 Simple example of reconfigurable computing *(from Ref. 10).*

In many cases, FPGA-based processors can take the form of plug-in accelerators in PC or workstation based systems, providing a low-cost, high-performance alternative to mainframes and supercomputers.[11] FPGA-based processors are implemented as attached processors or coprocessors, dedicated to off-loading the computationally intensive tasks from the host processors in these systems. Such reconfigurable coprocessors are viable platforms for a broad range of computationally intensive applications including scientific computing, design automation, cryptography, image processing, and real-time digital signal processing.

In many of these applications, multiple FPGAs are needed either to achieve the desired logic capacity or to obtain higher levels of parallelism.[12] The FPGAs are connected in either a linear or a matrix topology, depending on the application requirements. The more popular linear arrangements are best suited for implementation of sequential algorithms that operate on strings of data, e.g., audio and video processing, pattern matching, and encryption/decryption applications. The matrix arrangements have been applied to implement multiple sequential algorithms in parallel for applications in molecular biology and so forth.

Many of the current generation of FPGAs are oriented more toward general-purpose applications that require a mixture of combinatorial and sequential logic elements as the building blocks. These general-purpose architectures have some limitations. The interface to the main processor's bus must be implemented using significant amount of programmable logic resources in the FPGA. The internal FPGA architectures are not always optimal for data path algorithms, which are typical of coprocessing applications. Also, providing access to a large number of internal registers requires careful design. In many FPGAs, on-chip memory capacity may be too small for implementation of some algorithms. However, the evolving FPGA architectural improvements are also increasing their capabilities as configurable coprocessors.

The process of mapping algorithms into FPGAs is not fully automated and, therefore, specialized tools have been developed for each architecture. Typically, one needs to identify the algorithm (or portion of it) to be implemented in the hardware, so that the specialized software can convert the algorithm into a hardware description. Then, the commercially available *place-and-route* tools are used to physically implement the design logic into target FPGAs. Currently, there is no systematic way of determining how well an algorithm will map into the hardware. Therefore, an automatic mapping of the algorithms into FPGAs may require new compiler technology.

A common theme in FPGA-based processing is that the application specialization of a reconfigurable logic computing system provides orders of magnitude higher performance than general-purpose microprocessors. This higher performance results from the reconfigurable array manipulation of a large number of bits in parallel, whereas the conventional technology is focussed on fixed, word-wide data of up to 64 bits in current generation of microprocessors. When ALUs are implemented in the reconfigurable logic, they can be specialized to perform precisely the function required by the application.

Reconfigurable computing systems have been employed in wide range of applications including:[12]

- Binary and arithmetic operations
- Encryption/decryption and compression applications
- Sequence and string matching
- Physical system simulation
- Video and image processing

Although several of the reconfigurable computing systems recently developed show high performance, they do exhibit some common limitations, as follows:[13]

1. *Low bandwidth and high latency interface.* Many of the current reconfigurable computing systems are associated with host processors as peripherals and therefore generally attached to a peripheral bus with limited bandwidth and high latency to the processor. The low bandwidth makes communication expensive and limits the throughput achievable with the reconfigurable computing system.
2. *High reconfiguration overhead.* Since the reconfiguration costs are high in almost all currently available reconfigurable logic tech-

nologies, the reconfiguration time must be amortized over a large amount of processing power available to justify the cost of reconfiguration system development. In some cases, this may mean that a single configuration must be maintained throughout an application, even when different portions of the application might be accelerated with different kinds of specialized logic.

Therefore, a general-purpose reconfigurable computing system should have following attributes:

- Easy integration into user's application processor/computer system
- Compatibility with the existing standards and transparency to the end user
- Accessibility from within the common application program software
- Ease of programmability and software development

Several research groups have built reconfigurable computing engines for high application performance at low cost by specializing them for certain computation tasks. Some of these computing engines will be discussed in Sec. 6.5.

A new development has been the use of dynamically programmable gate arrays (DPGAs) in the reconfigurable computing system applications. In the FPGA-based processing, where the function of each array element is fixed between the relatively slow reconfiguration sequences, the DPGA array elements may switch rapidly among several preprogrammed configurations. This rapid reconfiguration allows DPGA array elements to be reused in time without significant overhead. The applications can preload multiple, specialized array personalities and switch rapidly from one configuration to another.

Many current-generation of SRAM-based FPGAs can be best modeled as the programmable look-up tables. The look-up table programming constitutes the configuration of each array element, whereas the programmable interconnect allows the array elements to be linked up, as required by the application. The interconnect in FPGAs is typically configured by programming pass gates and multiplexers. The DPGAs, like FPGAs, consist of a collection of simple computational array elements. Each array element can perform a simple logical function on several input bits, producing one or more output bits. Each DPGA array element uses a second look-up table to map a broadcast configuration selection into a local configuration. Figure 6.5 shows the conceptual diagram of a DPGA array element.[13]

The broadcast configuration directs each array element toward the type of function that is to be performed during the next clock cycle. Each DPGA array element can perform a different function in re-

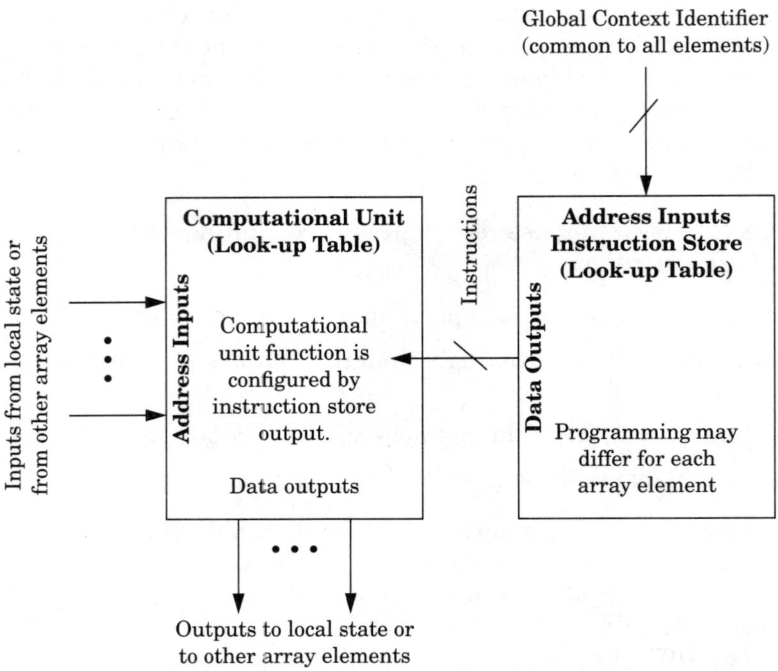

Figure 6.5 Conceptual diagram of DPGA array element *(from Ref. 13).*

sponse to the broadcast configuration identifier based on a preprogrammed context look-up table. Also, the configurable interconnect in a DPGA has a table of loaded configurations and selects between configurations, based on the current array text identifier. Therefore, within the space of preloaded contexts, the DPGA can switch personalities completely from one clock cycle to the next. The DPGA look-up table effectively serves as a cache of array element configurations. The DPGA can switch between configurations to accelerate different portions of an application. If the configurable array is too small to hold an entire acceleration logic block, the block can be partitioned across multiple context and evaluated in stages.

The DPGA-coupled processor has the potential to be employed as the key computational building block in general-purpose workstations and personal computers for application-specific acceleration. The operating system and applications can also specialize the DPGA logic to serve in coprocessing roles for computationally intensive functions such as signal processing, compression and encryption/decryption. The DPGAs may also be used to build special-purpose computing machines and multiprocessor systems. However, designing efficient and widely applicable DPGA-coupled microprocessors involves some major architectural challenges and key trade-offs. The interface between the fixed and reconfigurable logic will be one of the most important aspects of the composite architecture. This interface will have a critical impact on the communication overhead between the fixed and reconfigurable logic. The pin allocation and partitioning of available silicon area among on-chip memories (e.g., caches), fixed-functional units (e.g., ALUs, FPUs), fixed control, and reconfigurable logic will have to be optimized for maximum efficiency. In addition to any dedicated microprocessor bus-interface pins, some additional pins may be required for reconfigurable logic for use in system-specific interfacing and specialized control and communications.

6.4 ASIC Prototyping and Logic Emulation

6.4.1 Prototyping and logic emulation

The traditional approach to ASIC design prior to the availability of simulation techniques and tools was to build prototypes of their systems for design verification. More recently, ASIC designs have widely relied on software simulation tools prior to final design implementation in silicon. However, simulation has some inherent limitations, especially in cases where the design is normally simulated using test vectors based on specifications that may not be complete (or erroneous). Also, simulation is slow, since the software model of a logic design is inherently slower than the corresponding hardware. One methodology for ASIC prototyping to ensure highest probability of success is to use a hardware emulator during the design process. Although dedicated hardware emulators can be employed, generalized reconfigurable emulators are desirable to minimize cost and time. The hardware emulators allow downloading of a candidate netlist into a reconfigurable hardware engine where the functionality of a design may be evaluated in a faster, and often more thorough, manner than by using software simulation.

Several commercial FPGA kernel-based hardware emulators are used for prototyping by ASIC designers. However, some limitations that may constrain their performance need to be addressed.[14] The partitioning, placement, and routing algorithms for ASIC emulators are not fully optimized for timing or speed performance. The interconnection technology for FPGAs is not efficient, since significant resources are tied up in the interconnect routing. The timing constraints and limitations imposed upon the emulator operating speed are such that it is significantly slower than the target system.

The emulator performance is a function of both the FPGA kernel architecture and the system-level interconnection of FPGAs. Also, rapid configuration and reconfiguration of the emulator resources is necessary to support design changes. The timing constraints and limitations are directly affected by the internal architecture of the FPGA. The interconnection architectures for FPGAs used in emulators are more demanding than those for general-purpose applications. Therefore, for an efficient emulation and test generation, high controllability and observability of the internal circuit nodes is required, as compared to that for the general-purpose logic implementation.

The FPGAs consist of three major elements: (a) configurable logic blocks, (b) I/O blocks, and (c) interconnections. The FPGA architectural elements were discussed in Chap. 3. The FPGA performance is a direct function of its logic block and interconnect architecture. Many hardware emulators currently available use Xilinx SRAM look-up table (LUT) based FPGAs. There have been some studies performed in evaluating four different FPGA logic block architectures for performance: NAND gates, multiplexers, LUTs, and wide-input AND/OR gates.[15,16] The studies involved synthesis of selected benchmark circuits into FPGAs, using each of the four architectures, and then comparing the relative speed of the logic block implementation. The results showed that, although the performance of LUT- based and multiplexer-based architectures was comparable, the LUT architecture was a better choice in many cases, because the propagation delay is independent of the function being implemented, and the input combinations can be varied to accommodate optimum routing.

These studies also investigated the impact of logic block functionality and area trade-offs in the FPGA architectures. In LUT-based FPGAs, as the functionality of the logic block increases, the total number of logic blocks required to implement a circuit decreases. However, this increase in functionality comes at the cost of additional pins per block, which may cause significant increases in routing area. An area-efficient block is the one that has high functionality per connected pin. Therefore, the LUT-based logic block is preferable, because each pin can be connected to any logic function, providing high functionality per pin.

The interconnection technology requirements in emulators consist of a three-level hierarchy: (1) interconnection of logic blocks inside individual FPGAs using internal routing capabilities, (2) interconnection of multiple FPGAs into emulator modules, and (3) interconnection of multiple modules for building a large capacity ASIC emulator. Since the hard-wired interconnects are not reconfigurable under the software control, it limits the flexibility of the emulator operation. An optimal logic emulator design must offer the same flexibility as the software emulation.

An important consideration in a multi-FPGA-based logic emulator system is the selection of FPGA pins to use for routing inter-FPGA signals. While the FPGAs themselves can be routed and rerouted, the wires moving signals between FPGA pins are fixed by the routing structure on the implementation board. Although some impressive results have been achieved by the hand-mapping of the algorithms and circuits to FPGA systems, developing a completely automatic system for mapping of these structures is essential for their more widespread utility. In general, an automatic mapping approach goes through following phases: synthesis, partitioning/global placement, global routing, and FPGA place and route.[17]

During the synthesis step, the circuit to be implemented is converted from its source format into a netlist for appropriate implementation in FPGAs, possibly after several optimization steps. The partitioning and global placement breaks this mapping into subcircuits that will fit into individual FPGAs and determines which FPGAs a given subcircuit will occupy. Then the signals that connect logic in different FPGAs are routed during the global routing step, which determines the intermediate FPGAs that a signal will move through (if any) and identifies the FPGA I/O pins that will be used. The global routing phase of mapping to multi-FPGA system bears resemblance to routing for the individual FPGAs, and similar algorithms may be applicable to both the problems. After the logic assignment and inter-FPGA signals routing, standard FPGA place-and-route software can produce the programming files for the individual FPGAs.

6.4.2 Rapid prototyping and programmable interconnect architectures

A commonly used approach for using FPGAs for interconnect was by attaching them to fixed protocol devices such as the DSPs and memories so that the data flow between them could be programmed for many different interconnect patterns. Another approach has been to allocate FPGA to each fixed device, and the FPGAs are then connected together via a set of physical wires. This system provides the advantages of software-programmable pin protocols and allows for an unlimited number of devices to be physically connected. This interconnect system has been called a "virtual bus."[18]

The actual connectivity between FPGAs may take different forms of interconnects such as busses, pipelines, hypercubes, and trees. The simplest form of interconnect is the *global bus*. A major advantage of the bus is that all components on the bus share same connectivity, and it is well suited for systems that process blocks of data from a single I/O port. It also has the advantage of being incrementally expandable. However, the global bus has a fixed bandwidth that must be shared between all devices connected to it. The maximum data rates for global busses are limited to the 20 to 30 MHz range. Another form of interconnect is the pipeline that optimizes the data flow between the local elements. The pipelines require the number of processing elements to be known before the connectivity is established. If the pipelined elements can be added incrementally, the pipeline allows each new element to add bandwidth to the system, as well as processing power. It means that, for lightly loaded data links, the clock rates can be increased to above 50 MHz.

An interconnect form called the *segmented virtual bus* has been developed by Giga Operations Corp. This bus is similar to the interconnect found inside FPGAs. Each FPGA is connected to a neighbor with local connections, which can be enabled or disabled with devices called *quickswitches*. A quickswitch has two modes: the "on" position in, which it literally behaves like a 5 Ω resistor, and the "off" position, in which it behaves like an open circuit. If some quickswitches placed in series with a data bus are turned off, a bus can be created with segments that are isolated from each other. A segmented bus makes it possible to convert a global bus structure to a pipelined bus under software control. In addition, other interconnect forms, such as trees, also can be segmented. This segmented virtual bus architecture has been used in Giga Operations' G900 Reconfigurable Interface Card (RIC) for image and signal processing in a virtual hardware environment. Figure 6.6 shows the global, pipeline, and segmented virtual bus architectures.[18]

The initial concept of partitioning a large design into multiple SRAM-based FPGAs for rapid prototyping was based on an interconnection scheme that allowed each FPGA to have direct communication with all other FPGAs in the system.[19,20] The first-generation architec-

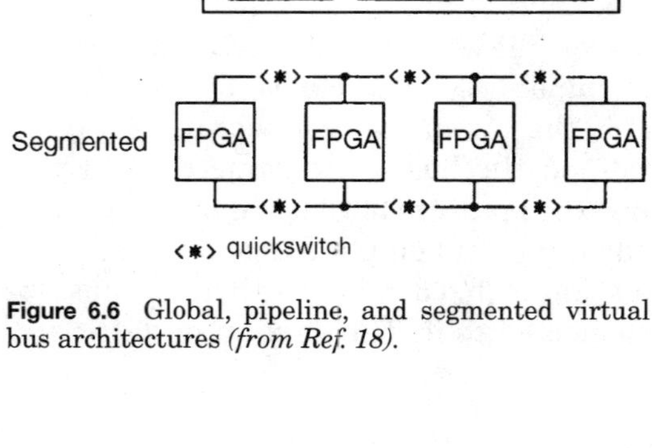

Figure 6.6 Global, pipeline, and segmented virtual bus architectures *(from Ref. 18)*.

ture provided communication between FPGAs in the adjacent neighborhood (e.g., up, down, left, right, and four diagonal connections), which was then extended to the four second-nearest neighbors and their edges connected back around to form a sphere. In this architecture, FPGAs serve both as the medium for implementing the logic, as well as the interconnection between the FPGAs.

In practice, many FPGA I/O pins are used for connecting I/O pins, which are not directly connected on the board but are a part of the same electrical net. Each such connection uses two additional FPGA pins, which leaves fewer pins for logic implementation and results in lower gate utilization. Also, FPGAs, when are used as interconnects, may result in poor timing characteristics. Because this architecture places no upper bound on the number of FPGAs in an interconnection path, this can lead to long and widely variable delays and timing problems such as set-up and hold-time violations.

A breakthrough was made through the concept of a partial crossbar interconnect architecture, a modified version of the Clos network, which has been widely used in many telecommunication systems.[21,22] Figure 6.7 shows the block diagram of partial crossbar architecture using field programmable interconnect devices.[19] A major innovation in this approach was the separation of interconnect functions from the logic functions, each to be served by a different type of device.

The interconnect function can be provided either by the FPGAs or by dedicated devices such as I-Cube's field programmable interconnect devices (FPIDs). In this architecture, all interconnections go through one crossbar, and therefore the delays are more uniform and bounded. The interconnection routing is a simple, repetitive, table-driven task. The architectural symmetry simplifies partitioning and placement

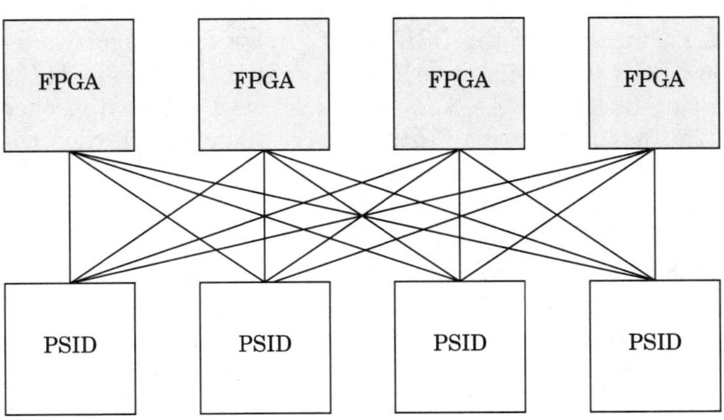

Figure 6.7 Partial crossbar interconnect architecture *(from Ref. 19 with permission of I-Cube Corp.).*

and provides overall better performance. I-Cube has enhanced the FPID architecture to provide a new product category called *programmable switching and interconnect devices* (PSIDs™). These are semiconductor devices that combine a nonblocking switch matrix with programmable I/O blocks for dynamic bit/bus switching as well as static routing and interconnect applications in the image/video processing, networking, and telecommunications industries.

In addition to the mapping of logic into FPGAs, the management of clock distribution and clock skew requires special attention. If the design includes memory, its emulation requires special consideration. It should be noted that the large delays that FPGAs and PSIDs incur (relative to the final design) will introduce set-up and hold-time violations in the prototype. Therefore, most commercial rapid-prototyping systems are designed with system-wide low-skew clock distribution network, and direct clock connections are made through the FPGAs fast buffers.

The interconnect network can be built using FPGAs, PSIDs, FPGA-style routing architectures, or custom-designed multiplexing chips. However, the selection criteria for an interconnect device should be based on factors such as amount of connectivity provided, minimum additional delays and very little skew, and low hardware/software costs.

Other architectural approaches are available for connecting logic and interconnect resources. One approach is to use the PSID as a pin switcher so that, for each FPGA, one PSID is used to switch pins from the FPGA before they are connected to other FPGAs on the board. This approach requires that the I/O port ratio of PSID to FPGA to be 2:1. In this scheme, each FPGA can be allowed to access all other FPGAs on the same board. This architecture allows the logic partitioning software to be decoupled from the FPGA automatic place-and-route (APR) program. Since use of the PSID as a pin-switcher removes all constraints on FPGA pin assignment, the average utilization of FPGA goes up, requiring fewer FPGAs. The main drawback of this approach compared to the partial crossbar interconnect structure is that the each node incurs at least two units of PSID delay.

An architecture that combines the partial crossbar interconnect structure and the PSID as pin-switcher approach may be advantageous. Figure 6.8*a* shows the block diagram of this architecture, which has the same benefits as the PSID as pin switcher, since the partition program may be decoupled from the FPGA APR program. The main drawback of this architecture is that, on average, two units of PSID delay are introduced for every node. However, this approach can free about 10 to 20 percent of the PSID I/Os for other purposes, such as connections to the target system or for diagnostic probing.

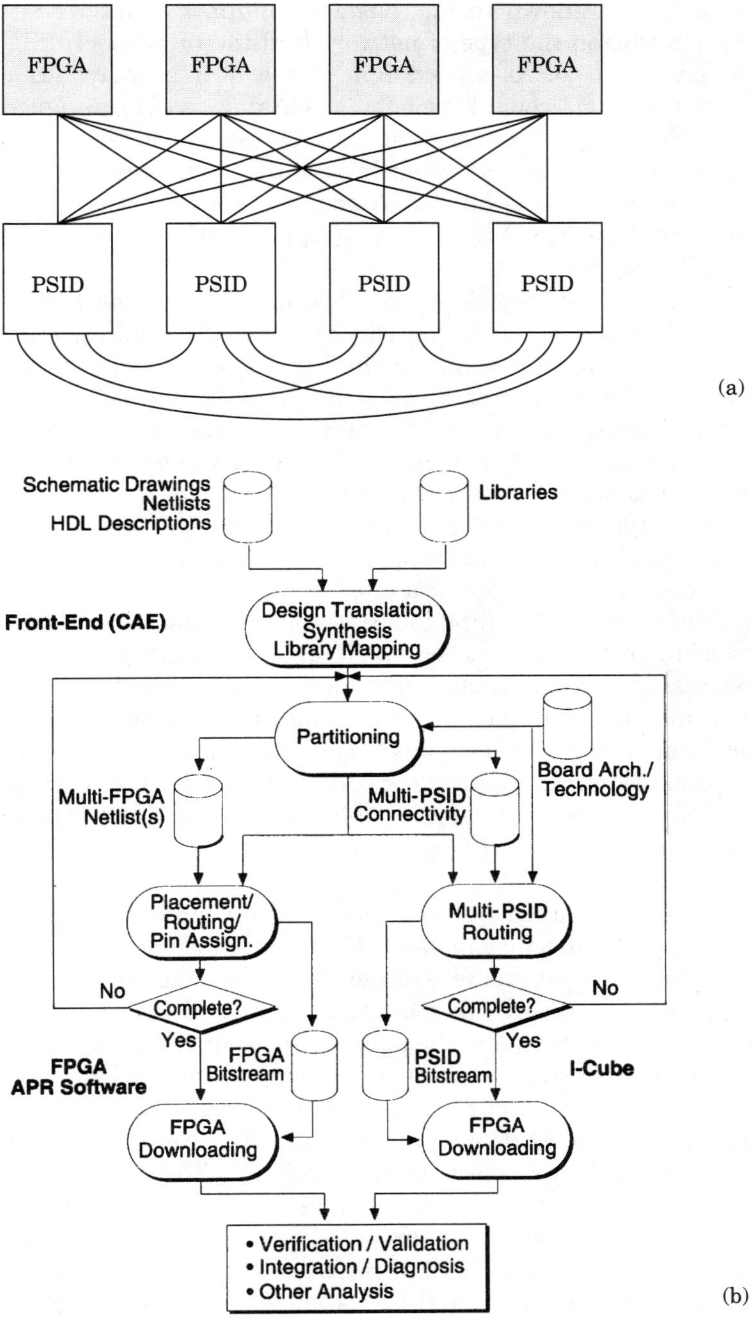

Figure 6.8 (a) Architecture combining partial crossbar interconnect and PSID as pin-switcher approaches and (b) generic design flow for FPGA and PSID-based prototyping *(from Ref. 19, with permission of I-Cube Corp.).*

In the architecture shown in Fig. 6.8*a*, the number of inter-PSID connections depends on the type of network and the number of PSID and FPGA devices used. As an example, for a design that uses 8 PSIDs, about 10 to 20 percent of the nets (FPGA pins) will be assigned to the same PSID. This means that at least 80 percent of the FPGA pins may have to be connected using inter-PSID connections unless the FPGA pin assignment is altered. A good rule of thumb is to use about 80 percent of the FPGA pins to get good completion results from the FPGA APR program.

Figure 6.8*b* shows the generic design flow for building the FPGA- and PSID-based prototypes. This flow can be modified to address critical nets by placing pin constraints on the FPGAs, as appropriate. As an example, FPGA pins for critical nets can be assigned to the same PSID, which reduces the delay for those nets. Since the number of nets involved in this special treatment is usually small, the overall FPGA gate utilization should not be much affected.

In rapid prototyping based on logic emulation, the network is mapped to a technology that allows operating speeds such that all interfaces to targeted applications can operate in real time. The rapid prototyping does enable full functional verification and can achieve system operating frequencies exceeding 20 MHz. This ability to run at real-time speeds gives rapid prototyping its major advantage over the simulation and emulation and can greatly reduce the design time for DSP applications. Logic synthesis has become an important tool that enables rapid migration from a targeted ASIC library to the FPGA library. Rapid prototyping is made possible by new developments in three technologies: FPGAs, programmable interconnect, and logic synthesis tools.

In 1992, a company called Aptix Corporation introduced the field-programmable interconnect component (FPIC™), which uses a routed (heuristic) architecture to create a large switch matrix between all FPIC pins.[23] The FPIC device has more than 1000 leads, of which 936 can be used for interconnect, and the additional 64 can be connected to proprietary debug interfaces. The remaining pins are used for programming, power, ground, and self-test. Figure 6.9*a* shows the architecture and layout of the FPIC interconnect chip. All interconnects are programmable, using simple pass-gates controlled by the memory elements. These switches present passive, bidirectional data path external signals (0 to 5V), with pin-to-pin propagation delays as low as 4 ns.

Aptix Corporation has used FPIC devices to create a field-programmable printed circuit board (FPCB™), as shown in Fig. 6.9*b*. Each hole in the printed circuit board connects directly to a pin on the FPIC device. Then the interconnects on the FPCB are made by completion of the proper signal paths inside the FPIC device. The FPIC is configured

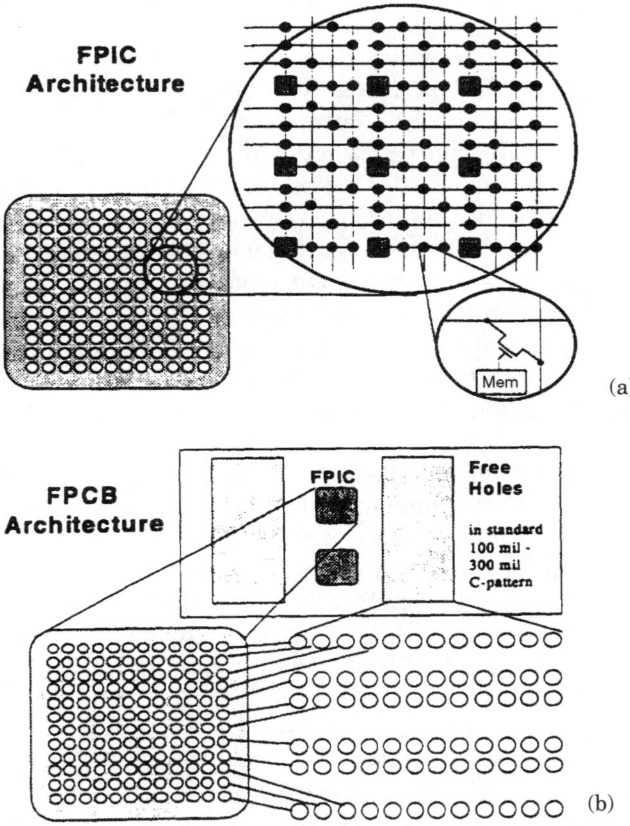

Figure 6.9 Aptix (a) FPIC chip and (b) FPCB *(from Ref. 23 with permission of Aptix Corporation).*

through a host interface module (HIM), which shifts in data from a workstation to program the FPIC memory. Both the FPICs and FPCBs use Aptix software system called Axess™. This software reads in a system-level netlist and, under the user control, translates it into the data-streams that are needed to configure both the FPIC devices and FPGAs.

Aptix Corporation has recently introduced the MP-3 system prototyping FPCB, which accepts a variety of FPGAs (mounted on special adapters) along with the standard components such as memory, DSP, and microcontrollers. To accommodate the various FPGA configurations, special FPGA adapters are used that plug into pin-hole areas that connect DIN, DOUT, CLK, programming, and power and ground pins of the FPGAs to dedicated circuits on the FPCB. This capability makes the MP-3 system a candidate for DSP system prototyping.

6.4.3 Design for emulation

A logic emulator may be considered as a system consisting of many FPGAs connected through the programmable interconnect arrays (PIAs) and software to automatically program the hardware, according to the gate-level or HDL design. The function of this logic emulator hardware and software is to support the operation of the emulated design as an ultrafast simulation engine, or real-time operation connected in circuit to real hardware.[22] A good logic emulator design compiler is one of the most sophisticated EDA tools. Its function is to take large design netlists from various sources, map them into target FPGA technologies, partition them across tens of boards and hundreds of FPGAs, providing all necessary interconnections, while maintaining high and predictable utilization of the hardware and maximum emulation speed.

The greatest future development direction for logic emulation is its continued adoption as a core development tool and improvement of the efficiency of compiler software technology. Emulator-specific partitioning research seeks to combine multiple algorithms to partition very large logic networks into many hierarchically connected FPGAs, and to observe circuit functionality as well as connectivity to optimize partitions for timing as well as pinout. The HDL-driven logic emulation is also becoming more popular.

The improvements in logic synthesis tools has enabled designers to create technology-independent descriptions of a system-level design, emulate the design in FPGAs and, when the design is stable, retarget it to selected technology. In system emulation, the entire system, including any ASICs, are retargeted to a programmable hardware environment. The ASIC function is combined with microprocessors, memories, other system components, and system interfaces in the programmable hardware environment. Software tools automate the technology mapping and configuration of the system emulation hardware.

In emulation, the network is verified by retargeting all or parts of the system to a different technology such as FPGAs, and the verification takes place in a hardware environment rather than in the virtual environment of a simulator. Therefore, the ability to run the emulation hardware at speed is essential. In an FPGA-based emulation, it is important to realize that, although each FPGA can perform its operation at or above the system-level speed, there is no guarantee that the multiple-FPGA implementation of the entire system can do the same. To achieve *at-speed* system level operation, the following speed constraints need to be taken into consideration:

- Minimize the number of FPGAs needed to reduce the number of boundary crossings.

- Analyze the timing characteristics of the signals traveling between the FPGAs, and use optimization techniques such as pipeline design, latching of signals, or reduced logic depth when generating these signals.

In ASIC emulation, the ASIC netlist is retargeted from the ASIC vendor's library, typically to an array of FPGAs, using software tools and dedicated hardware developed for automating this process. Once the design has been retargeted, the ASIC emulation hardware can be exercised by the designer's simulation vectors. However, a designer has to be aware of targeted FPGA device limitations (gate count, pinout) at the earliest cycle of the design. The following are some of the basic rules for efficient prototyping and ASIC design implementation:

- Synchronous design
- Use of pipeline design techniques
- Short arithmetic functions (minimize logic levels)
- Minimal bit width
- Use of I/O flip-flops (latches) where possible
- Careful mapping between functions and FPGAs
- Minimization of high fan-out net or use of available buses
- Use of small logic blocks (20 to 40k gates)
- Limited I/O module count

A compete FPGA-based emulation software system using Virtual Wires technology has been developed at MIT Laboratory for Computer Science at Cambridge, MA.[24] The MIT Virtual Wires Softwire compiler accepts a netlist description of the system to be emulated and produces programming information for the FPGA hardware, consisting of an inexpensive board designed for the Virtual Wires in-circuit emulation. The compiler also provides an interface to the standard logic simulator tools for hardware accelerated simulation. It differs from the existing systems that dedicate each FPGA pin to a single emulated signal in using a multiplexing scheme that allows each pin to support multiple emulated signals. Virtual Wires can improve gate utilization by over 50 percent and can speed emulation by reducing the size of FPGA array and the length of communication paths.

The MIT Virtual Wires Softwire compiler has been built using internally developed modules and off-the-shelf commercial CAD products. A netlist description of the system to be emulated is first divided into multiple FPGA-sized partitions. The core of the software compiler, referred to as the Virtual Wires Kernel, produces a pipelined, statically-routed communication network to connect the partitions. This is followed by technology mapping and FPGA place-and-route software, which reduces the networked partitions to FPGA programming data.

The core of the Virtual Wires Softwire compiler is a set of software modules that read a partitioned netlist and produce an interpartition communication network. After the compiler takes the netlist description of the user's system and hardware description files, the software goes through following sets of procedures:

- Technology mapping of system procedures to FPGA primitives
- Partitioning the netlist into per-FPGA sections (partitions)
- Visualization of inter-FPGA connections using the Virtual Wires kernel
- Back-end compilation of the virtualized netlist into FPGA programming data

Additional files and information are generated to support the logic simulator interface to the Virtual Wires hardware to allow debugging of the Virtual Wires system and to provide the compilation statistics. The conversion of emulated system description to the target FPGA technology is performed in two stages: replacement (or encapsulation) of complex structures and mapping of the low-level primitives. High-level structures, such as small RAM/ROM (typically 16 to 256 bits) can be mapped directly into the look-up tables used in SRAM-based FPGAs. However, the larger memory arrays in the input list are mapped to SRAM chips attached to each FPGA in the Virtual Wires hardware. Once all high-level transformations of the emulated system have been performed, generic technology mapping of the primitive gates to FPGA technology is typically done. After preliminary technology mapping, the emulated netlist is partitioned into per-FPGA sections using algorithms such as clustering and min-cut methods. The remaining stages of Virtual Wires compilation process are mostly FPGA technology specific.

The Virtual Wires system has been used for both the hardware-accelerated simulation and full hardware emulation of the Sparcle chip, an 18k-gate ASIC implementation of the Sparc microprocessor with enhanced multiprocessor support.[25]

In general, the emulation systems also have a number of drawbacks, including their prohibitive cost, limited density, and large form-factor for board-level implementation. The hardware emulation systems rely on the programmable hardware to implement logic functions, but the mapping of the custom design into the programmable hardware is usually not very efficient. For example, a hardware emulation system designed using 20 of 10k gate FPGAs may only be able to provide 50,000 usable gates after the design has been mapped into the programmable architecture and routed.

However, there are examples of low-cost ASIC prototyping PLD systems based on high-density multichip module (MCM) technology. The following sections describes two examples of such field programmable systems (FPSs).

ASIC prototyping system EPF8050M. An example of this ASIC prototyping system is EPF8050M architecture, which uses four Altera FLEX 8000 family devices (EPF81188) of 12k gate PLDs each, and an Aptix field programmable interconnect (FPIC) device on an MCM substrate to provide 50,000 usable gates.[26] All communication within the device and the I/O pins is routed through the FPIC, providing a transparent decoupling of the internal logic from external pin assignments. There are 160 I/O channels between each FLEX device and the FPIC, and 360 I/O channels between the FPIC and the MCM I/O pins. This ratio of 360 I/O pins for 50,000 gates fits into Rent's rule, ensuring that the designs should be able to utilize available logic resources without running out of the I/O pins.

Figure 6.10 shows the architectural block diagram and side view of the EPF8050M MCM.[26] All four FLEX 8000 family PLDs and one FPIC are SRAM-based and require configuration upon power-up. The configuration and test channel for all five devices are routed to a total of 52 unique MCM pins, allowing independent testing and configuration. The remaining 150 MCM pins are used for V_{CC} and GND connections. The four FLEX 8000 devices are housed in ceramic BGA packages and mounted to the top of the MCM. The FPIC device is attached to the bottom of the MCM using controlled collapse chip connection (C4) solder bumps process for mounting. The combination of BGA package and C4 attach process permits MCMs to be easily reworked for element replacement.

The ASIC designs targeted at the EPF8050M must be partitioned into the four FLEX 8000 devices. Design entry for the EPF8050M may be initiated in any standard CAE environment that exports EDIF, VHDL, or Verilog. The MAX+PLUS II software that supports the EPF8050M can perform logic synthesis, optimization, and minimization for mapping the design into the FLEX 8000 architecture. Once the design has been optimized, a min-cut algorithm is used during automatic partitioning to minimize the required interconnect. The timing simulation files generated can be simulated in MAX+PLUS II simulator or imported into any standard CAE emulator for use in board-level simulation.

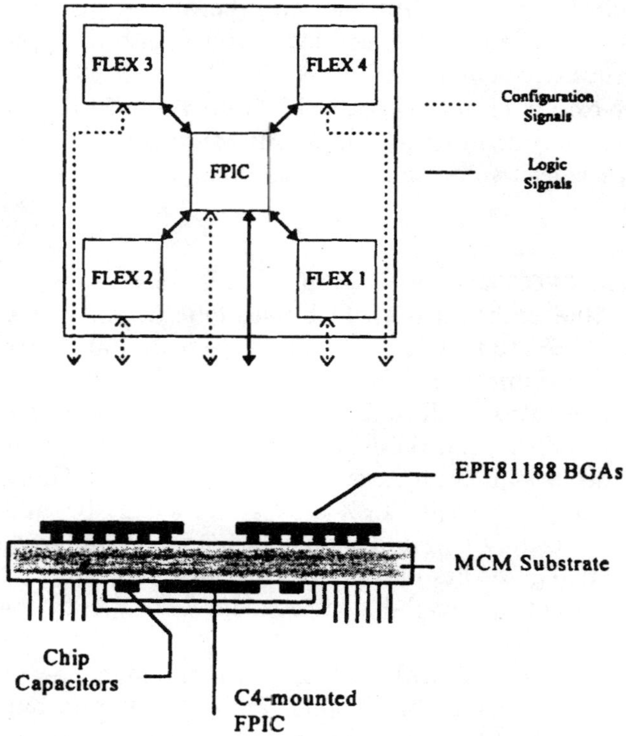

Figure 6.10 ASIC prototyping system EPF8050M MCM architecture and side view *(from Ref. 26 with permission of IEEE).*

Field programmable MCM (FPMCM) prototyping system. A prototype has been built for silicon-on-silicon field programmable MCM with a capacity of 40k gates and 256 user I/Os that demonstrated an increase in capacity by a factor of four over the FPLD family from which it was designed.[27] This MCM was designed using AT&T silicon-on-silicon, flip-chip process (μ-SMT) in conjunction with their 3000 series FPGAs, which are functionally identical to Xilinx 3000 series. The basic architecture consisted of arranging FPGAs in a star configuration around a single large field programmable interconnect (FPIC) device to simplify the partitioning software.

The device selected for implementation from the 3000 series was the 3042, which showed significantly easier routability and optimal dollar cost per usable gate than the larger devices. Figure 6.11 shows the architecture of a first-generation FPMCM that consists of twelve 3042 chips, a large FPIC and four chip capacitors on a 30.6 × 30.6 mm size substrate. All of the lines connect to the central FPIC to simplify test-

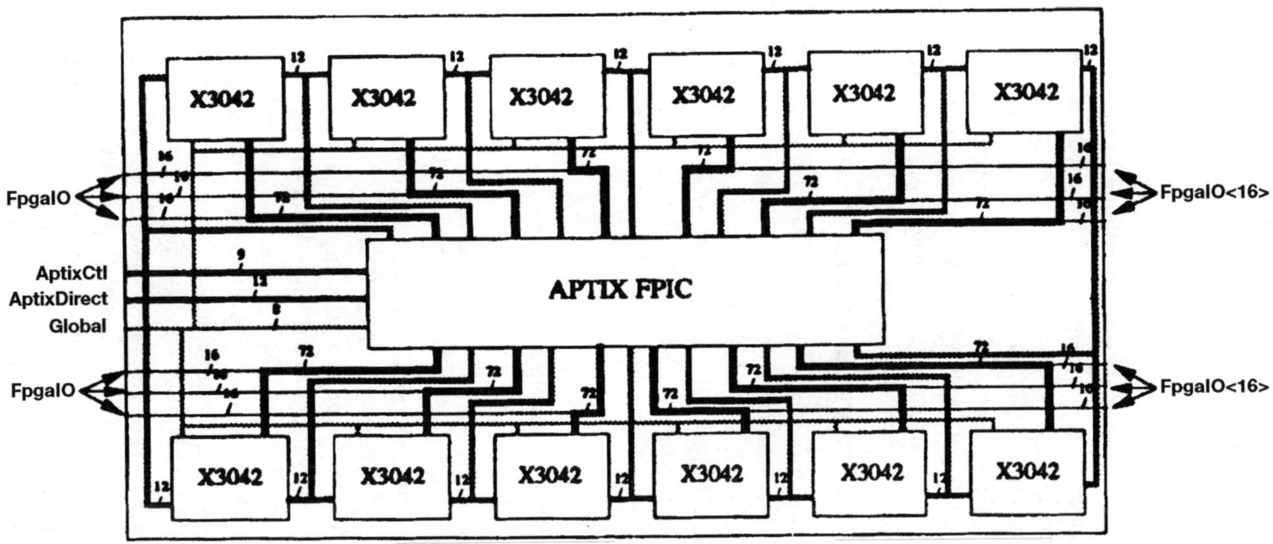

Figure 6.11 Block diagram of first-generation FPMCM architecture *(from Ref. 27 with permission of IEEE).*

ing and diagnosis, although some direct connections are provided between the neighboring FPLDs to reduce the number of FPIC pins needed. Also, dedicated lines are provided for global signals such as system clocks. The module design allows each of the FPLDs and the FPIC to be configured separately and in parallel. The MCM was packaged in a 327 pin grid array (PGA) using electroless gold on a four-layer polyimide board.

An application development system allowed the FPMCM to be used as a high-performance coprocessor in a Sun workstation. The system was used to provide an efficient implementation of the data path circuits from a C-like description of the hardware, which is used for both simulation and logic synthesis. The design is then automatically partitioned and placed inside the FPMCM. It was shown to deliver 200 MOPs of computing power for image processing applications.

6.5 Custom Computing Machines (Examples)

Custom computing machines are high-performance FPGA-based reconfigurable logic systems targeted to computation-intensive applications. There have been scores of such systems developed over last few years and are commercially available. Many more are being researched and are in "proof-of-concept" development stages. This section describes some of the custom computing machines that have been developed.

6.5.1 Configurable hardware algorithm mappable preprocessor (CHAMP)

The configurable hardware algorithm mappable preprocessor (CHAMP) technology being developed by U.S. Air Force and Lockheed Sanders is reconfigurable hardware targeted toward the implementation of preprocessor algorithms.[13] The processing of images such as for Infrared Missile Warning (IRMW) systems require highly parallel data processing to handle very large amounts of real-time data. As an example, for a sensor that is sampling data at 1000 frames per second and for 128×128 images, resulting in 16M pixels per second, many IRMW algorithms may typically require a throughput of more than 1 billion operations per second (BOPS). This high throughput may not be feasible with many general-purpose processors. The goal of the CHAMP program was to create a preprocessor that is directly mappable from the algorithms and would be reconfigurable even during a mission. The mapping was to be performed, eventually, with fully automated tools.

The selection of the implementing technology was combined with the selection of the high-level architecture. The technology targeted was Xilinx 4000 series FPGAs, based on the available devices

(XC4013), and the efficiency of implementing pipelines with the configurable logic blocks (CLBs) used as RAM. A compromise between the fixed and crossbar interconnect scheme was selected based on the examination of algorithms to be implemented. Some sections of the algorithm required frame-to-frame interaction, resulting in large amount of data for storage, which required the use of an external RAM. Figure 6.12*a* shows the CHAMP 1 processing element, which consisted of two XC4013 FPGAs and a dual-port RAM of size 16k × 32 bits associated with each of these processing elements.[28]

The control scheme included counters for each address for the dual-port RAMs, control for I/O with external boards and sensors, and a 32-bit counter for starting "events" relative to a single start signal. The FPGA technology was selected for the interconnect task based on the access time of the dual-port RAMs being the driving factor in the system performance. Figure 6.12*b* shows the hardware implementation of

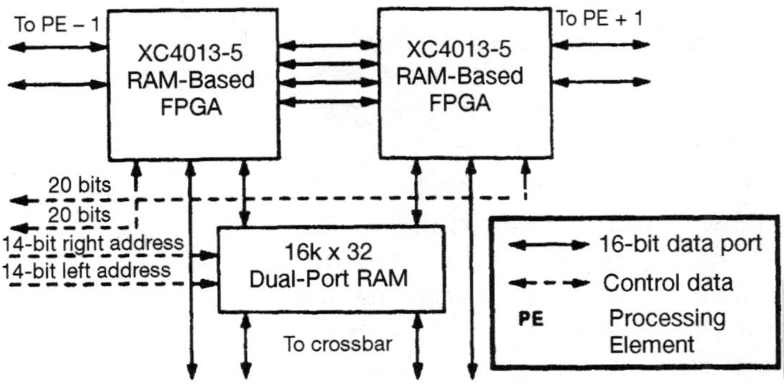

(a)

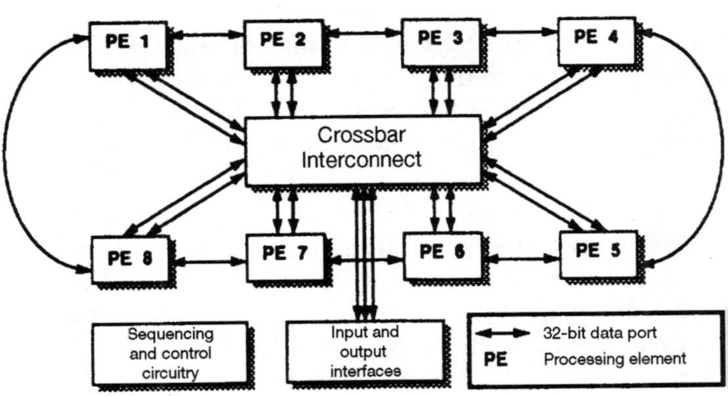

(b)

Figure 6.12 Configurable hardware algorithm mappable processor (CHAMP 1): (a) processing element (PE) and (b) architecture *(from Refs. 28 and 29 with permission of IEEE).*

CHAMP 1, consisting of a 6U VME board with 8 PEs.[29] It includes a large three-port RAM (256K × 16), a 16-bit high-speed input interface, a 16-bit high-speed output interface, a VME interface, and four 32-bit sensor interfaces. There is also logic provided for downloading of all FPGAs using the VME interface, except for the VME interface FPGA, which is loaded from a serial PROM.

The key factor in architectural implementation was the actual FPGA design process, including routing of the XC4013s using NeoCad or Xilinx automated routing tools. For the XC4013 to perform the necessary functions at required speed, an internal feature known as *fast-carry logic* was used. The CHAMP architecture and preliminary design was validated by the mapping of an IRMW algorithm, which required a highpass, spectral, and median filters; a background normalizer; and threshold detection. The mapping involved running the executable C code implementation of the algorithm to produce outputs that corresponded with the input data provided. The mapping results for the IRMW algorithm implemented showed 1 BOPs of real-time throughput while utilizing 75 percent of the CHAMP 1 board's processing resources.

The CHAMP 1 board was completed in 1994. The use of CHAMP 1 in several applications showed shortcomings in (a) inadequate RAM, (b) not enough CLBs, resulting in shortage of logic capacity, (c) shortage of connections from data to control, and (d) difficulty in programming of global control. A follow-up on this activity was to address these shortcomings and increase processing density and performance by implementing the CHAMP 2 hardware through the multichip module (MCM) approach.

The CHAMP 1 was designed to accept data from an IR camera, which had 128 × 128 frame size and utilized 16k × 32 dual-port RAMs in each PE. For CHAMP 2, the ability to accommodate 256 × 256 (64k) frames was considered a minimum and 512 × 512 (256k) as desirable. Therefore, 256k × 32 dual-port RAMs were used. The total number of CLBs available and the logic capacity were increased by using Xilinx XC4025 FPGAs.

In 1994, a benchmark evaluation of CHAMP 1 performed against a TMS320C40 processor board showed an advantage of approximately 25:1 in terms of processing per unit area. The CHAMP 2 MCM is expected to make a 3.33:1 improvement in CLBs per unit area (and a 20:1 improvement in memory per unit area versus CHAMP 1).

6.5.2 Programmable active memories (PAMs)

In 1988, Digital Equipment Corporation's Paris Research Laboratory (PRL) began investigating the use of SRAM-based FPGAs for building reconfigurable hardware processors. These coprocessors were called *programmable active memories* (PAMs).[30] A PAM is a virtual machine controlled by a standard microprocessor that can be dynamically and indefinitely reconfigured into a large number of application-specific circuits. An experimental device, DECPeRLe-1, was built in 1992. Subsequently, four generations of PAM and its programming environment have been built for use in applications that have high computing requirements (e.g., Viterbi decoder, Heat Equation solver), high I/O bandwidth and real-time constraints, hardware interface capability such as for a digital camera interface, and data formats for which the standard CPU hardware is not optimized. The following are examples of some of the applications for PAM:

- A stereo vision system used to compute range maps from simultaneous images obtained by two cameras. A software implementation of this system computes the correlation of a pair of images in 59 seconds on a SPARCstation II, as compared to a specialized board with four DSPs performing the same task in 9.6 seconds, and the DECPeRLe-1 in 0.28 seconds.
- A 3-D Heat Equation solver, running at 25 MHz and computing 40 BOPs.
- A digital audio 256-channel professional quality sound synthesizer.
- Implementation of benchmark for the second-level triggers of planned Large Hadron Collider at CERN, Geneva.

Figure 6.13 shows the architecture of a specific PAM, called DECPeRLe-1, that was built in 1992.[31] Each of the 23 squares shown represents a Xilinx XC3090 FPGA, whereas each of the 4 rectangles with letters "R" denotes 1 MB of SRAM. Each line represents 32 wires, physically laid out on the printed circuit board (PCB) of PAM. A PAM is connected—through the in and out links—to a host processor for downloading configuration bitstreams into the PAM. After configuration, the PAM behaves, electrically and logically, like the ASIC defined by the specific bitstream. It may operate as a coprocessor under a host control, specialized to speed up some critical computation, or it may operate as both and connect the host to some external system such as an audio or a video device or another PAM.

The computational core of this PAM is a 4×4 matrix of XC3090 FPGAs shown by letter "M" in Fig. 6.13. Each of these FPGAs has 16 direct connections to each of its four neighbors. The four FPGAs in each row and column share two common 16-bit buses, so that there are four 64-bit buses traversing the array, one per geographical direction N, S, E, and W. The size of cache RAM is 4 MB, made up of four independent

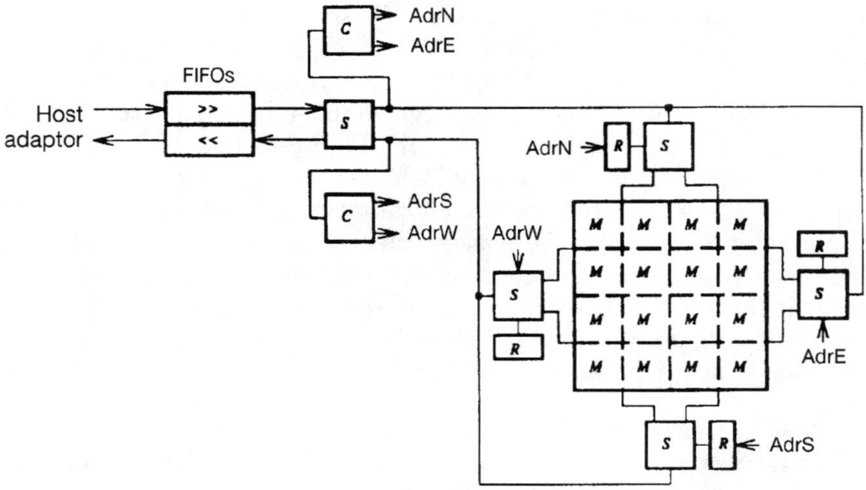

Figure 6.13 Digital Equipment Corporation's DECPeRLe-1 program active memory (PAM) architecture *(from Ref. 31 with permission of IEEE).*

32-bit-wide banks. The 18-bit addresses and read/write signals for each RAM are generated within one of two controller FPGAs, shown by letter "C." The data to and from each RAM goes to the corresponding switch FPGA, shown by the letter "S."

The PAM has four 32-bit-wide external connectors, of which three (not shown in Fig. 6.13) link edges of the FPGA matrix to the external connectors. These are used for establishing real-time links, at up to 33 MHz, between PAM and the external devices such as audio, video, and detector interfaces. Their aggregated bandwidth exceeds 400 MB/s. The fourth external connection links to the host interface of the PAM is a 100 MB/s TURBOchannel adapter. To avoid having to synchronize the host and PAM clocks, the host data goes through two FIFOs, for input and output, respectively. On the PAM side of the FIFO is another switch FPGA, which shares two 32-bit buses with the other switches and controllers. One extra FPGA on PAM is not configurable by the user, and its function may be considered analogous to the ROM, since it provides control over the states of the PAM via software from the host.

A PAM program consists of three parts: (a) the driving software, which runs on the host and controls the PAM hardware, (b) the logic equations describing the synchronous hardware implemented on the PAM board, and (c) placement and routing directives that guide the implementation of logic equations onto the PAM board.

The driving software is written in C or C++ and is linked to a runtime library encapsulating a device driver. The logic equations and placement and routing directives are generated algorithmically by a C++ program.

It was calculated that, at 25 MHz, the PAM had a virtual computing power of 368 billion (denoted by *giga*) boolean operations per second (GBOPs). This was compared to a 32-bit standard microprocessor operating at 100 MHz (100 MIPs) that had a virtual computing power of 3.2 GBOPs, and a 200 MHz, 64-bit processor featuring 12.8 GBOPs. A 100 MHz, 64-bit floating point multiplier delivering one operation per cycle (100 MFLOPs) would rate 281 GBOPs. A feasibility study was performed in 1994 on three possible PAM implementation technologies for a high-performance system: (1) programming a parallel machine, (2) designing a specific PAM configuration, and (3) building a custom system. The results of this study can be summarized as follows:[31]

- For many high-speed computational problems, PAM technology proved to be superior, both in performance and cost, to current forms of general-purpose processing systems such as the pipelined machines and massively parallel processor systems. The cost of PAM was comparable to that of a high-end work station, and much lower than the cost of a supercomputer.
- PAM technology was best applicable to low-level, massively repetitive tasks such as image or signal processing. PAM was not suitable for many supercomputer applications because of their software complexity.
- For many real-time problems with low-volume applications, PAM implementations showed performance and cost equal to those of specific, custom-designed systems. By tuning a specific application for a PAM, quite high-performance implementations are possible.
- PAM technology is particularly suitable for a specific group of applications: high bandwidth interfaces to the external world, with a fully reprogrammable, real-time capability. These include high-bandwidth networks, audio/video input and output devices, and data acquisition systems.

6.5.3 SPLASH attached processor systems

The Splash 1 and Splash 2 attached processor systems were designed and built at the Supercomputing Research Center (SRC) as custom computing platforms. The Splash-2 is an FPGA-based attached processor that can be configured to perform wide variety of tasks. Its architecture is well suited for repetitive computations and high data transfer rates that characterize most low-level image processing problems. To keep up with the high data rates in real time, the processing engine must provide specialized data paths, application-specific operators, creative data management, and careful sequencing and pipelining. These machines are sufficiently flexible that new algorithms can be implemented on existing hardware, and they are fast enough to enable real-time operation for many image processing tasks.[32] The Splash 2 is classified as an *attached processor* (not a coprocessor), because it does not reside directly on the host processor bus.

The basic building block of Splash 2 is Xilinx XC4010 FPGA. Each of XC4010 contains a 20×20 array of configurable logic blocks (CLBs). The CLB contains three look-up tables (LUTs) and two flip-flops. The two LUTs can each implement any Boolean function of up to four inputs, and their outputs can be combined with a ninth input to form a single Boolean function of nine inputs. It also has an additional fast-carry logic, which allows a single CLB to implement a two-bit full adder. An additional wire allows the carry-out of one CLB to be connected directly to the carry-in of an adjacent CLB. The CLBs are connected by three types of routing resources: single-length interconnect between adjacent switch boxes, a double-length interconnect between alternate switch boxes, and a set of long lines that span the width and height of the chip.

Figure 6.14*a* shows the overall Splash 2 system architecture, which has been designed with an SBus interface and serves a Sun SPARCstation 2 as a host.[33] An SBus adapter card is placed on the host and connects via a cable to the Interface Board of Splash 2. The Interface Board and the Splash 2 Array Boards reside separately in a cabinet on a Futurebus+ backplane. Splash 2 can operate either synchronously with the host or asynchronously as an attached processor. The programs for Splash 2 are loaded onto the system by the host through SBus connection. In some applications, the processing on Splash 2 is then driven by a clock on the interface board, and the data is delivered to and taken from Splash 2 by DMA channels on the interface board. The interface board can also be configured to accept clock and data from an external source for applications in which the role of the host is only to load programs and provide high-level control.

A Splash 2 system containing eight array boards has been built and tested, although the architecture can support up to 16 array boards. Each array board contains 17 Xilinx XC4014 FPGA chips as its processing elements, as shown in Fig. 6.14*b*. Sixteen of these, X1 through X16, form a processing array and are connected with a 36-bit-wide data path linearly and via a crossbar. The seventeenth chip, labeled X0, provides a broadcast capability for the data received on the SIMD bus to the other 16 chips. Each of these 17 chips has 512 KB memory attached. The crossbar for Splash 2 is a unique feature for the implementation of a 36-bit-wide data path by aggregating nine 4-bit Texas Instruments SN74ACT8841 crossbar chips. Each of these chips can be

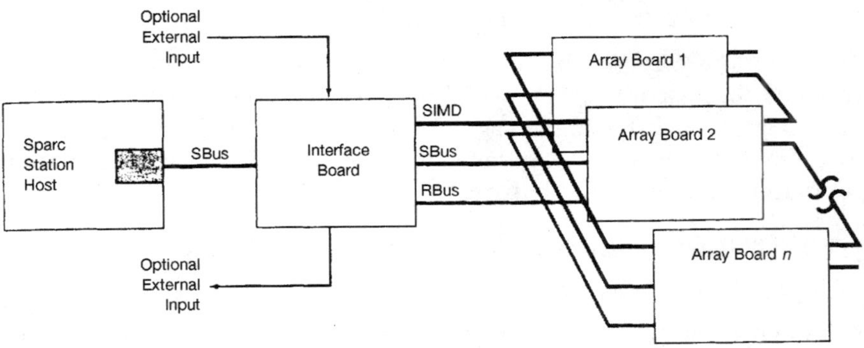

(a)

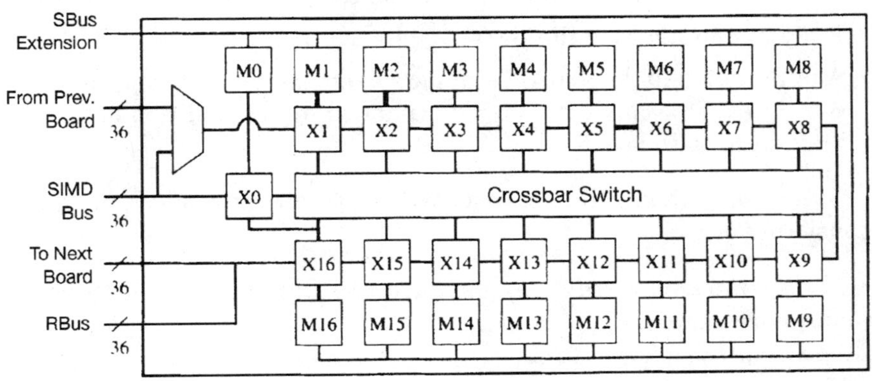

(b)

Figure 6.14 Splash 2 architecture: (a) system level and (b) array board *(from Ref. 33 with permission of IEEE).*

loaded at start-up with as many as eight different configurations. Three bits from X0, controlled by a program that must be loaded into X0, select which of the eight configurations of the crossbar are in effect at any given point in time.

The Splash 2 can be viewed as a machine with a linear data path in which the SIMD bus is used to transmit data from the interface board to the first FPGA on the first array board. The data then can be moved through the linear data path on that board, then to the first FPGA on the second array board, and so on. The data from the last FPGA on the last array board returns to the interface board via the RBus. The linear path is also 36-bit wide and bidirectional (except for the initial segment along the SIMD bus) so that the data can flow in both directions for correlation computations. The range of applications include image processing, computational science benchmarks, data compression and encryption benchmarks, and molecular biology.

6.5.4 Teramac custom computing system

Teramac is a configurable custom computing system developed as a test bed for research into special-purpose parallel architectures and applications such as image segmentation and finding the best match in a DNA database for a given data sequence. The Teramac custom computer includes these features:[34]

- 1,000,000 gate capacity for synchronous logic circuits.
- Up to 1 MHz clock rate.
- A memory storage of 0.5 GB, organized into 64 independent, 32-bit-wide banks, each with independent read and write ports. The banks may be combined horizontally and vertically to form large memories.
- Fully automatic compilation.
- A minimum (single-board) Teramac system supports designs of up to 64k gates. This capacity may be expanded by the addition of up to a maximum of 16 boards.

Figure 6.15 shows the Teramac system architecture block diagram, consisting of

1. Programmable hardware configured to functionally reproduce a user's circuit
2. RAM for the user's designs requiring memory
3. A controller, which is responsible for controlling the programmable hardware as well as exchanging configuration and state data with the host
4. A host workstation that provides the control center for user interface, compiler and debug environment[35]

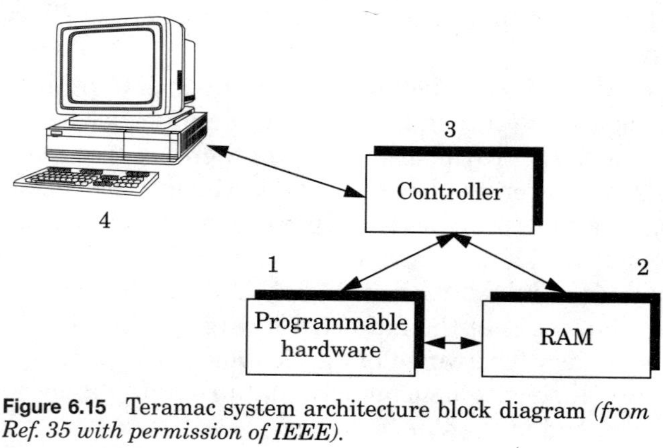

Figure 6.15 Teramac system architecture block diagram *(from Ref. 35 with permission of IEEE).*

Teramac's hardware implementation consists of 16 identical PC boards that are interconnected with cables, with each board carrying four identical multichip modules (MCMs). Each of these MCMs carries 27 identical FPGAs so that a fully configured Teramac contains 1,728 FPGAs. Teramac was designed, based on Rent's rule (see Sec. 5.1.3), which resulted in creation of a custom FPGA called PLASMA to ensure that enough routing resources (wires) are available on-chip and enough I/Os off the chip. Rent's rule was also used at the system level for a balanced distribution of 300,000 interchip wires among the MCM wiring, board-level interconnections, and interboard connections.

The basic building block for Teramac is programmable logic and switch matrix (PLASMA), a hierarchical structured and designed FPGA that is 16.2 mm × 16.2 mm in size and has over 3,000,000 transistors. The PLASMA chip has programmable atomic logic element (PALE) as the basic primitive for implementing logic. Figure 6.16*a* shows a PALE structure consisting of a six-input, two-output universal gate, with each output followed by a pair of latches that can be configured as an edge-triggered flip-flop.[36] The LUT can be configured to implement any six-input Boolean function. Each LUT output drives a PALE output block, which drives the PALE output. The output block can be configured to pass its input directly to its output or pass the input to the output via a level- or edge-triggered register.

The PLASMA chip contains 256 PALEs that are organized into sixteen groups. All PALE inputs and outputs connect to the crossbar. Each of these groups of 16 PALEs, plus a crossbar, are combined to form a hextant. Eight PALEs bound each side of the hextant, and eight hextants bound each side of a central crossbar. These two levels of crossbars allow (1) interconnection of PALEs within the same set, (2) interconnection of PALEs in different sets, and (3) interconnection of PALEs and I/O signal pins. The I/O signal pins are connected to the top and bottom of the central crossbar as shown in Fig. 6.16*b*.

The PLASMA contains nearly 400,000 bits of configuration memory integrated throughout the chip for storage of configuration information of both the PALE LUTs and the crossbar routing resources. The memory cell design is a 5-T memory cell. Some of the PALEs can be configured to perform as register files instead of as normal logic. Each register file contains 64 two-bit words and has eight read and eight write ports. Special circuit design techniques allow the look-up tables to function as multiplexers and demultiplexers for the register files. All user data including register file and I/O pin values may be set and examined via a scan chain. The programming of the PLASMA chip is accomplished using a configuration scan chain. This scan chain is separate from the user scan chain so that fast reading and writing of the user state can take place.

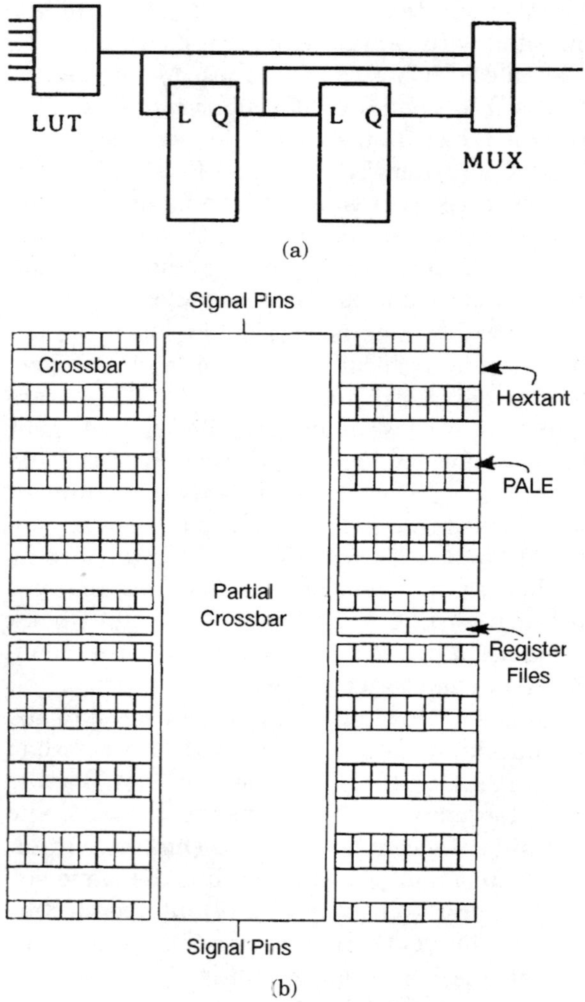

Figure 6.16 PLASMA chip: (a) 1/2 PALE logic structure and (b) interconnect structure *(from Ref. 36 with permission of IEEE).*

The compiler receives a netlist of simple gates as input and produces a Teramac configuration to implement the netlist. The compiler runs in three phases. The first phase begins by mapping the simple gates in the netlist into the types of resources available on Teramac (e.g., register files and RAMs). It then partitions the netlist across the sea of PLASMA chips, assigning each PALE or a register file slice to a specific chip, and each I/O pin on each chip to a specified signal. In the second phase, each chip netlist is placed and routed onto the PLASMA chip to which it has been assigned. The third phase analyzes the resulting configuration to determine the maximum clock frequency at which it can run.

Custom computing machines such as Teramac can be used as search engines for matching and retrieval of information on the DNA sequences archived in biomedical databases. It can also be used for the content based retrieval of images that will require massively parallel search engines.

6.5.5 Virtual Computer

The Virtual Computer from Virtual Computer Corporation is based on a form of virtual computing that allows an algorithm to be implemented in the hardware. Its design is based on the analysis of an algorithm to identify its computationally intensive inner loops and then implementing those inner loops in completely reconfigurable hardware.[37] The Virtual Computer features 52 Xilinx 4010 FPGAs; 24 I-Cube field programmable interconnect devices (FPIDs); three 64-bit I/O ports, out of which one is for hardware configuration/read-back (CBus), and two for general-purpose I/O (VBus); 8 MB of 25 ns RAM; and sixteen 8k × 16-bit of 25 ns dual-port RAMs. The Virtual Computer has 520,000 gates of reconfigurable logic per board and can be completely reconfigured in under 25 ms. A special area of the board designed to serve as the main computing engine is called the *virtual array* and consists of 40 Xilinx 4010 along with 24 I-Cube FPIDs surrounded by 256k of dual-port RAM.

The Virtual Computer uses a form of interconnect called *virtual pipeline,* as shown in Fig. 6.17*a*.[37] It consists of two rows of logic terminated by two dual-port RAMs on either side. The other port of the dual-port RAM on each side connects to a control Xilinx FPGA that also has a 1 MB SRAM. This control Xilinx FPGA is also connected to a 36-bit local bus and acts as smart memory/pipeline interface. There are four of these *virtual pipes* on each board, and each of these pipes has 100,000 gates of logic in between the dual-port RAM and 20,000 gates that allow for smart memory and local bus interface. The board bus interface is made independent by running buffered control and data lines off to three 96-pin connectors on the backside of the board. The first of these interfaces is to a Sun SparcStation II through the SBus. The other interfaces can be supported as required. Figure 6.17*b* shows the Virtual Computer board-level layout.

The control section of the Virtual Computer, called *kernel,* is supported by the general environment resource manager (GERM), which handles the kernel's data I/O. The Virtual Computer supports two 64-bit general-purpose I/O ports (VBus) that can be interfaced to almost

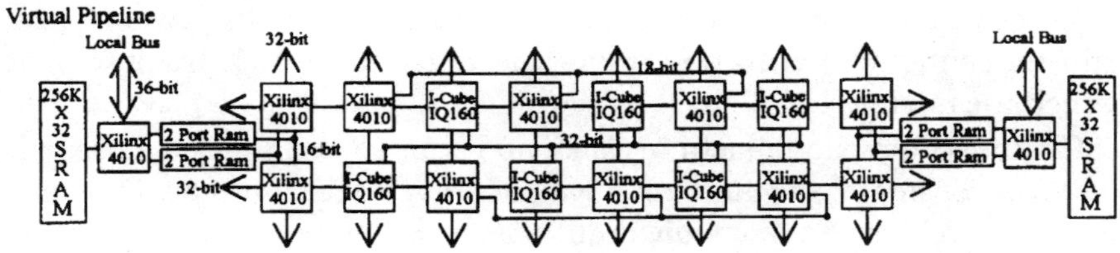

(a)

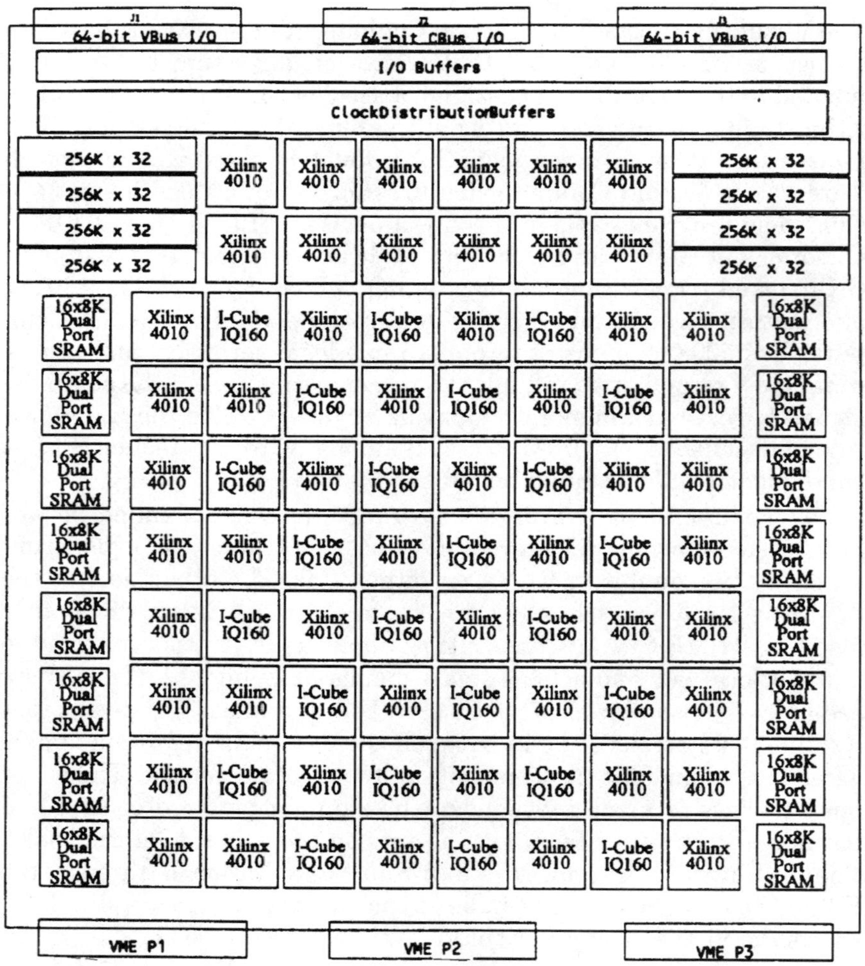

(b)

Figure 6.17 Virtual Computer's (a) virtual pipeline and (b) board-level layout *(from Ref. 37 with permission from Virtual Computer Corp.).*

any system through the GERM. Since the GERM is reconfigurable, different bus protocols can be supported by loading the appropriate configuration files.

The configuration bus (CBus) is an independent 64-bit bus used for configuration of the board. The configuration and read-back control (CRC) section is responsible for the configuration, read-back and clock control for single stepping and debugging. This allows the state of any set of registers to be examined.

The Virtual Computer has different programming models associated with it. At the highest level, a run-time function library is provided. One level down from this, function would have to be written in some object-oriented language such as Verilog or C++ code and linked with a large library of hard macros. The lowest level would be full custom, hand placement and routing.

6.5.6 MOdular and reprogrammable real-time processing hardware

The *MOdular and reprogrammable real-time processing hardware* (MORRPH) architecture is a general-purpose reconfigurable processing unit primarily designed to solve real-time two-dimensional image processing problems and other applications such as one-dimensional signal processing, two-dimensional cellular automata problems, and three-dimensional image processing.[38] In the MORRPH concept, the modular open-ended (incompletely specified) architecture consists of an $M \times N$ rectangular mesh of processing elements (PEs) called the *processing array.* Each processing element contains a single FPGA chip and interconnections for several support chips. The FPGA chips within the PEs provide an array of logic resources, consisting of combinational logic functions, flip-flops, and internal chip routing resources. The types of support chips, which are included in the PEs, are not fixed; they are determined by the requirements of computational task performed by the MORRPH. Currently, the PEs can accept several types of Xilinx 4000 series FPGAs. Figure 6.18*a* shows the PE structure.

Between every set of PEs there is a bidirectional communication bus. The communication buses at the north, east, and west edges of the array are combined into a single bus to be used for high-bandwidth communication with data sources or collection devices. These three external connections are defined as the three I/O buses. The communication buses at the south edge of the array are also combined into a single bus that is used for communication with the host computer and programming of the FPGAs.

A prototype board that utilizes Xilinx 4000 series FPGAs in a 2×3 element processing array has been developed and verified. The proto-

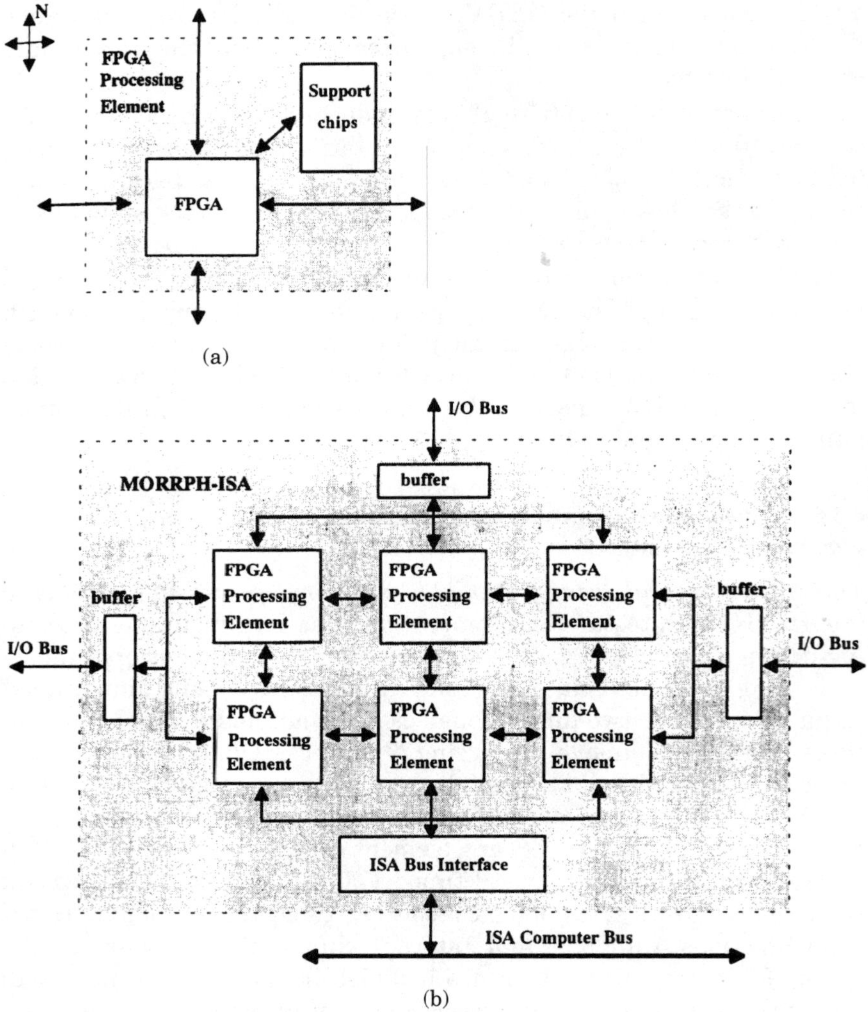

Figure 6.18 (a) MORRPH processing element (PE) structure and (b) MORRPH-ISA card architecture *(from Ref. 38 with permission of IEEE).*

type is implemented as an industry-standard architecture (ISA) adapter card, called MORRPH-ISA, as shown in Fig. 6.18*b*.

In developing a solution for a particular real-time processing task, the algorithm is first developed and tested on a software platform, which may use a high-level language or a schematic capture tool. Currently, for the MORRPH board, the real-time image processing applications are developed using the Powerview schematic capture software from Viewsim Corporation. A library of frequently used image processing functions is under development. An interface to the

Powerview program has been developed to verify the image processing modules, allowing simulation of an image processing hardware design using test image data.

Once the hardware design for a real-time processing task is verified using simulation, its software description is then partitioned and compiled into the resources of one or more FPGA chips. Currently, the design logic is manually partitioned into several groups, which are created to satisfy the logic and I/O resource constraints of each individual FPGA chips. After partitioning, the process of global routing assigns the input and output nets, the nets connected to support chips, and the nets that cross FPGA chip boundaries, to the interconnect resources of the MORRPH board. After partitioning and global routing, the hardware design must be translated into the resources of the destination architecture. The logic in each group of the partition is individually compiled into the resources of each distinct FPGA chip of the MORRPH board using the Xilinx Design Manager (XDM) software. This creates the bitstream files used for programming the individual FPGA chips. Finally, the required FPGA and support chips are assembled on to one or more MORRPH boards for the targeted application.

References

1. Bradley K. Fawcett, "Taking advantage of reconfigurable logic," *IEEE ASIC Conf. Proc.,* 1994, pp. 227–230.
2. Bradley K. Fawcett, "Field programmable gate arrays and reconfigurable computing," *SPIE,* vol. 2607, pp. 159–166.
3. Clive "MAX" Maxfield, "Logic that mutates while-u-wait," *EDN,* Nov. 7, 1996, pp. 137–142.
4. Doug Conner, "Reconfigurable logic: Hardware speed with software flexibility," *EDN,* March 28, 1996, pp. 53–64.
5. John Boquet, et al., "Design tools for dynamic reconfigurability," *Electronic Products,* Sept. 1996, pp. 31–33.
6. Bournemouth University, WWW site for Dynamically Reconfigurable Hardware.
7. *Atmel Reconfigurable Logic Data Book,* Application Note: Implementing Cache Logic with FPGAs.
8. Scott Gould, et al., "An SRAM-based FPGA architecture," *Proc. IEEE 1996 Custom Integrated Circuits Conference,* pp. 12.2.1–12.2.4.
9. *Xilinx Data Book 1996:* XC6200 Field Programmable Gate Arrays.
10. John H. Mayer, "Reconfigurable computing redefines design flexibility," *Computer Design,* Feb. 1997, pp. 49–52.
11. Bradley K. Fawcett, "Reconfigurable processing with FPGAs," WWW presentation and discussion forum.
12. Rodger Hosking," Reconfigurable hardware acceleration for embedded DSPs," WWW presentation and discussion forum.
13. Andre DeHon, "DPGA-Coupled Microprocessors: Commodity ICs for the early 21st Century," MIT Transit Project, Transit Note # 100 at their WWW site.
14. Ubaid R. Khan, et al., "FPGA architectures for ASIC hardware emulators," *Proc. IEEE ASIC 1993 Conference,* pp. 336–340.
15. J. Rose, et al., "Architecture of field-programmable gate arrays: The effect of logic block functionality on area efficiency." *IEEE Journal Solid State Circuits,* vol.25, no.5, Oct. 1990, pp. 1217–1225.

16. S. Singh, et al., "The effect of logic block architecture on FPGA performance," *IEEE Journal Solid State Circuits,* vol.27, no.3, March 1992, pp. 281–287.
17. S. Hauck, et al., "Pin assignment for multi-FPGA systems" (extended abstract), IEEE Workshop on FPGAs for Custom Computing Machines.
18. Brad Taylor, et al., "Segmented busses deliver 15x memory bandwidth increase for image and signal processing in a virtual hardware environment," Paper supplied by Giga Operations Corporation.
19. I-Cube Application Note: Using FPID devices in FPGA-based prototyping.
20. Stephen Sample, et al., "Apparatus for emulation of electronic hardware system," Quickturn Design Systems, Inc., U.S. Patent # 5,109,353, April 28, 1992.
21. Michael Butts, et al., "Method of using electronically reconfigurable logic circuits," Mentor Graphics Corp., U.S. Patent # 5,036,473, July 30, 1991.
22. Michael Butts, "Future directions of dynamically reprogrammable systems," *Proc. IEEE Custom Integrated Circuits Conf. 1995,* pp. 24.1.1–24.1.8.
23. Henry Verheyen, et al., "Rapid prototyping using system emulation technology for DSP design validation," *EDA&T* '95, Aug. 21–26, 1995 (preprint paper supplied by Aptix Corporation).
24. M. Dahl, et al., "Emulation of the Sparcle microprocessor with the MIT Virtual Wires emulation systems," *Proc. IEEE Workshop on FPGAs for Custom Computing Machines*, pp. 14–22.
25. A. Aggarwal, et al., "Sparcle: An evolutionary processor design for multiprocessors," *IEEE Micro,* 13(3):48–61, June 1993.
26. Richard Terrill, "50,000 gate ASIC prototyping PLD using four FLEX 8000 devices and a programmable interconnect," *IEEE International ASIC Conf. Proc. 1994,* pp. 348–352.
27. John Darneur, et al., "A silicon-on-silicon field programmable multichip module (FP-MCM)—Integrating FPGAs and MCM technologies," *IEEE Trans. Comp., Packaging and Manufacturing Technology,* Part B, vol.18, no.4, Nov. 1995, pp. 601–608.
28. Brian Cox, "Field programmable gate array based reconfigurable preprocessor," *Proc. IEEE 1994 Workshop on FPGAs for Custom Computing Machines,* pp. 40–48.
29. Brian Cox and John Nieznanski," Common processor element packaging for CHAMP," *Proc. IEEE Symposium on FPGAs for Custom Computing Machines,* pp. 39–44.
30. Patrice Bertin, et al., "PAM programming environments: Practice and Experience," *Proc. IEEE 1994 Workshop on FPGAs for Custom Computing Machines,* pp. 133–137.
31. Jean A. Vuillemin, et al., "Programmable Active Memories: Reconfigurable systems come of Age," *IEEE Transactions on Very Large Scale Integration (VLSI) Systems,* vol.4, no.1, March 1996, pp. 56–69.
32. A. Lynn Abbott, et al., "Accelerating image filters using custom computing machine," SPIE, vol. 2607, *Proceedings, 1995* (FPGAs for Fast Board Development and Reconfigurable Computing), pp. 62–70.
33. Duncan A. Buell, et al., "Splash 2 FPGAs in a Custom Computing Machine," Chapter 1, in *Custom Computing Machines: An Introduction,* IEEE Computer Society Press.
34. W. Bruce Culbertson, et al., "The Teramac configurable custom computer," *SPIE, vol. 2607 Proceedings,* 1995 (FPGAs for Fast Board Development and Reconfigurable Computing), pp. 201–209.
35. Rick Amerson, et al., "Teramac—Configurable custom computing," *Proc. IEEE 1995 Workshop on FPGAs for Custom Computing Machines,* pp. 32–38.
36. Rick Amerson, et al., "An FPGA for multi-chip reconfigurable logic," *Proc. IEEE 1996 Custom Integrated Circuits Conf.*, pp. 137–143.
37. Steven Casselman, et al., "Virtual computing and the virtual computer," Paper supplied by Virtual Computer Corporation.
38. Thomas Drayer, et al., "A Modular and reprogrammable real-time processing hardware, MORRPH," *Proc. IEEE Symposium on Custom Computing Machines 1995,* pp. 11–19.

Chapter

7

PLDs for Military and Space Applications

7.1 Introduction to Radiation Effects

Programmable logic devices are increasingly used in military and space applications because of their attractive attributes such as low cost, quick turn-around time, and technological advancements that have resulted in availability of device families with more than 100,000 gates. In military and space applications, it has become important to reduce both cost and turn-around time for board- and system-level integration into the payloads. Although the custom ASICs and gate arrays that have been traditionally used in military and space applications have a higher level of performance and greater radiation tolerance, they also have a large NRE cost and relatively longer design cycles for prototyping and delivery of flight units. The rapid advancements in FPGA technology have made available devices that meet the density and performance goals of many space-level design requirements.

A major hurdle in the wider acceptance and usage of the PLDs has been the concerns about their radiation tolerance in space environments. The space radiation environment, radiation effects including total dose and single event phenomenon (SEP), radiation hardness assurance testing (RHA), and the radiation effects mitigation through the use of redundancy techniques such as shielded packages or radiation-hardened (RH) devices is discussed in the following sections.

7.1.1 Space Radiation Environments

The space environment presents a certain radiation risk to all electronic components, including field programmable logic devices used in

Earth-orbiting satellites and planetary mission spacecrafts. The irradiating particles in this environment consist primarily of high-energy electrons, protons, alpha particles, and cosmic rays. The military weapons environment, such as a nuclear explosion (often referred to as the *gamma dot*), is characterized by X rays, gamma neutrons, and other reaction debris constituents occurring within a short time span. This can cause latchup and transient upsets in integrated circuits such as PLDs. Although the natural space environment does not contain the high-dose-rate pulse characteristics of a nuclear weapon, exposed electronic systems can accumulate a significant amount of total dose from the electrons and protons over a period of several years, depending upon the satellite inclination and orbital parameters.[1]

The radiation effects of charged particles in the space environment are dominated by *ionization,* which refers to any type of high energy that creates electron-hole (*e-h*) pairs when passing through a material. In addition to creating *e-h* pairs, the radiation can cause displacement damage in the crystal lattice by breaking the atomic bonds and creating trapping recombination centers. Both of these damage mechanisms can lead to degradation of the device performance.

The important ionizing electromagnetic radiation types are X rays and gamma rays. Ionizing particulate radiation can be light uncharged particles such as the neutrons; light charged particles such as electrons, protons, alpha, and beta particles; and heavy charged particles (heavy ions) such as iron, bromine, krypton, Xenon, and so on, which are present in the cosmic ray fluences. The cosmic rays originate from two sources: the sun (solar) and galactic sources outside the solar system. Table 7.1 provides a general overview of the various types of radiation and parameter degradation or circuit failures caused by them.[2]

Gamma rays (or X rays) basically produce a similar kind of damage as light charged particles, since the dominant mechanism is charge interaction with the material. Neutrons have no charge and primarily react with the nucleus, causing crystal lattice damage. Heavy charged particles create "tracks" or a volume of ionization within the substrate of the semiconductor devices. This charge can collect on circuit nodes and cause data loss through single-event upset (SEU) or transient upset. The major contributor to ionizing particle fluences are the Van Allen radiation belts in the Earth's geomagnetic environments, solar flares, and cosmic rays. The energy of these charged particles can range from few hundred kilovolts to 10 GeV. Therefore, the designers using PLDs for space applications have to be concerned with two types of radiation damage caused by charged particle ionization: total ionizing dose (TID) effects and single-event phenomenon (SEP), which is also referred to as single-event effect (SEE).

TABLE 7.1 Overview of Various Types of Radiation and Corresponding Transistor/Circuit-Level Parameter Degradation and Failure ***(from Ref. 2)***

Type of radiation	Type of degradation or failure	
	Transistor level	Circuit level
	Photons (X rays and gamma rays)	
Total dose (Low dose rate) (High total dose)	▪ Threshold voltage shifts ▪ Drive current shifts ▪ Carrier mobility reduction ▪ Parasitic leakage current	▪ Performance degradation ▪ I/O parametric shift ▪ Lost functionality ▪ Data upset/loss
Transient (High dose rate) (Low total dose)	▪ Collapse of depletion region ▪ Burn-out of metal interconnects ▪ Damage to junction region	▪ Latchup ▪ Catastrophic failure ▪ I/O glitches ▪ High photocurrents
	Charged particles and cosmic rays	
Electrons and protons	▪ Threshold voltage shifts ▪ Drive current shifts ▪ Carrier mobility reduction ▪ Parasitic leakage currents ▪ Carrier recombination and trapping ▪ Collapse of depletion region (photons only)	▪ Performance degradation or improvement ▪ I/O parametric shifts ▪ Lost functionality ▪ Data upset (single bit/node, protons only)
Alpha particles	▪ Collapse of depletion region	▪ Data upset (single bit/node)
Heavy ions	▪ Collapse of depletion region	▪ Data upset (single or multibit/node) ▪ Hard oxide failures ▪ Latchup
	Uncharged particles	
Neutrons	▪ Junction leakage ▪ Bipolar beta reduction ▪ Carrier recombination and trapping	▪ No performance effects on CMOS ▪ Improves latchup tolerance on CMOS circuits ▪ Performance degradation on bipolar circuits

The total dose effect is the cumulative ionization damage expressed in rads (material) that is caused by all ionizing particles passing through a semiconductor device. For the MOS devices, the ionization traps positive charge in the gate oxide called the oxide traps and produces interface states at the Si–SiO_2 interfaces. These effects typically exhibit both a time and field dependence characteristic, resulting in threshold voltage shift, reduced speed, higher leakage currents, and possible loss of device functionality. In digital bipolar devices that exhibit relatively higher radiation tolerance compared to MOS technology, the ionizing radiation may cause decrease in gain and breakdown voltage, and higher leakage currents. The magnitude of these changes depends on a number of factors such as the device type and process technology, total radiation dose and its energy, dose rate, applied bias and temperature during the irradiation, and post-annealing period.

The SEP is caused by high-energy particles, such as those present in cosmic rays and solar flares, and can result in two types of errors: soft and hard. The soft errors, often referred to as SEUs, can cause a flip-flop (e.g., in Actel FPGAs) or configuration bit of SRAM-based look-up tables (LUTs) such as those used in Xilinx devices to change from "1" to "0" (bit-flip) and vice-versa. These soft errors cause no permanent damage, and the device can be reprogrammed. However, the heavy ions can also cause hard errors due to the parasitic p-n-p-n devices present in bulk CMOS technology. This can result in a latchup and device burnout unless the current is limited or the power is cycled off and on again. Therefore, unlike the TID degradation, SEE rates are not evaluated in terms of a time or dose until failure, but a probability that a SEE will occur within a known span of time. These total dose effects and SEP are discussed in the following sections.

The natural space environments in which most of the orbiting satellites operate ranges in altitude from low Earth orbits (150 to 600 km) to geosynchronous orbits (roughly 35,880 km). The radiation environment of greatest interest is the near Earth region, about 1 to 12 Earth radii (where R_e = 6380 km), which is mainly dominated by electrically charged particles trapped in Earth's magnetosphere and, to a lesser extent, by the heavy ions from cosmic rays (solar and galactic). Interplanetary space probes such as the Voyager (and Galileo to Jupiter) encounter ionizing particles trapped in the magnetosphere of other planets as well as the solar flares and heavy ions from cosmic rays.

The main sources of energetic particles that are of concern to the spacecraft designers are as follows:

- Protons and electrons trapped in Van Allen radiation belts
- Heavy ions trapped in the magnetosphere
- Cosmic ray protons and heavy ions
- Protons and heavy ions from solar flares

The electrons in the Earth's atmosphere have energies ranging from low keV (kilo-electronvolts) to about 7 MeV and are trapped in the roughly toroidal region, which is centered on the geomagnetic equator and extends to about 1 to 12 Earth radii. These trapped electrons are differentiated by *inner zone* (<5MeV) and *outer zone* (≈7 MeV) electron populations. The trapped protons originating mostly from the solar and galactic cosmic rays have energies ranging from a few million electron volts to about 800 MeV. The motion of the trapped charged particles is complex, as they gyrate and bounce along the magnetic field lines and are reflected back and forth between the pairs of conjugate mirror points (regions of maximum magnetic field strength along their trajectories). Figure 7.1 illustrates the spiral, bounce, and drift motion of the trapped particles.[3]

The actual electron and proton flux encountered by a satellite is strongly dependent on the orbital parameters, mission launch time and mission duration. Electrons and protons from the trapped radiation belts on the interaction with spacecraft materials produce secondary radiation (e.g., *bremsstrahlung,* or braking radiation from the deceleration of electrons). This secondary radiation can extend the penetration range of primary radiation and lead to an increase in dose deposition. These incident electron and proton fluxes are typically calculated from the trapped radiation environment models developed by the U.S. National Space Sciences Data Center (NSSDC).

In addition to the trapped geomagnetic radiation, another contribution to the incident particle flux for an orbiting satellite is the transiting radiation from the solar flares and galactic cosmic rays. These solar energy particle events, usually occurring in association with solar flares, consist mainly of protons, some alpha particles (5 to 10 percent), heavy ions, and electrons. This solar flare phenomenon is categorized as an ordinary (OR) event or an anomalously large (AL) event. The particles fluxes from the solar flares can last from a few hours to several days. These AL events, although occurring rarely, can cause serious damage (e.g., latchup, SEU) to nonhardened IC chips

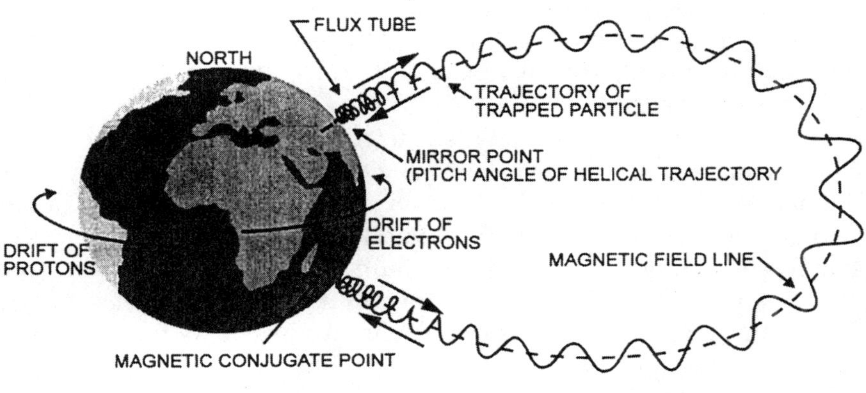

Figure 7.1 The spiral, bounce, and drift motion of trapped particles in Earth's environment *(from Ref. 3).*

such as PLDs. Another significant contribution to the transiting radiation is from the cosmic rays originating outside the solar system and consisting of 85 percent protons, 14 percent alpha particles, and 1 percent heavier ions. These galactic cosmic rays range in energy from a few million electronvolts to over 10 GeV per nucleon.

The particles from energetic solar flares (OR events) are heavily attenuated by the geomagnetic shield at low altitude and low inclination orbits, such as U.S. space shuttle orbits (28.5° inclination). In a 500 km, 57° inclination orbit, some particle fluxes do penetrate. A characteristic of the geomagnetic field, which is particularly significant to SEP is the South Atlantic Anomaly (SAA), referring to an apparent depression of the magnetic field over the coast of Brazil, where the Van Allen radiation belts dip low into the Earth's atmosphere. This SAA is responsible for most of the trapped radiation in low Earth orbits (LEOs).

The most important characteristic of the environment encountered by satellites in low Earth orbits is that, several times each day, they pass through the proton and electron particles trapped in the Van Allen radiation belts. The levels of fluxes seen during these passes varies greatly with orbit inclination and altitude. The amount of protection that the geomagnetic field provides a satellite from the cosmic ray and solar flare particles is also dependent on the inclination and, to a smaller degree, the altitude of the orbit. As altitude increases, the exposure to cosmic ray and solar flare particles gradually increases. However, the effect of satellite orbit inclination on the exposure to these particles is more important. As the inclination reaches polar regions, which are outside the closed magnetic field lines, it is fully exposed to cosmic ray and solar flare particles for a significant portion of the orbit.[4]

The highly elliptical orbits (HEOs) are similar to LEOs, because they pass through the Van Allen radiation belts each day. However, because of their high apogee altitude (typically greater than 30,000 km), they also face long exposures to the cosmic ray and solar flare environments, regardless of their inclination. At geostationary orbits (GEOs), the satellites are almost fully exposed to the galactic cosmic rays and solar flare particles.

The total dose absorbed by a semiconductor component such as a PLD during a space mission depends not only on the radiation environment encountered but also on the amount of shielding between the device and the incident flux. For small shield thicknesses (e.g., the spacecraft skin of 100 mils of aluminum), the transmitted electrons and protons contribute roughly equally to the absorbed dose whereas, for the thicker shields, the dose accumulates primarily from high energy protons.

The number of single-event induced upsets (SEUs) experienced by a semiconductor component such as a PLD in a given radiation environment depends primarily on its threshold for upsets (usually given by the critical charge Q_c or critical LET) and the total device volume sensitive to ionic interaction. There are several models available for the prediction of upset/error rates due to cosmic ray events. The trapped proton levels are typically calculated using NASA AP8 model.[5] In this model, the flux intensities can be ordered according to the field magnitude (B) and the dipole shell parameter (L). The AP8 model has both solar minimum and solar maximum versions, making it possible to account for solar variations by selecting the appropriate model version.

An example of another model is the popular Cosmic Ray Effects on MicroElectronics (CRÉME) model developed at the Naval Research Labs, which predicts a device upset/error rate based on description of the radiation environment.[6] Since its release, CRÉME has become a standard and widely used in SEE calculations. However, studies of the CRÈME model and comparison of its predictions with direct measurements of the heavy ion environment near Earth have demonstrated some deficiencies. The flight of the combined release and radiation effects satellite (CRRES),[7] which includes state-of-the-art instrumentation for measurement of heavy ions and an experiment called *microelectronics package* (MEP) for on-orbit testing of single-event upsets,[8] has resulted in improved model of the near-Earth heavy ion environment. This model is called CRRES/SPACERAD heavy ion model of the environment (CHIME).[9]

The CRÉME and CHIME models include solar cycle variations and magnetospheric attenuation calculations. The CRÈME model calculates LET for simple shield geometry for aluminum shields and targets. CHIME improves the LET calculations by permitting the user to select a shield material density and target material density.

This section provided a general overview of the natural space environment, its modeling, and radiation effects for the Earth orbiting satellites and planetary missions. The nuclear weapon radiation environment is of concern primarily for the strategic and tactical nuclear weapons designers working on “operate-through” (nuclear event survivable) systems and will not be discussed here.

7.1.2 Total ionizing dose effects

In general, bipolar technology devices are more tolerant to total ionizing dose (TID) radiation damage effects than are bulk MOS technology parts. In bipolar transistors, the principal factors responsible for performance degradation are trapped positive charges that build up in the passive oxides surrounding silicon surfaces. This excess positive charge results in the creation of surface states at these interfaces. Because a bipolar transistor depends on a junction for its operation, the effect of radiation degradation is not severe as for the MOS transistor, which depends on a surface and corresponding interfaces for operation.

The MOS technology has been preferred by designers for military and space applications because of its lower static power requirements and better noise margins. Also, the MOS devices, being majority carrier devices, are inherently hard to neutron irradiation to 10^{16} neutrons/cm^2. However, the nonhardened MOS devices are quite sensitive to total dose effects. For example, commercial NMOS and CMOS devices may fail at doses as low as 1 to 2 krad(Si). For CMOS devices, the radiation damage is related to ionization effects and consists of the buildup of trapped charge in the oxide that may cause an increase in the number of bulk oxide traps and interface traps.

Ionization radiation (charged particles) traversing through the bulk of material generate electron-hole (*e-h*) pairs within the silicon dioxide. Typically, the amount of energy required to create an *e-h* pair is roughly two to three times the bandgap energy. In silicon, approximately 3.5 eV is required to create an *e-h* pair. A major effect of these *e-h* pairs created uniformly throughout the solid is the increase in bulk conductivity of the material. In a MOS structure, the insulating oxide (SiO_2) layer is most sensitive to ionizing radiation. As the radiation passes through the oxide, the energy deposited creates *e-h* pairs.

After the *e-h* pairs' creation and some instantaneous recombination, these electrons and holes in the presence of an electric field are free to diffuse and drift away from their points of generation. The electrons, being more mobile than the holes, are swept out of the oxide to be collected at the gate electrode. Some of the holes that escape initial recombination stay near their point of generation, acting as *oxide traps,* causing a net positive charge. There are other holes that undergo a *trap-hopping* process through the oxide. In the presence of positive applied gate bias, when the holes reach the SiO_2/Si interface, a fraction of them are captured in the long-term trapping sites, referred to as the *interface traps.*

The interface traps, which are localized states with energy levels within the silicon bandgap, can cause small negative voltage shifts that may persist in time for few hours to several years. They may also cause a degradation in the mobility of the carriers in the channel of MOS transistors that can lead to a reduction in the channel conductance. Figure 7.2 illustrates the process of charge generation and initial recombination, hole transport, and long-term trapping near the SiO_2 surface.[10]

The amount of energy absorbed by the device material increases from the cumulative effect of radiation exposure. Since the number of *e-h* pairs generated is directly proportional to the amount of energy

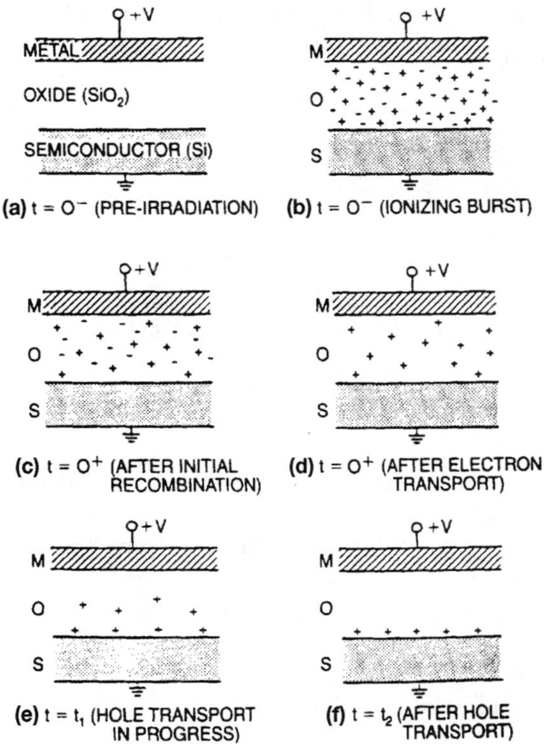

Figure 7.2 Schematic illustrating processes of charge generation and initial recombination, hole transport, and long-term trapping near SiO_2–Si interface *(from Ref. 4).*

absorbed, the ionization damage in the form of oxide traps and interface traps also monotonically increases. Thus, the total dose received by a programmable device can be measured in units of rads(Si) or rads (SiO_2), a unit equal to 100 ergs absorbed per gram of Si (or other material specified), defined by the relationship given below.

$$1 \text{ rad(Si)} = 100 \text{ ergs absorbed/gram of Si}$$

$$= 10^5 \text{ J absorbed/gram of Si}$$

The radiation hardness of a MOS programmable logic device is determined by the rate at which the two types of traps (oxide and interface) build up with increasing levels of total dose radiation. To summarize, the parametric degradation in threshold voltage and operating margins, leakage currents, and variations in timing parameters for commercial (unhardened) programmable logic devices in response to total dose radiation depend on various factors, such as:

- Manufacturing processes and fabrication variables
- Total dose irradiation levels
- Dose rates
- Circuit irradiation biasing conditions
- Time between completion of the irradiation and measurement of circuit parameters (annealing period)
- Test temperatures (during irradiation and testing)

Section 7.1.4 describes total dose effects measurement test procedures.

7.1.3 Single-event phenomenon

The single-event phenomenon (SEP) induced by the charged particles penetrating through the microelectronic circuits is a major concern in space environments. The SEP refers to the ionization effects due to a single high-energy particle as it strikes the sensitive nodes (or sensitive volume) within the electronic device. Satellites at GEO orbits and in regions outside the Earth's radiation belts can experience upsets due to heavy ions from galactic cosmic rays or solar flares. SEP can cause three different types of effects, as follows:

- Single-event upset (SEU) is an ionizing radiation-induced logic change in the programmable logic device flip-flops or RAM configuration bit. This can cause a bit-flip from logic "1" to "0" and vice-versa because of the high current level or excess charge collected at a sensitive bit node in the memory circuit. This can be a *soft error,* which means that the logic reversal is temporary, and the cell may become functional again on reset.
- Single-event latchup (SEL), or a *hard error,* may occur when a single charged particle causes a latched low-impedance state by turning on the parasitic SCR structure (p-n-p-n or n-p-n-p) in bulk CMOS devices. It can also result in a *snap-back* effect, which is a result of photocurrent-induced initiation of avalanching in the drain junction of an n-channel MOS transistor. If the induced parasitic current levels are sufficiently high, they can cause permanent device failures such as a junction burnout. These parasitic current states are highly dependent on circuit layout and device geometry.
- Transient radiation effects are those that depend on the ionizing dose rate delivered to a circuit rather than the total amount of dose delivered. A higher dose rate creates a greater number of *e-h* pairs, both in the insulator layer and the silicon substrate. High photocur-

rents generated can also alter the contents of storage element to cause soft errors, without permanent damage to the circuit. This effect caused by a burst of ionizing radiation from an extended source, is referred to as a *transient upset,* as compared to an SEU caused by a single energetic ion.

The SEUs can have significant effects within the heart of proton belts, which may roughly have 10^7 to 10^9 protons/cm^2/day with energies above 30 MeV (which is approximately the minimum energy required to penetrate a typical spacecraft). The rate of energy loss, and therefore the density of generated charge, depends on several factors such as energy, mass, and charge of ionizing particle. The energy deposition rate (dE/dX), the stopping power, or the linear energy transfer (LET) are measured in units of MeV/[cm^2/mg] (or MeV/μm). An important consideration in device SEU susceptibility is the critical charge, which is defined as the minimum charge that must be deposited in a device to cause an upset. It is equivalent to the number of stored electrons representing the difference between a stored "0" and a stored "1".[11] In general, for unhardened programmable logic devices, the critical charge decreases with reduced feature sizes.

The heavy ions or transient radiation (high dose rate burst) can induce latchup at certain internal circuit nodes, which can produce high photocurrent, resulting in large voltage drops in the isolation wells, substrate, and metallization. The cosmic ray environment is specified in terms of particles with a given energy loss (LET spectrum). This LET spectrum can be combined with device dimensions to find the probabilities of different amounts of energy deposition and then to calculate the number of events that will deposit a charge greater than Q_c. A lower LET threshold implies that the device is more sensitive to an SEU or latchup. A recent study also showed that the devices are more susceptible to proton-induced upsets if the heavy ion induced LET threshold is less than about 8 MeV/(cm^2/mg).[12]

The SEU threshold (or vulnerability) is commonly measured in terms of critical charge (Q_c) and the critical LET. The Q_c is primarily correlated to circuit design characteristics, whereas the critical LET is a function of both circuit design and technology. Q_c can be determined from computer simulation, whereas the critical LET is found experimentally by bombarding the device with various ion species (e.g., in a cyclotron). The LET is basically a measure of how much energy (or charge) is deposited per unit distance within the device for the first micrometer of track length. The critical LET values can be different for devices fabricated with identical designs and processes but on different substrates. Section 7.1.4 discusses SEP measurement test procedures.

7.1.4 Radiation hardness assurance (RHA) testing

Semiconductor devices have a radiation response that is a function of technology, fabrication process, and design characteristics. Therefore, the radiation sensitivity, especially of unhardened commercial programmable logic devices, can vary from lot to lot. As such, actual radiation testing is necessary to characterize the device response to ensure that it meets mission requirements. However, radiation testing on ground is complicated by the fact that it differs significantly from the actual space environment, which is quite complex, as discussed in Section 7.1.1. In space, there is a large variability in environmental parameters: particle types, energies, and their time dependency. An ionizing dose is delivered steadily over a long period of time, with mission periods usually ranging from few months to over ten years.

An ideal situation would be to put the test devices in an actual space environment and measure their radiation response. However, this approach is not practical and limited to research types of satellites. Therefore, for all practical purposes, the space environment has to be simulated on the ground for testing semiconductor devices. The major objectives of ground-based radiation testing are as follows:

- Identify the mechanism of the radiation interaction with the device material and correlate these effects with the device failure.
- Characterize the response of specific device types and technologies for use in space applications.
- Evaluate the device susceptibility to radiation and conformance to mission (application) requirements.

Radiation sources include radioactive isotopes such as Co-60, X-ray tubes, and particle accelerators. In addition to radiation sources, the radiation response of semiconductor devices strongly depends on the biasing conditions and temperature during irradiation. Therefore, device modeling and testing are performed for the worst-case biasing conditions. This radiation testing for programmable logic devices can be for (1) total dose effects, which is the damage caused by ionizing energy through cumulative absorption by the sensitive material, (2) SEU (or soft errors) simulated by the high-energy, single-particle environment, and single-event latchup (SEL), (3) transient response for dose-rate effects (latchup), and (4) neutrons. The transient dose rate effects and neutrons testing are primarily the concern for military and nuclear weapons design environment. The following sections discuss test methods for the first two: total dose effects and SEU/SEL phenomena.

Total dose effects testing. Total dose testing is performed to measure the damage caused by ionization energy absorbed by the sensitive material. Radiation induces gradual parametric degradation (changes in threshold voltage and leakage currents) and, eventually, functional failures. The rate of degradation may depend on the dose rate (flux). The total dose testing can be performed by exposing a group of programmable logic devices to a number of fixed, incremental dose level steps and, after each radiation step, making electrical measurements before returning the devices to irradiation facility for the next dose level.

The irradiation sources commonly used for the total dose effects due to ionization investigation through ground simulation testing can be listed as follows:

- Gamma rays through radioactive isotopes such as Co-60 and Cesium-139 (Cs-139)
- X rays
- Electrons
- Protons

Each of these sources exhibits a specific energy spectrum and affects the test devices differently. Co-60 is the most commonly used gamma-ray irradiation source for the measurement of ionization effects on silicon devices. It emits photons of energy at 1.173226 and 1.332483 MeV, and it has a half-life of 5.27 years. In a typical irradiator, the Co-60 source is sealed in a steel jacket and placed inside a thick lead shield chamber. The devices to be irradiated are plugged into sockets on the fixture board, which is connected with wires to the power supply and biasing source located outside the chamber. The distance of the source or source exposure window can be varied to get different dose rates such as 10^5 to 100 rad(Si)/h or less.

As an example, a typical high-radiation orbit mission of 10 years may average 200 krad total dose (including a 2:1 radiation design margin, RDM). Then, a "one-shot" total dose exposure to that level (200 krad) at a dose rate of 100 krad/h may take only 2 hours. However, the total dose testing is usually performed at various steps (cumulative), followed by electrical measurements at each of these steps. Figure 7.3 shows the schematic of a typical arrangement for Co-60 gamma ray source testing setup, test levels, and so on.[13]

X rays (even of low energy) can also be used to simulate the space environment. The main advantage of using X rays as an irradiation source is the low cost, wider distribution, and higher safety standards available for X-ray equipment. For commercial applications, 10 keV X-ray sources are often used for wafer-level testing and in-line monitoring of radiation tolerant products before their assembly and packaging.

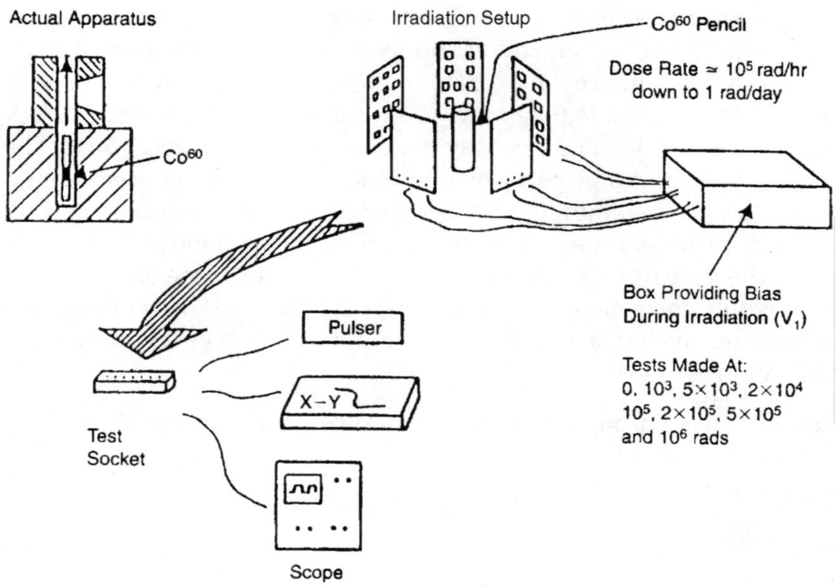

Figure 7.3 Schematic of a typical arrangement for Co-60 gamma ray source test setup *(from Ref. 7).*

However, analysis of X-ray test results should take into consideration effects such as dose enhancements and the absorption coefficient in comparing them to higher energy radiation source test results.

Electron beams can act as sources of ionization, and a useful electron source is the Van de Graaff accelerator, which may be typically designed to operate at particle energies between 0.1 and 5 MeV. The dose rates can be varied by altering the beam current and beam focus, or sweeping the beam to yield particle fluxes from about 2×10^{10}/cm^2/s to 2×10^{13}/cm^2/s, which roughly corresponds to the dose rates ranging from 600 to 600,000 rad/s. These dose rates are several orders of magnitude higher than the typical space dose rates of 10^{-5} to 10^{-3} rad/s. Linear accelerators (LINACs) provide electrons of higher energies, typically 4 to 40 MeV (or even higher), in rapid square pulses. However, these are quite expensive machines and have a high test time overhead. The protons in space environment (heart of radiation belts) cause ionization damage primarily through nuclear interactions. The ground testing with protons is often done mainly for displacement damage studies, using cyclotrons.

Ground testing for total dose radiation is usually performed at dose rates that are typically several orders of magnitude higher than the actual space-like dose rates. In the low dose rate space environment, a significant amount of hole detrapping occurs due to annealing effects.

Therefore, the higher dose rate ground-based testing is usually considered the worst case for failure mechanisms dependent on the oxide traps, although they do not accurately simulate the interface traps (time-dependent) effects. However, it has been recently observed that, for some linear devices such as operational amplifiers, higher degradation is observed in parameters such as offset voltages and bias currents during ground testing at low dose rates a compared to test results at higher dose rates.

Total dose testing on programmable logic devices is performed per MIL-STD-883, Method 1019, which defines the requirements for effects from a Co-60 gamma ray source.[14] This procedure also provides an accelerated aging test for estimating low dose rate effects, which may be important for some device types that exhibit time-dependent effects. This test procedure includes specifications for the following:

- Test apparatus including radiation source, dosimetry system, electrical test instruments, and test circuit board
- Test procedures including sample selection and handling, dosimetry measurements, radiation levels and dose rates, and temperatures
- Electrical performance measurements, test conditions including bias and loading, and post-irradiation procedures
- MOS accelerated aging test procedures and requirements

Single-event phenomenon (SEP) testing. As discussed earlier, the SEP (SEU and SEL) natural space environment concerns are related to (1) protons from the Earth's trapped radiation belts and solar flares, and (2) heavy ions from solar flares and galactic cosmic rays. The SEP testing is performed in accordance with the general guidelines established by the EIA/JEDEC standard EIA/JESD57.[15] This test method defines the requirements and procedures for Earth-based single-event effects testing of integrated circuits and is valid only when using a Van de Graaff or cyclotron accelerator.

A SEP test consists of irradiating a given device with a prescribed heavy ion beam of known energy and flux in such a way that the number of SEUs or latchup can be detected as a function of the beam fluence (particles/cm^2). For both SEU and latchup, the beam LET, which is equivalent to the ion's stopping power dE/dx (energy/distance), is the basic measurement variable. The SEU error rate is expressed in terms of critical charge (Q_c), which is related to the circuit design and device cross section (i.e., geometry of the sensitive volume). Q_c can be represented by a threshold LET, which is minimum charge per unit length that must be deposited in the sensitive volume to cause upset or latchup.

A full device characterization requires irradiation with beams of several different LETs, which requires changing the ion species, energy, or angle of incidence with respect to the chip surface. The end product of SEP testing is a plot of the device upset rate (or cross section) as a function of the beam LET.

The SEU testing is performed using the in-flux test method and requires a source of energetic heavy ions. The cost, availability, and ion/energy capabilities are all important considerations when selecting a test facility. The following are some of the commonly used heavy ion sources for conducting SEP testing:

- Cyclotrons
- Van de Graaff accelerators
- Synchrotrons
- Fission sources
- Alpha emitters

Cyclotrons provide the greatest flexibility, since they can supply a number of different ions (including alpha particles) at a finite number of different energies. This allows the selection of ions with adequate penetration range in the device. The cyclotrons used for SEU testing may have accelerating potentials of up to 300 MeV (or even higher) and use gaseous ion species such as krypton, argon, oxygen, and neon. The major disadvantages of the cyclotrons are expensive test setup and downtime associated with ions source replacement and changes of ion energy.

Van de Graaff accelerators offer relatively faster change of ion species and are less expensive to operate than the cyclotrons. However, they have limited energy range and function. A less expensive available simulator uses fission products from californium Cf-252 that have LET distributed primarily in the range of 41 to 45 MeV/[cm^2/mg]. These simulators are primarily used to determine the device limiting cross sections, and test data gathered is used as a characterization tool to establish upset/latchup thresholds for circuit design. There are other experimental techniques that use short pulses of highly focused laser beams as a research tool for the investigation of single-particle ionization effects.[16,17]

In a cyclotron, the SEU testing is performed in an evacuated chamber. The delidded part, such as a programmable logic device in a test socket, is mounted on a platform that can be rotated so that the angle of incidence between the ion beam and the chip surface can be changed. The device under test (DUT) mounted on the test fixture is connected through interfacing cables to a power supply and tester outside the chamber. Figure 7.4*a* shows a schematic version of an SEU test setup, and Fig. 7.4*b* shows the typical vacuum chamber.[15]

Listed below are some of the definitions and terminology involved in SEP testing and bit error rate calculations:

- *Fluence.* Fluence is the ion flux integrated over the run time and is expressed in units of ions/cm^2. Therefore, flux is the number of ions passing through one cm^2 area normal to the beam per unit time, measured in (ions/cm^2) · s.
- *LET.* LET is the amount of energy lost per unit length as the ion travels through a material. The most commonly expressed unit is MeV/[cm^2/mg]. LET$_{eff}$ is the LET modified to account for change in path length of an ion through the sensitive volume when the angle of incidence of the ion is not normal to the surface.

$$LET_{eff} = LET\ (0)^{\circ}\ X\ Sec\ \theta$$

where θ is the angle of beam with respect to the normal (to the DUT surface). LET$_{th}$ is the LET required to cause a SEE. LET$_{sat}$ is the minimum observable cross section and, in a non-RH part, appears as the asymptotic upper section of the LET versus cross section curve.

- *Cross section (σ).* Cross section is the device SEE response to ionizing radiation. A cross section is defined as the number of events per unit fluence. Assuming the depth of sensitive volume is small, compared to the lateral dimensions, the SEE cross section (σ) can be expressed by following relationship:

$$\sigma = \text{number of errors (bit upsets)/fluence} \times \cos$$

The cross section is usually measured in cm^2/device or μm^2/bit. The saturated (or asymptotic) cross section will have an area equal to the sum of all sensitive areas of the device, provided one heavy ion induces one upset per strike in the device sensitive volume.

A programmable logic device is tested for SEU by populating it with certain design functions using the available programming elements such as antifuses in Actel FPGAs and RAM configuration bits in look-up tables (LUTs) based Xilinx devices, and then monitoring them to detect any bit errors. LET can be varied either by changing the ion species or the angle of incidence. The particle flux (or beam current) is varied such that the bit-error rate (BER) is compatible with the speed of the tester to ensure that all errors are counted. A common practice is to increase the LET through a sensitive volume by varying the angle of incidence so that LET$_{eff}$ = LET × Sec θ. Then, at each LET$_{eff}$

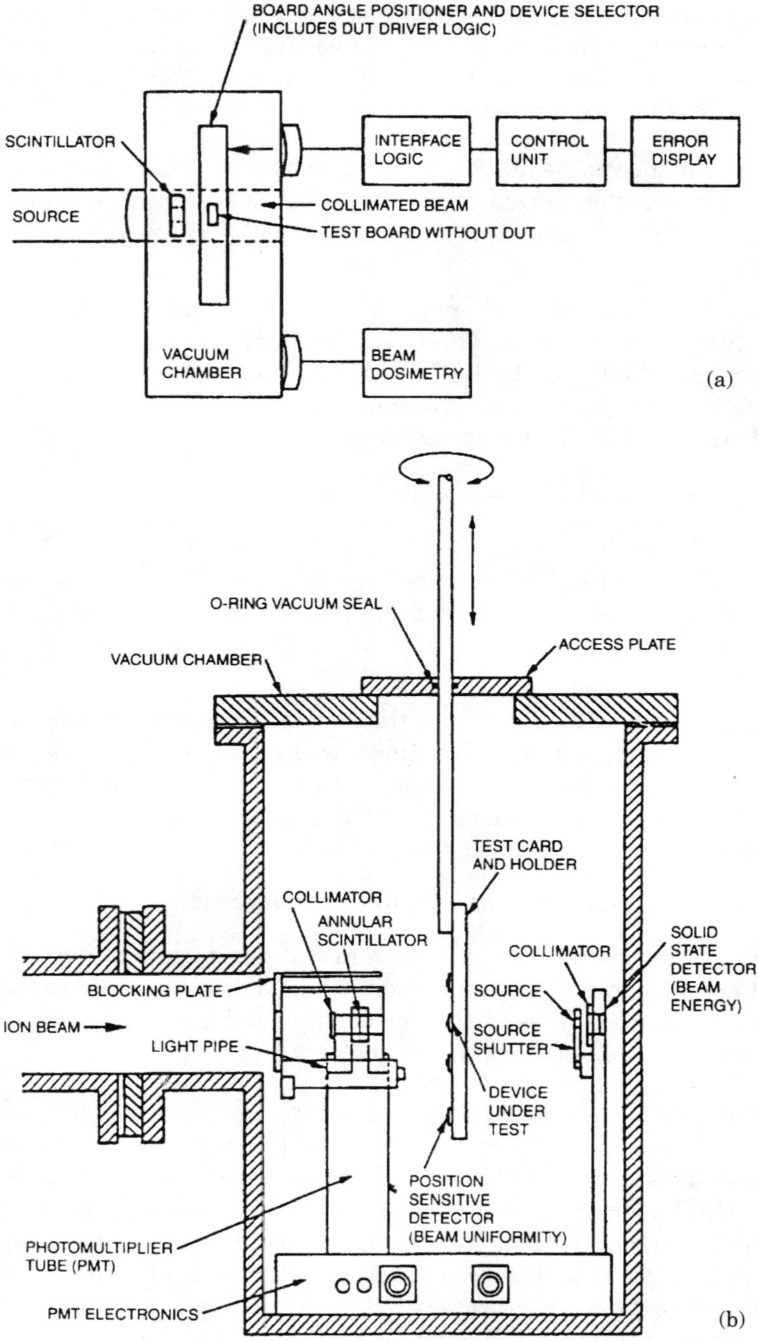

Figure 7.4 (a) Schematic overview of SEU test setup and (b) typical test chamber *(from Ref. 13)*.

value, the BER is measured by counting the statistically significant number of errors (bit upsets). Then, the error cross section is the total number of errors divided by particle fluence.

For heavy ion-induced latchup testing of programmable logic devices, the test circuit is designed to prevent latchup-induced burnout, and the power is interrupted after each latching event. The range of heavy ion fluxes is established to provide the valid dosimetry and typical flux range is 10E2 to 10E5 (ions/cm^2) · s. The total number of particles (fluence) must be adequate to establish with high statistical confidence that all sensitive volume in the DUT has been irradiated. Typically, a fluence of 10E6 ions/cm^2 for soft devices and 10E7 ions/cm^2 for hard devices is used. The range of ion species selected to test a given device is determined by the LET threshold of the device.

In general, the heavier ions produce larger LET and are typically used first to determine the upper section of LET curve. Then, the particles with lower LETs are selected to determine the shape and threshold of the curve. Also, the ion specie is selected such that the penetration depth of the ion is large, compared to the depth of the charge collection region, to ensure that the beam LET is nearly constant while the ion traverses the charge collection region of the DUT. If possible, the data should be taken to 2× the LET required for the cross section to saturate to ensure that the cross section is saturated. If this level is not possible, data should be taken up to a LET_{eff} of 120 MeV [cm^2/mg]. When the DUT upset occurs during the initial run, the following are some of the steps typically recommended during testing of PLDs:

- Vary flux/fluence to get a meaningful number of device upsets (typically at least 100) or a fluence of 10E6 ions/cm2, whichever comes first.
- Repeat runs (two or three times, as required) to establish repeatability and beam stability.
- Vary the beam angle to get additional data points on the cross section versus LET plot.
- Change DUT biasing configuration and/or operating parameters such as clocking rate, input data patterns, and so on.
- Change DUT temperature (if applicable).
- Select another ion specie to introduce a new range of LETs.

When the device does not upset in the initial run, some of the parameters varied to obtain upsets are

- Fluence increase (typically 10E7 particles/cm^2 maximum)
- Beam angle increase

- Variation of bias and/or operating parameters
- Selection of higher LET ion species

Figure 7.5 shows a sample heavy ion upset curve with cross section expressed as a function of effective LET.[18]

7.2 Radiation Test Data for FPGAs

Actel FPGAs have been particularly attractive to the designers for space applications because of their antifuse based technology, which is relatively more radiation tolerant than the SRAM- or EPROM-based devices. Therefore, Actel FPGAs have been extensively characterized for both total dose and SEP by various organizations including NASA and the aerospace industry. Actel FPGAs technology and architecture were discussed in Chap. 3. Actel first introduced ACT 1 family and device A1020 with 2000 gates density that used a 2.0 µm process. This was followed by the enhanced versions A1020A with 1.2 µm and A1020B with 1.0 µm process technologies. Similarly, the higher-density ACT 2 family device A1280 with 8000 gates was originally a 1.2 µm design that went through a die shrink to provide A1280A version with 1.0 µm minimum feature size. The 1280XL integrator series devices have gone through additional die shrinks to provide minimum feature sizes of 0.8 µm and 0.6 µm. ACT 3 family devices ranging from A1460A with 6000 gates, to device A14100A with 10,000 gates, have minimum feature sizes of 0.8 µm.

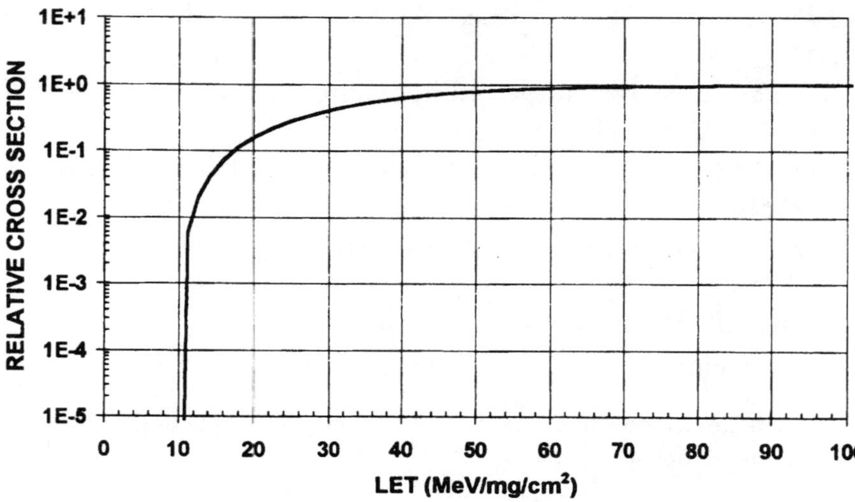

Figure 7.5 A sample heavy ion upset curve with cross section expressed as a function of effective LET *(from Ref. 18)*.

Actel has been using mainly two wafer fabrication foundries for its ACT 1 and ACT 2 FPGA product families: TI and Matsushita Electronic Corp. (MEC) of Japan. However, the newer Actel products such as ACT 3 devices, XL, and 3200 DX families are using multiple wafer source foundries. A large variation in radiation tolerance has been observed for Actel devices, both in total dose as well as SEU sensitivity, depending on the device family, product type, feature size, and the wafer foundry origin. Both total dose and SEE testing has been performed on various FPGA families from Actel, and some limited testing on Atmel, Lucent Technologies (formerly AT&T Microelectronics), and Xilinx devices. The total dose and SEE test results only for Actel FPGAs will be presented in the following sections.

7.2.1 Total ionizing dose test results

Total ionizing dose (TID) characterization testing has been performed on both Actel ACT 1 A1020B and ACT 2 family A1280A devices for various NASA and aerospace industry programs using Co-60 source. It should be noted that these parts are from manufacturer's standard military/space and commercial process flows, which are not guaranteed to be radiation hard. The results for two of these series of tests are reported below.[19] The term rads or krads used are abbreviated forms of rads(Si) and krads(Si).

Test set #1. This TID evaluation was performed on both Actel A1020B and A1280A parts, each part type from two different wafer lots. Some of these parts had been burned-in during screening, whereas the others were non-burned-in parts. During each radiation test, six parts (consisting of three burned-in and three non-burned-in) were irradiated under static bias, and two parts were used as control samples. The irradiation levels ranged from 5 to 75 krads, with intermediate exposures at 10, 15, 20, 30, and 50 krads, depending on the device type and level of degradation experienced. The dose rates were setup between 0.06 to 1.5 krads/hour, depending on the total dose exposure level. After the final exposure, the parts were annealed at room temperature (25° C) for 168 hours.

The blank FPGAs were programmed with test chip pattern consisting of (1) combinatorial logic, (2) flip-flops, (3) input/output latches, and (4) shift register and counter. In some cases, special measurements of power supply currents were made during irradiation, using a strip chart recorder. The start-up currents were measured between irradiation steps using a current probe and oscilloscope.

Electrical measurements included various dc and ac parametric tests such as input leakage currents (ICCL, ICCH), output voltages

(VOL, VOH), propagation delays (TPLH, TPHL), and so forth. A functional test vector set was developed from design and simulation files, using automated test pattern generation (ATPG) software to run on a Sentry (S-50) ATE system. Six functional tests (FUNCT1 through FUNCT6) were performed at V_{CC} = 4.5, 5.0, and 5.5 V—three of these at 1 MHZ and other three at 5 MHz. These electrical tests were performed initially and during each interim step following irradiation as well as for post-annealing final measurements.

The total dose results on A1020B showed the parts passing functionally up to 10 krads and failing between 15 to 30 krads, depending on the device wafer lot date code. At 50 krads exposure, there were significant increases in ICCL and ICCH. These parts showed significant recovery upon annealing at room temperature. The A1280A parts passed all functional tests up to 10 krads and exhibited first functional failures at 15 krads and above. However, the A1280A showed significantly higher increases in ICCL and ICCH (3 to 10 times more than the specification limit of 20 mA) at 10 krads and above and, unlike A1020B parts, they showed no significant recovery upon annealing at room temperature. No significant differences in functional failure levels and/or increases in ICCL/ICCH leakage currents were found for burned-in versus non-burned-in parts. Figure 7.6 shows the typical ICCL and ICCH curves for one of the A1020B wafer lots.

Test set #2. This total dose evaluation was performed only on A1280A FPGAs, four device types with SN 2, 3, 4, and 5. The total dose irradiation levels were 3, 4, 5, 6, and 7 krads, and dose rate was 0.059 krads/hour. The test pattern and electrical measurements performed including functional tests were similar to that for Test set #1.

After the 3, 4, and 5 krads irradiation, significant increases were observed in both ICCL and ICCH, although all irradiated parts continued to remain within specification limits. After 6 krads and the following exposure, all four parts exceeded ICCL and ICCH specification limits, as shown in Fig. 7.7. The dc/ac parameters other than supply current were relatively less sensitive to the total dose and usually remained within specification limits. These included input leakage currents, output voltages, and various propagation delays. All irradiated parts showed some recovery during subsequent annealing steps.

In summary, during some tests, the older version of A1020 (2.0 μm feature size) has shown significantly better radiation tolerance to 100 krads exposure and above. However, the radiation tolerance of Actel A1020B (1.0 μm feature size) devices tested varied from 5 to 30 krads, depending on the device lot date code. Also, the failure mode for the A1020B was functional whereas, for the A1280A devices, the first fail-

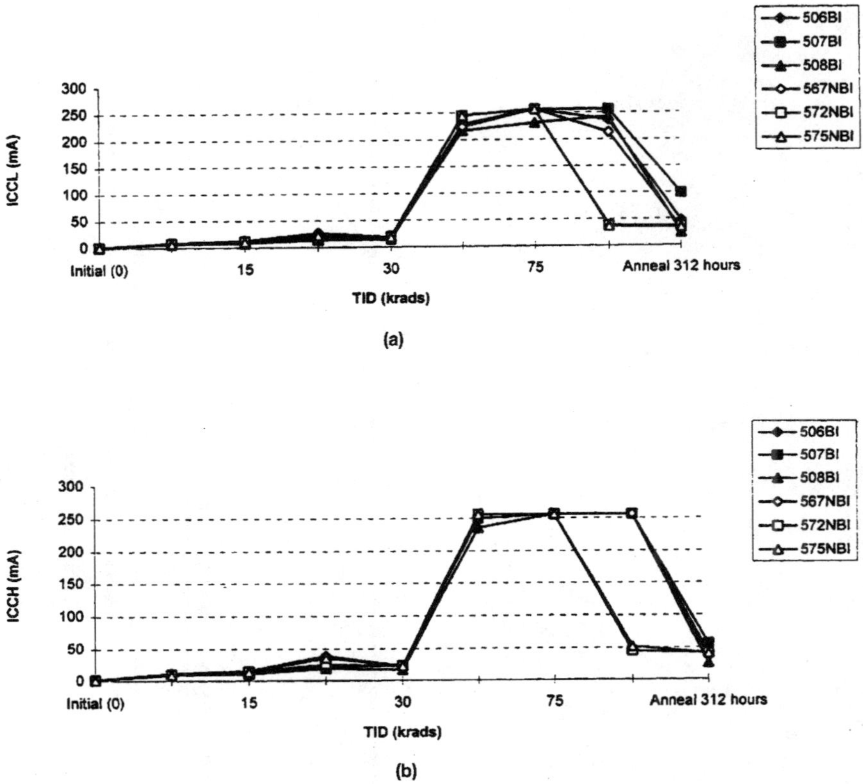

Figure 7.6 Typical TID versus (a) ICCL and (b) ICCH curves for one of the A1020B wafer lots tested *(from Ref. 19). Note:* "BI" suffix after part serial number indicates that the part was burned-in during screening, whereas "NBI" indicates non-burned-in parts.

ures occurred due to large increases in ICCL and ICCH. A study has shown that total dose degradation affects the charge pump circuit in A1280A, causing large increases in power supply current and raising the minimum operating voltage.[20]

Thus, commercial non-RH FPGAs may be used is space applications where the total dose radiation requirements are quite low. However, due to significant lot-to-lot variations, it is recommended that the sample parts from the flight lot wafer procurement be characterized for their radiation tolerance.

7.2.2 Single-event effects (SEE) test data

The SEU testing performed on Actel A1020 FPGAs have shown the sequential (S) modules logic building blocks as being more sensitive in comparison to combinatorial (C) modules. Therefore, the flip-flops made from C-modules are relatively harder to SEU effects than those

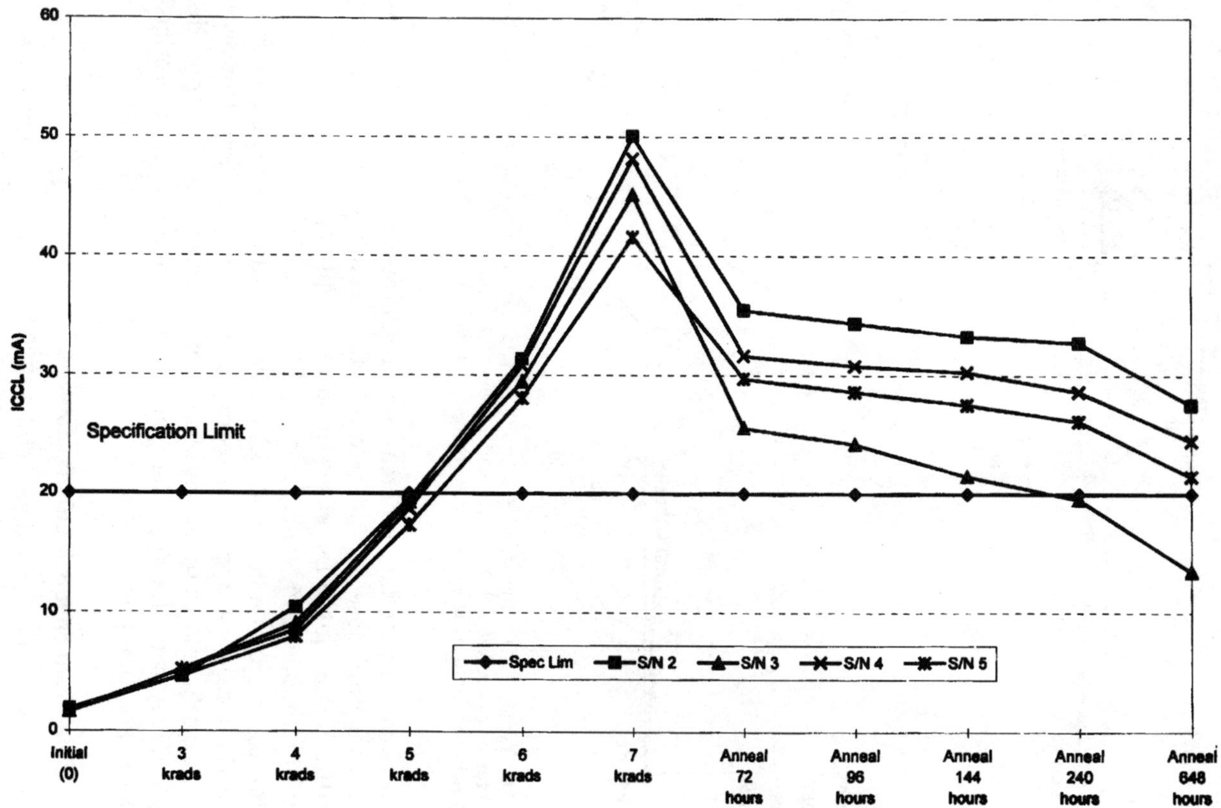

Figure 7.7a Typical TID versus ICCL curves for A1280A devices tested.

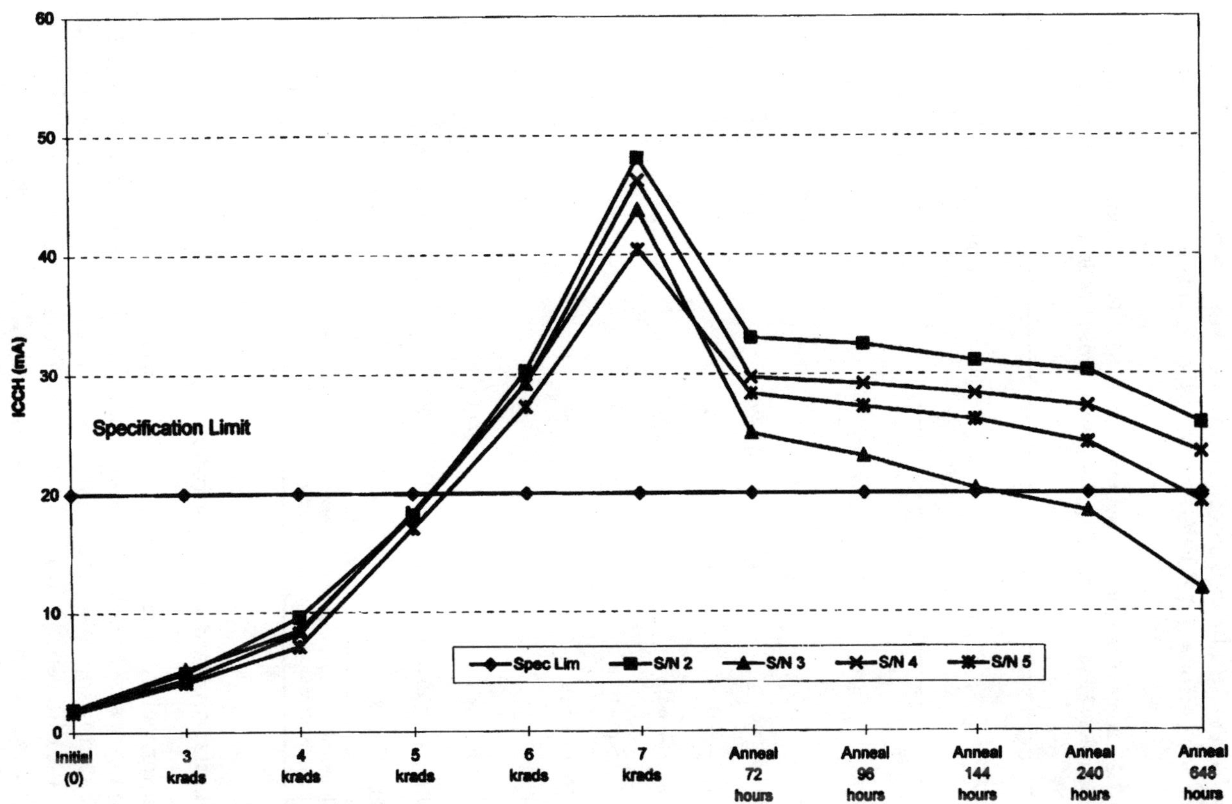

Figure 7.7b Typical TID versus ICCH curves for A1280A devices tested.

made from S-modules. In general, the review of existing SEE test data literature on Actel FPGAs shows that latchup has not been a problem, except for some A1020B test lots that showed moderate SEL susceptibility. Table 7.2 shows some examples of representative SEU performance for various Actel FPGA families, device types and test data.[20]

TABLE 7.2 Some Examples of Representative SEU Performance of Various Actel FPGA Families, Feature Sizes, LET_{th}, Saturation Cross Section, and Test Temperatures *(from Ref. 20)*

Device type	Feature size (μm)	SEU LET_{th} MeV/[cm^2/mg]	Sat. cross section cm^2/flip-flop or RAM bit	Test temp.
A1020	2.0	25	5×10^{-6}	R → 100° C
A1020A	1.2	25	3×10^{-6}	R
A1020B	1.0	28	2×10^{-6}	R
A1280 (C-mod)	1.2	23	3×10^{-6}	R → 100° C
A1280 (S-mod)	1.2	5	8×10^{-6}	R → 100° C
A1280A (C-mod)	1.0	28	2×10^{-6}	R
A1280A (S-mod)	1.0	5	8×10^{-6}	R
A1460A (C-mod)	0.8	~30	$\sim 2 \times 10^{-7}$	R
A1460A (S-mod)	0.8	~8	1×10^{-6}	R
A1460A (I/O mod)	0.8	~8	2×10^{-6}	R
A1460A (C-mod/3.3 V)	0.8	~25	8×10^{-7}	R
A1460A (S-mod/3.3 V)	0.8	~6	2×10^{-6}	R

Note: R indicates that the testing was performed at room temperature (25° C), and R → 100° C implies a testing range of room temperature to 100° C.

7.3 Radiation Hardness Assurance (RHA) for FPGAs

One approach for reducing the SEU susceptibility of ASICs as well as certain types of FPGAs is based on the fact that memory elements can be created using many different types of D flip-flops and the observation that different D flip-flop types display a wide range of SEU susceptibilities. Once the susceptibilities of various flip-flop types have been determined, the more SEU-tolerant flip-flop types can be selected for use in critical areas. Also, simple error detection and correction (EDAC) circuits can be incorporated into FPGA based designs.

Another method for enhancing the SEU tolerance of flip-flops is through a circuit topology that uses triple modular redundancy (TMR), feedback, and a compact voter system for EDAC. Figure 7.8 shows the basic D flip-flop TMR configuration, which consists of three flip-flops (FF1, FF2, and FF3) that redundantly store the input for D.[21] The output Q is a majority function of three flip-flops, i.e., "the best two out of three wins." Therefore, it two select inputs agree, then these inputs represent the answer. Thus, D0(00) is wired to ground, and select D3 is wired to VCC(11). If two select inputs disagree, then the third flip-flop breaks the tie. The select D1(01) and select D2(10) are wired to the output of the third flip-flop.

An optional error monitoring circuit (also shown in Figure 7.8) uses gates G2 and G3 to provide an error signal. The principle of operation of this circuit is similar to that of the voter. Thus, if the first two outputs differ, then the output of the multiplexor is hard wired to a "1", signifying an error condition. If the two outputs agree, then there may be an error only if there is a miscomparison.

Some other RHA approaches for PLDs in space applications include the use of radiation hardened PALs and FPGAs or the use of a shielded package. These are discussed in the following sections.

7.3.1 Radiation hardened PALs and FPGAs

Programmable array logic (PAL) devices such as the popular 22V10 were discussed in Chap. 2 in terms of general architecture, features, and vendor families. Most of the commercial PALs and FPGA available are not designed to be radiation hardened for space applications. However, some bipolar PALs, being inherently more radiation tolerant than their CMOS counterparts, have been used in space applications.

An example of PAL specifically designed for space applications is United Technologies Microelectronics Center (UTMC) Systems, Inc., UT22VP10 Universal RADPAL™. These PALs are built on a triple-level metal 1.2 µm CMOS process, with the key technology being the nonvolatile programming element. The UT22VP10 uses a planar amorphous silicon antifuse to provide a low-impedance connection between the metal layers. It is based on the familiar sum-of-products (AND-OR) logic structure that is complemented with a programmable macrocell. It has 12 dedicated inputs and 10 I/Os to provide up to 22 inputs and 10 outputs for creating logic functions. The core architecture is based on a one-time programmable antifuse AND array that drives a fixed OR array. Figure 7.9 shows a block diagram of UT22VP10 architecture, and macrocell.[22]

The UT22VP10 architecture allows implementation of up to 10 sum-of-products expressions. The user specifies whether each of the potential outputs is registered or combinatorial. Also, output polarity

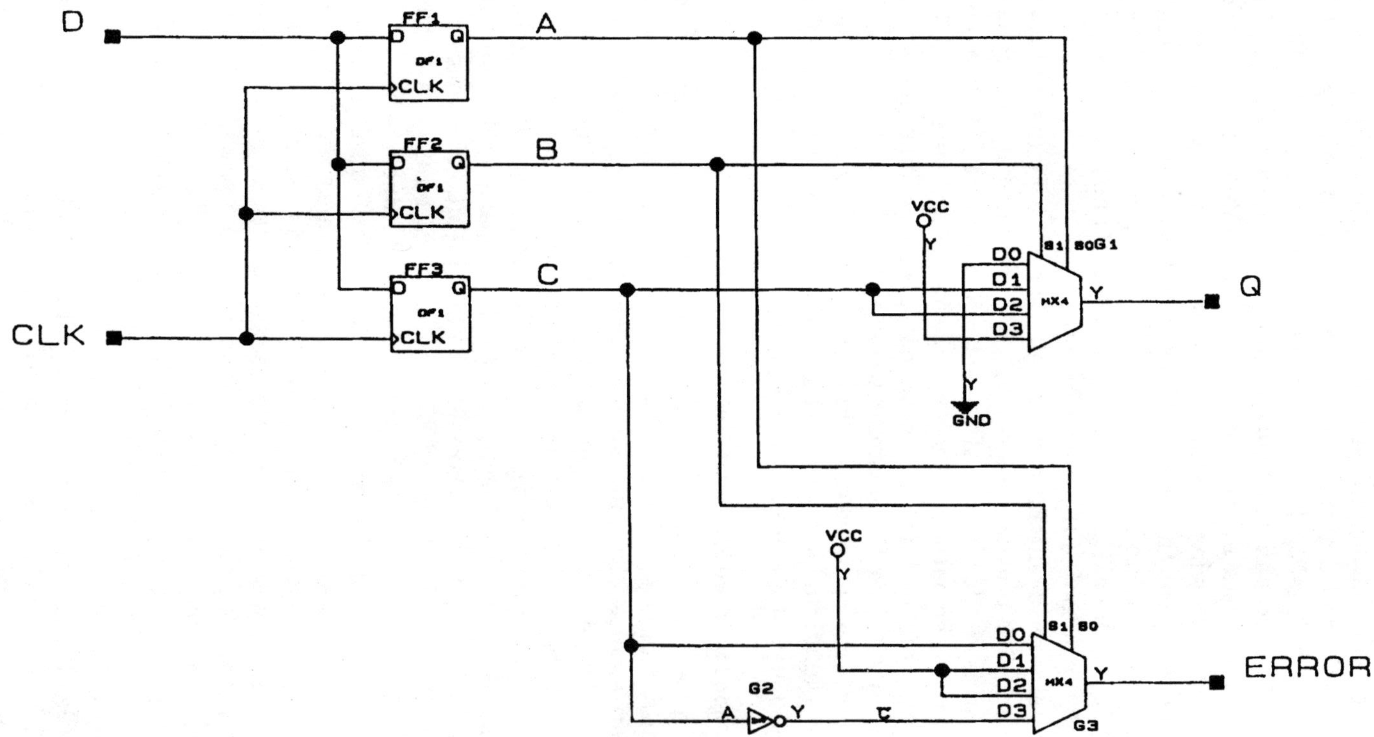

Figure 7.8 Basic D flip-flop triple modular redundancy (TMR) configuration including an optional error monitoring circuit *(from Ref. 21)*.

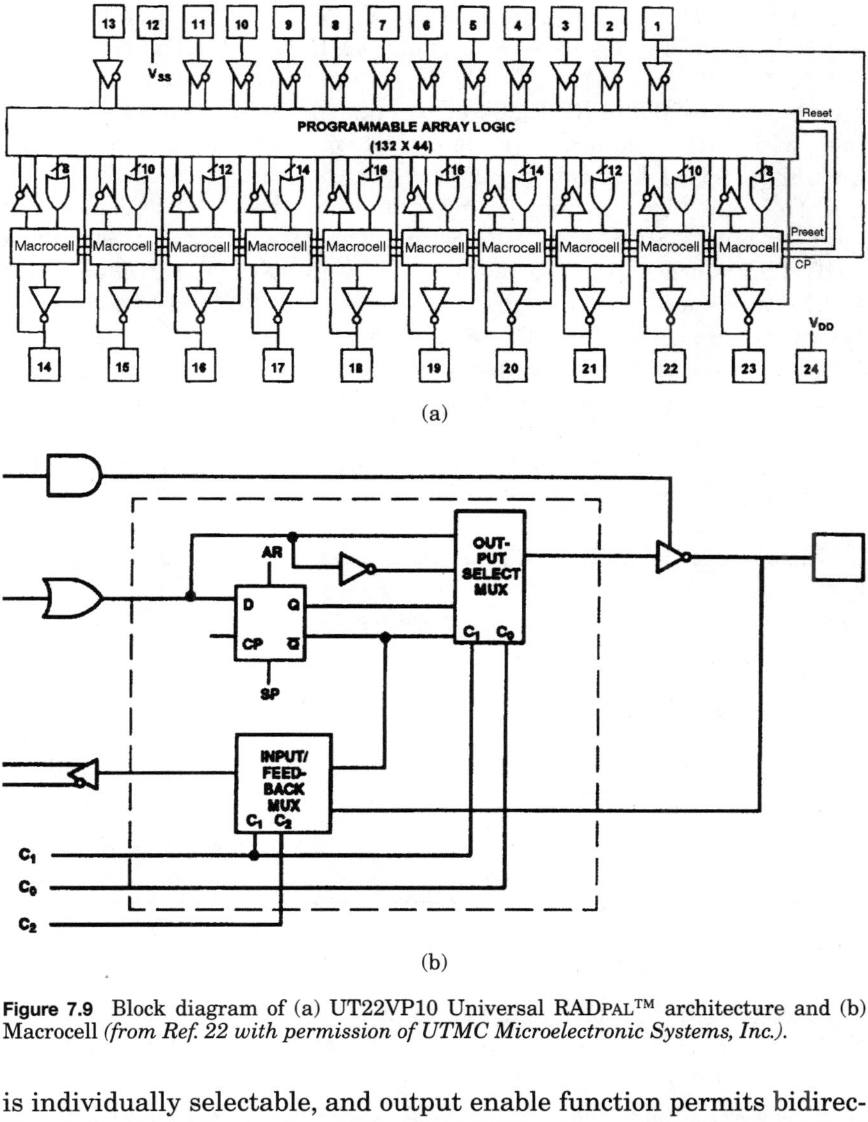

Figure 7.9 Block diagram of (a) UT22VP10 Universal RADPAL™ architecture and (b) Macrocell *(from Ref. 22 with permission of UTMC Microelectronic Systems, Inc.).*

is individually selectable, and output enable function permits bidirectional configuration on an individual basis. A macrocell is associated with each of the 10 OR functions and provides control over each of the outputs. These devices are specified for the following key radiation tolerance characteristics:

Total dose:	1×10^6 rads(Si)	
SEU LET threshold:	>50 MeV/[cm^2/mg]	@V_{CC} = 4.5 V
Latchup:	LET > 109 MeV/[cm^2/mg]	@V_{CC} = 5.5 V; T = 125° C

LockHeed Martin Corp. (Manassas, Virginia), in partnership with Actel Corporation, has developed radiation hardened (RH) versions of Actel A1020B (1.0 μm) and A1280XL (0.8 μm). These are being designated as RH1020 and RH1280 and are pin-to-pin functionally compatible with Actel non-RH versions. All manufacturing operations, including device fabrication, assembly, and testing, are done at the LockHeed facility, and the parts are built on the radiation-hardened epilayer bulk CMOS process (0.8 μm). The base layers are standard RHCMOS process with the addition of high-voltage transistors and ONO antifuses of Actel designs. These parts are supplied by Actel Corporation and are available to standard military drawings (SMDs).

The RH1280 devices have been characterized for total dose as well as single-event latchup (SEL) testing over a temperature range (T_c) of 55 to 125° C using V_{cc} = 5.5, and for single-event upset (SEU) over T_c of 55 to 125° C using V_{cc} = 4.5 V. These parts are specified for following key radiation tolerance characteristics:

Total dose:	300 krads (maximum)	
SEL:	No latchup at LET_{eff} of 177 MeV/[cm^2/mg]	
SEU:	Combinatorial (C-mod)	
	LET_{eff} (no upset)	17 MeV/[cm^2/mg]
	Max device cross section	320 μm^2/bit
	(LET = 120)	
	Sequential (S-mod)	
	LET_{eff} (no upset)	4 MeV/[cm^2/mg]
	Max. device cross section	110 μm^2/bit

7.3.2 RAD-PAK shielding technology

RAD-PAK® is an IC packaging concept provided by a company called Space Electronics Inc. (SEI) and used for shielding of semiconductor chips from total ionizing dose (TID) effects.[23] This approach often allows use of commercial devices (typically radiation *soft*) in applications that specifically require hardened devices.

A key aspect of RAD-PAK technology for shielding from electrons is the use of high-Z materials (where Z is the atomic number) such as a tungsten layer integrated into the package lid. High-Z materials are significantly more effective (per gm/cm^2) than low-Z materials, due to the large angle scattering that occurs in the high-Z material.

SEI RAD-PAK approach has been used for Actel A1020 and A1280 FPGAs in some space hardware flown for NASA programs. However, it should be noted that, while the RAD-PAK may provide an effective solution in a total dose environment, it does not provide any shielding against high-energy protons and heavy ions that may cause SEUs and latchup. In an application where critical factors include total dose, SEU, and latchup considerations, radiation-hardened FPGAs such as the ones produced by LockHeed Martin/Actel Corp. (RH1280/RH1020) are more suitable alternatives.

References

1. Ashok K. Sharma, *Semiconductor Memories: Technology, Testing and Reliability,* Chap. 7, Semiconductor Memory Radiation Effects, IEEE Press, 1996.
2. UTMC Technical Description Notes.
3. W. N. Spjeldvik and P. L. Rothwell, "The Earth's radiation belts," AFGL-TR-83-0240, Air Force Geophysics Lab., Hanscom AFB, MA, Sept. 29, 1983.
4. J. Barth and E.G. Stassinopoulos, "Ionizing Radiation Environment Concerns," Section 3 in *Single Event Effect Criticality Analysis (SEECA)* document, version 1.0: Feb. 15, 1996.
5. D.M. Sawyer and J.I. Vette, "AP-8 trapped environment for solar maximum and solar minimum," *NSSDC / WDC-A-R&S,* 76-06, Dec. 1976.
6. J.H. Adams, et al., "Cosmic ray effects on microelectronics, Part I: The near Earth particle environment," NRL Memorandum Report 4506, Naval Research Laboratory, Washington D.C., 1981.
7. J.A. Simpson, et al., "The experiment for high energy nuclei composition (ONR-604), in CRRES/SPACERAD Experiment Descriptions", M.S. Gussenhoven, E.G. Mullen, and R.C. Sagalyn, eds., *Air Force Geophysics Laboratory Report AFGL-TR-85-0017,* Hanscom AFB, MA., p. 163, 1985.
8. E.G. Mullen, et al., "CRRES microelectronics test package (MEP)," *IEEE Trans. Nucl. Sci.,* NS-40, p. 288, 1993.
9. D. Chennete, et al., "The CRRES/SPACERAD heavy ion model of the environment (CHIME) for cosmic ray and solar particle effects on electronic and biological systems in space," *IEEE Trans. Nucl. Sci.,* vol. 41, no. 6, Dec. 1994, pp. 2332–2339.
10. J.R. Srour, "Basic mechanism of radiation effects on electronic materials, devices and integrated circuits," DNA Tech. Report: DNA-TR-82-80, Aug. 1982.
11. J. F. Ziegler and W. Langford, *Science,* vol. 206, p.776, 1979.
12. J. G. Rollins, "Estimation of proton upset rates from heavy ion test data," *IEEE Trans. Nucl. Sci.,* vol. NS-37, pp. 1961–1965, Dec. 1990.
13. *ESA Radiation Design Handbook* ESA-PSS-01-609, Issue 1, Jan. 1987.
14. MIL-STD-883, "Test Methods and Procedures for Microcircuits."
15. EIA/JEDEC Standard EIA/JESD57: "Test Procedures for the Measurement of Single-Event Effects in Semiconductor Devices from Heavy Ion Irradiation."
16. S. P. Buchner, et al., "Laser simulation of single event upsets," *IEEE Trans. Nucl. Sci.,* vol. NS-34, p. 1228, Dec. 1987.
17. A. K. Richter, et al., "Laser simulation of heavy charged particle tracks using focussed laser beams," *IEEE Trans. Nucl. Sci.,* vol. NS-34, p. 1234, Dec. 1987.
18. Ed Peterson, "Single-Event Analysis and Prediction," Section III of *1997 IEEE NSREC Conference Tutorial.*
19. Ashok K. Sharma, et al.,"Evaluation of Actel A1020 and A1280 Field Programmable Gate Arrays," *NASA Parts Technology Report 64198,* February 1996.
20. Richard Katz, "Programmable Logic Application Notes," Goddard Space Flight Center (GSFC) Electronic Packaging & Space Parts News (EEE Links) 1996–1997.
21. Richard Katz, et al., "SEU hardening of field programmable gate arrays (FPGAs) for space applications and device characterization," *IEEE Trans. Nucl. Sci.,* vol.41,no.6, Dec.1994, pp. 2179–2186.
22. UTMC Microelectronic Systems UT22VP10 Universal RADPAL product description data sheet.
23. Space Electronics Inc. (SEI), *Space Radiation Effects Handbook for RAD-PAK® Microelectronics.*

Chapter

8

Advanced PLD Technologies and Architectures

8.1 Intellectual Property (IP) Cores and Megafunctions

8.1.1 Intellectual property (IP) cores

In general, cores are predesigned and preverified functional blocks that can accomplish a well defined task in an ASIC- or PLD-based system. There are companies, known as third-party intellectual property (IP) providers, that supply users with the cores that are usually designed for ASICs but are becoming available now for PLDs as well. A major reason for this trend is the rapid increase in logic density of programmable logic devices that can allow on-chip integration of an IP block and, at the same time, leave enough room for other custom-designed logic implementation. It means that an IP core has become a higher-level building block in a more complex design.

The form in which the core is provided determines the entire design methodology.[1] The simplest and least risky from the user's point of view are *hard cores,* which have the advantage of proven layout coupled with accurate timing specifications that can be easily *dropped* into the chip. Hard cores are fully diffused blocks—the actual physical layout of logic functions such as a microcontroller, DSP, and so on. These devices are identical to the packaged semiconductors, except that they lack I/O pins and a package. These cores have already been optimized for compactness and density by their successful use in other high-performance applications. The well defined timing parameters of hard cores makes them more applicable for timing critical applications such as high-performance processor engines and high-speed I/O functions.

A hard core or a block cannot be changed. It should fit into an existing design the same way an IC fits onto a printed circuit board. During simulation, the hard core appears as a *black-box* with stimulus-in and response-out. The cores are typically the property of large semiconductor companies such as Motorola, TI, and others, who either supply them directly or make them available from a licensed second source. However, a number of smaller companies are emerging as independent suppliers of hard macros.

In contrast to the hard cores, *soft cores* are more flexible and the *soft* refers to a lack of geometric boundaries for a logic function.[2] A soft core description is commonly at a behavioral or register-transfer level (RTL) and almost always in a hardware description language (HDL) such as VHDL or Verilog. Some vendors offer soft cores in gate-level or netlist descriptions, which are encrypted to protect the intellectual property (IP) rights. The soft cores are designed to meet minimum performance specifications over a range of technology implementations. The files also come with test benches. Since the file contains HDL statements, the designers can synthesize the code along with other logic in the circuit being designed. However, the drawback to testing the core after it has been synthesized with other logic in the design is that the accompanying test bench is applicable only to the core-level testing. This means that, after synthesis, the test bench must be modified to verify the new logic as well.

A soft core's technology independence provides the flexibility to have the devices fabricated at any foundry. In contrast, if the logic is implemented as a hard core, one needs to redesign and reverify the logic for each targeted technology. A soft core requires just simulation and timing verification after synthesis to a target technology.

Figure 8.1 is flowchart of the relationship between the cores, which represent the actual design implementation and the models used for design verification. In a soft core based design, one needs not only a behavioral or an RTL description of the desired logic, but also a core model to simulate the logic's behavior and a test program (or test bench) for core verification after the implementation in a given technology. A core described in VHDL or Verilog is not automatically synthesizable. If one is using a soft core at the behavioral level or RTL, a synthesis script is needed from the core vendor for synthesizing the core to a gate level. The base line requirement for a synthesis script is that it should read all needed RTL files, perform some basic optimizations, and provide the resultant netlist. A good synthesis script does not require repeated synthesis runs or manual modification of the netlist to insert delay elements or fix other timing problems. In addition, many core vendors provide synthesis scripts with following features:

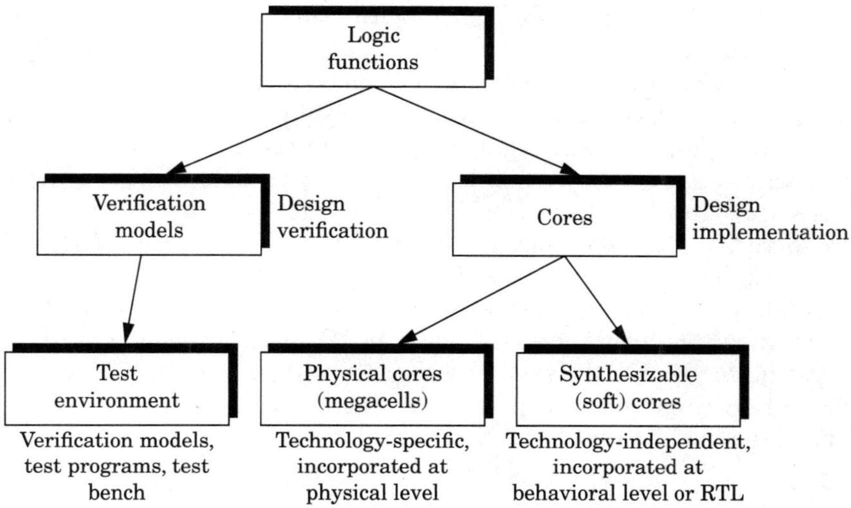

Figure 8.1 Flowchart showing relationship between the cores used for actual design implementation and design verification models *(from Ref. 2).*

- Allowable use of variables or parameters for attributes that a designer may need to modify, such as target library name and timing requirements on the interface between the core and supporting logic
- Full specification of timing requirements on standard interfaces
- Synthesis commands to satisfy all setup and hold-time requirements on standard interfaces
- Automatic synthesis to check file dependencies and rerun synthesis tools only when input files change

A soft-core model lacks interconnect parasitic delay information and, thus, the accurate timing information. Some vendors provide a bus-functional model that simulates the core's behavior at the pins without modeling the core's internal configuration. This bus-functional portion can be used to model interaction between the core logic and the external interface circuitry. The behavioral or RTL simulation verifies functionality but not performance. The timing performance can only be verified after synthesizing the core to the gate level for a target technology.

A *firm core* is a cross between a hard and a soft design. It is a synthesized gate-level description of some function. Therefore, like a hard core, the designer cannot change the circuit. However, the device is similar to a soft core in that it can be simulated, placed, and routed with the surrounding logic. The firm core also provides the flexibility of fabrication foundry selection.

Table 8.1 shows the various IP categories (including the algorithmic, soft, firm, hard) and their general characteristics.[3] The system designers require IP building blocks to have been fully tested and verified before they can integrate them into their designs. Since the IP testing and verification process is complex and time consuming, the vendors are beginning to create system-level evaluation models (SEMs). These SEMs are algorithmic/behavioral versions of IP that the designers can use to verify the system-level performance and functionality within their designs. A major advantage of using algorithmic/behavioral representation of IP components is that they can be more easily verified within a system-level context. Currently, IP and EDA vendors are cooperating to provide the system designers with increased IP test and verification capabilities, especially in the domain of SEMs for soft, firm, and hard IP cores.

TABLE 8.1 Various IP Categories, Including Algorithmic, Soft, Firm, Hard, and Their General Characteristics *(from Ref. 4)*

IP category	Representation	Characteristics
Algorithmic	C/C++ algorithms	Most flexible; technology and architecture independent Fast system-level simulation Describes component functionality only High IP security risk
Soft	Microprocessor-targeted C code RTL HDL netlists	Flexible, technology-independent mapping Slow system-level simulation Describes hardware architecture function but not low-level performance High IP security risk
Firm	Assembly-code algorithms Gate-level HDL netlists Generic technology library Timing models, floorplan	Less flexible Very slow system-level simulation Reliably predicts hardware low-level performance High IP security risk
Hard	Object-code/machine-code Gate-level netlists Technology-specific, process-specific library Timing models, routing Polygon data	Least flexible, technology-specific fab-process mapping Very slow and/or difficult system-level simulation Predicts hardware performance most reliably Low IP security risk

The predefined cores can enable very high-performance and high-density designs. However, a core has to be carefully optimized to the target architecture and subjected to extensive verification testing before it can be released in the market. One of the major goals in IP packaging is to maximize design security while providing flexibility for user customization. A main hurdle in this area has been the lack of open standards in IP packaging/interface protocols. A widely accepted approach to this problem is for the IP vendors and system designers to work with different level representations of IP at different stages in the design. One representation, which contains all the modifiable design elements, is at higher level of abstraction and omits the design information that the IP vendor wants to protect. Another representation, which is synthesized from the first, is protected and not modifiable. This representation contains the IP vendor's protected design information, modified according to the parameter values specified by the system designer using the first representation.[3]

The EDA industry is addressing the IP packaging/interface standards through two major initiatives: the Virtual Socket Interface (VSI) and the Open Model Forum (OMF). In the VSI initiative, which includes most of the major EDA companies and IP vendors, a baseline standard has been proposed that specifies the requirements for *virtual sockets* as a means of integrating the *virtual components* such as system-level macros, cores, and megacells. This proposal also includes the guidelines for delivering design data, enabling IP vendors to produce and maintain a standard set of deliverable documentation along with their IP product. The OMF was established late in 1994 with the goal of developing an industry-standard procedural interface across all system simulators. The interface called the Open Model Interface (OMI) will permit any OMF-compiled IP model to run on any OMF-compliant simulator, regardless of the language in which the IP model is developed (C++, VHDL, or Verilog).

The baseline VSI standard outlines requirements for three different types of IP models: processor, bus functional, and behavioral. The processor models are helpful in hardware/software co-verification. The bus-functional models (BFMs) are mainly used to verify correct component behavior at the block-level interfaces instead of block-internal functionality. The IP behavioral models are used for all classes of IP cores (soft, firm, and hard) to verify nonprocessor (custom logic) IP functionality.

Another key issue in the IP-based designs is the access to IP documentation. This documentation includes all the information system designers need to identify and locate IP blocks for integration into their designs. For example, most DSP core vendors provide information at the functional block level, whereas designers trying to integrate IP from multiple sources into complete systems often need details about the gate-level architecture and timing as well as the core internal design and verification methodology. The VSI data-delivery standards will be helpful in this area, since they will enable designers to compare different IP components using similar data in similar formats. The comparison will enable the designer to identify and evaluate candidate IP building blocks to meet application design requirements.

In a new generation *system-on-a-chip* design approach, the designers will employ a third-party processor core to serve as the CPU and then surround this core with additional third-party IP that represents standard system functions such as universal asynchronous receiver-transmitter (UART), universal serial bus (USB), Viterbi decoders, or ISDN interfaces.[4] Finally, this will be integrated with an additional set of functional blocks that represent the proprietary portion of the hardware design. The software portion of this approach is similar. The proprietary portions of the design is represented by the application layer and firmware drivers, whereas additional firmware drivers and the OS may be provided as IP by third parties.

The IP standards will play an important role in the overall design process, which can be broken down into three basic levels. The first level is the design and verification of hardware at the block level and of software at the module level. The second level is the assembly of these blocks into a virtual prototype, where the entire system's functionality is represented in software for the purpose of analysis and verification. The third level is the creation of physical prototype through hardware emulation techniques that reflect on-chip functionality of the target system.

This three-level approach allows a large and complex design process to be broken down into manageable tasks. At the first level of the process, hardware and software blocks are designed and verified using conventional techniques such as HDL- and gate-level simulation for hardware and host-based debugging operations for the software. At the next level of virtual prototyping, early in the design cycle, one needs to verify that the embedded software under design will interact in a predictable manner with the system hardware—even though the chip-level hardware does not yet exist. Therefore, the entire system must be modeled and must go through the hardware/software co-verification process. This level, which is a hybrid of emulation and simulation, allows the design's various blocks of IP to be exercised using stimulus that represents a real system operation.

The virtual prototyping does not provide a complete picture of how the final system will operate. This requires the ability to exercise the system at speed (or at reduced speed) to reflect the real-world application environment, and accomplished by installing the hardware portion of the design in a hardware emulation system. Both the virtual and physical prototype design levels require standards because of the proliferation of tools to support them.

Figure 8.2 shows the generic flowchart for IP-based design process virtual prototyping environment and availability of existing standards to address specific segments, as well as development of number of other standards that would be required.[4] A key component in the hardware/software co-verification technology is the full functional module (FFM), which executes the firmware instructions sequences in the same manner as the target processor and passes the results to a functional model of the system hardware. There are two types of FFM-behavioral (traditional) and optimized model (OFFM). The OFFM increases simulation performance by separating the function from the timing.

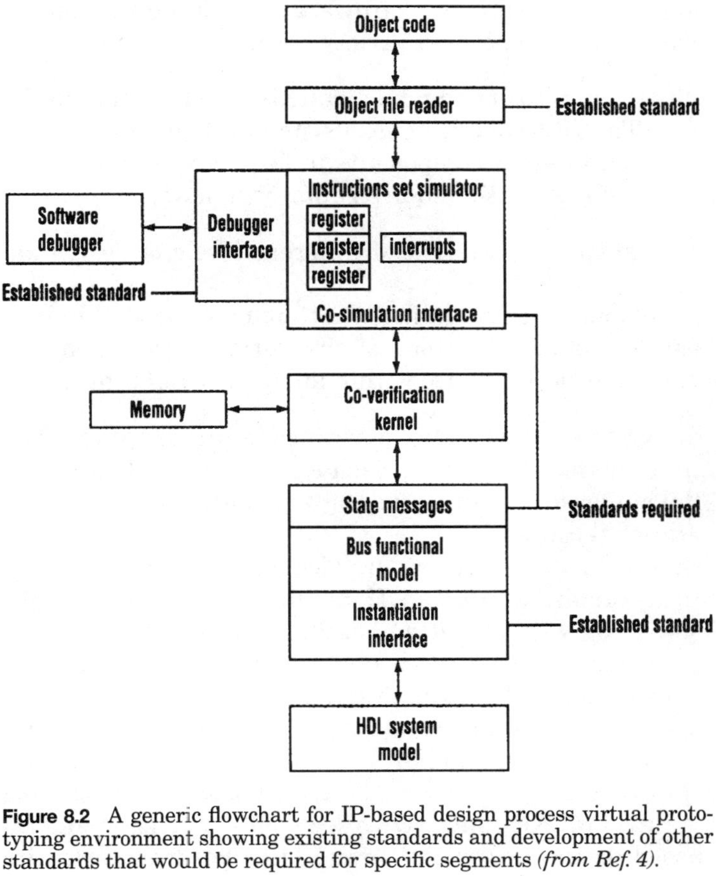

Figure 8.2 A generic flowchart for IP-based design process virtual prototyping environment showing existing standards and development of other standards that would be required for specific segments *(from Ref. 4)*.

The OFFM has following components:

1. An instruction set model (ISM) that is able to process instruction stream in a manner consistent with the internal registers, as well as logic units of the target processor
2. A bus-functional module (BFM), which is able to convert internal processor states into appropriate signals on the processor's external pins

In the case of ISM, there are three possible areas for standardization, as follows:

1. Reading of the object code file representing the target processor instructions and data
2. The debugger interface, which handles command and control of the ISM during debugging operation
3. Interface between the co-simulation kernel, which manages the interface between the software and hardware sides of the simulation, and ISM, which may come from a variety of sources

The availability of standards can make a valuable contribution to the IP-based designs. The IP-based design consists of integrating executable specifications of different components to form complete systems. This approach leads to two basic requirements, as follows:

1. Design component representation at different levels of design abstraction
2. A system development environment that enables the designer to integrate executable specifications of the various system components and test their performance within an operational context

Once the overall system functionality is verified, a process of continuous design improvement is applied to each component, adding implementation detail while continually verifying system-level performance within the operational context.

Figure 8.3 shows the flowchart for Alta Group's Convergence simulation architecture application to an IP-based DSP system design.[3] The designers starting with a written specification use SPW to integrate algorithmic/behavioral IP components and verify their overall system-level performance. Then, the designers use the software and hardware-component executable specifications generated by SPW's code generation system (CGS) and hardware design system (HDS), respectively, to produce C/C++ software or RTL hardware executable specifications. The SPW, CGS, and HDS are capable of reusing and integrating IP-based design in various forms. After hardware/software

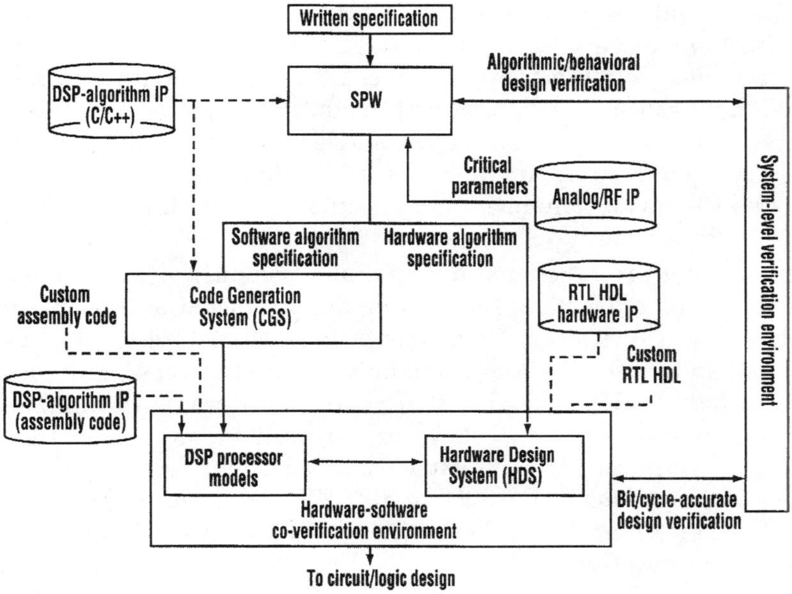

Figure 8.3 Example of an IP-based DSP system design supported by Alta Group's Convergence Simulation Architecture *(from Ref. 3)*.

co-verification, the designer automatically generates RTL Verilog or VHDL for logic and datapath synthesis. The custom logic, which results from logic/datapath synthesis, is finally integrated with a firm or hard model of the DSP core.

Actel recently announced the availability of six cores, out of which three are for the telecommunication industry, designed by its partner, Inicore AG. The first six cores have been optimized to Actel's ACT 3, ACT 3 PCI, 1200XL, and 3200DX families. The selected cores will also be ported to Actel's new ES family of system-programmable gate arrays (SPGAs). The optimized cores will be available as a compiled netlist, VHDL, or Verilog RTL source code. Lucent Technologies is targeting following three areas for cores: bus interfaces, DSP functions, and communication blocks. Gatefield Corporation offers a PCI target-interface core aimed at the company's new GF250F family. The PCI core is available as a netlist with layout constraints to meet the timing requirements.

8.1.2 Altera MegaCore™ functions

Altera's FLEX 10K family embedded array architecture development was driven by the requirements for large blocks of RAM and integration of other megafunctions. The FLEX 10K architecture that combines traditional logic arrays blocks (LABs) with embedded array blocks (EABs) can be used both for large RAM implementations as well as complex logic functions (see Sec. 2.3.2). Therefore, Altera has taken an approach to provide an architecture tailored for megafunction integration but not specific to any single application. Altera's own MegaCore functions program and Altera Megafunction Partners Program (AMPP) provide a number of intellectual property functions for a variety of applications.[5]

Megafunctions are functional blocks that accomplish a well defined task in a system. They may replace standard product ICs, or they can be entirely new functions for implementation such as processors, bus controllers, and interfaces. Megafunctions are built in text-based languages, including VHDL, Verilog HDL, and the Altera Hardware Description Language (AHDL). They can be used in any design file, including schematics, EDIF netlist files, and HDLs. Most megafunctions are optimized for FLEX 8000 and FLEX 10K device families. As initial offerings, Altera is developing MegaCore functions library with following eight megafunctions:

- 8237 DMA controller
- 6502 8-bit processor
- 6402, 16450, and 8251 UART processors
- 6850 Asynchronous communications interface adapter
- 8259 Interrupt controller
- 8255 Parallel I/O controller

Each function in the MegaCore function library consists of optimized postsynthesis AHDL netlist files that can be instantiated into a user's design. Altera will also include simulation capable Verilog HDL and VHDL design files to facilitate integration within the user's selected EDA tools.

The AMPP, an alliance between Altera and developers of synthesizable megafunctions, was created to encourage and increase the use of megafunctions in PLD designs. This program is focusing on three areas of interest: digital signal processing (DSP), bus interfaces such as PCI and USB, and communications. Altera has announced AMPP program with a company called Integrated Silicon Systems Ltd. (ISS) that specializes in advanced DSP ASICs and DSP ASIC megafunctions. DSP megafunctions will support fast Fourier transforms (FFTs) and adaptive finite impulse response (FIR) functions. PCI megafunctions are available through an AMPP company called Eureka Technology, which has delivered target megafunctions running at 33 MHz. An alliance with CAE Technology and Sand Microelectronics provide USB megafunctions, including drivers, firmware, and test/verification.

In the communication area, Altera's digital modulator megafunction from Nova Engineering contains a parameterized complex multiplier/mixer and quadrature output NCO. The NCO is a look-up table (LUT) that has a quadrature output, phase accumulator, and phase offset input port. The complex multiplier/mixer multiplies two user-defined inputs with the NCO outputs for amplitude modulation or frequency down conversion. The adaptive pulse code modulation (ADPCM) megafunction from Integrated Silicon Systems (ISS) performs multichannel duplex ADPCM coding in telecommunications applications.

Another AMPP communications application is the Reed-Solomon CODEC megafunction for data encoding and decoding, provided by Object Oriented Hardware, U.K. The Reed-Solomon encoder receives raw data and adds check symbols. The codeword length and the number of check symbols are assigned using the foundry independent standard product (FISP) bus interface or via direct inputs to the data block. The output produces the original data with the check symbols appended. The Reed-Solomon decoder can correct up to n symbol errors or $2n$ erasures in each block of data. The time required to perform the correction depends on the codeword length and the number of checkbits. The megafunction also provides statistical information about the number of correctable and uncorrectable errors that occur over the decoder channel. It is implemented in VHDL and optimized for the Altera FLEX 10K family architecture.

A major consideration in IP-based design is the selection of a core that meets the intended application requirements.[6] A PCI core designed for FPGA may fail to provide acceptable performance in some applications. Therefore, PCI cores need careful evaluation, especially, if they are to be run at full speed. The price one has to pay for an FPGA core depends on whether the core is being provided as a netlist or as source code. The shrink-wrapped netlist designs help suppliers save on support costs and allow the users to more easily guarantee functionality, timing, and compliance with the specifications. Many designers select cores because of the time-to-market demands, and a netlist is often the simplest and fastest way to incorporate a core into an FPGA design. The netlist users do not need synthesis tools. One drawback of netlist is that it is less portable to other devices than is the source code.

Some designers prefer source code rather than the netlist to maintain full control and customization to their requirements. However, modifying a core has its own drawbacks, since one might have trouble meeting tough timing requirements after making a design change. Some vendors attempt to reduce the number of modifications that the core users require by offering several versions of their design, or by parameterizing them. The core may come as a netlist, source code, or even schematics; the deliverables for all these formats include documentation and test benches. The major role of test benches is to verify that the core will operate properly with the associated logic design, or even when modifications are made to the core by changing the source code.

Altera has the OpenCore evaluation system, which allows the user to try a core before procurement. An encrypted file is provided that can be instantiated and simulated for verifying the interaction and performance of a core in one's design environment. However, it will not allow the generation of programming files, nor output the VHDL or Verilog simulation files that are needed as input to a synthesizer, until the core is actually purchased. The suitability of implementing a particular core in a CPLD or FPGA device depends on the capacity and performance requirements of the core. To provide a complete solution for IP-based designs, the PLD vendor needs to deliver a combination of following:

- Synthesis-efficient architectures with superior support for logic synthesis design methodologies
- High-capacity and high-performance PLDs capable of implementing complex IP cores
- An extensive library of IP cores that have been previously optimized for a target architecture

In summary, it is important from a user's perspective to look beyond the actual core when evaluating a solution for a high-performance core-based design by taking into consideration the target silicon architecture, software tools, and service. The target FPGA must have the appropriate density and performance required for the particular application. The tools must support usage of predefined cores, be easy to use, and integrate seamlessly with the preferred EDA environment.

8.2 System-Programmable Gate Arrays

In the last few years, several PLD vendors have made significant improvements in architectures and densities, making them serious contenders for low- and medium-scale integration ASIC design applications. The popularity of PLDs has grown because of their design flexibility, lack of NRE costs, and fast time-to-market advantages compared to the traditional ASIC design based methodologies. However, programmable logic devices such as FPGAs do not have the inherent predictability and routability to implement programmable systems-on-silicon. In addition, improved routability is a requirement to integrate the ASIC cores with programmable logic. Therefore, PLDs have not been considered appropriate design vehicles for high-performance applications such as implementation of DSP algorithms or a 66 MHz PCI core. A new development has been the concept of system programmable gate arrays (SPGAs) which, on a single chip, combine the performance advantages of gate array or standard cell architecture and programmable logic design flexibility. The following are some of the planned characteristics of SPGAs under development:

- Gate capacity beginning at 50,000 gates, with maximum capacity extending to 1 million gates and beyond
- Features that support single-chip systems design
- Efficient silicon target for either soft HDL blocks or embedded application-specific cores
- Predictable performance from HDL descriptions

SPGAs offer high logic capacity beginning at about 50,000 gates and are optimized targets for a wide range of design reuse applications. The future versions of these SPGAs will offer features that are common on board-level designs such as clock control using phase locked loops, large blocks of RAMs, special output drivers, and JTAG support. An important area for SPGA design would involve using them as development platforms for the IP cores and associated logic. Some of the promising areas for SPGA applications include high-speed networking front-ends, gate-intensive graphics processing, hardware emulation, and customizable single-chip accelerators for embedded processing.

A first SPGA product has been announced jointly by Actel Corporation (Sunnyvale, CA) and EDA vendor Synopsys (Mountain View, CA). According to the agreement between the two companies, Actel will license Synopsys's cell based array (CBA) architecture and adapt it to its own technology, supporting FPGAs. The SPGA contains a mix of both ASIC and FPGA resources on a single chip. Therefore, in Actel's SPGAs, the IP core, which cannot be implemented in their FPGA architecture, can be accommodated in the CBA technology (licensed from Synopsys). The CBA is a metal-masked gate array structure, which is programmable only in the upper two metal layers. Its cells have three tiny transistors for each larger driver transistor. The small inputs have lower capacitance than normal sea-of-gates transistors and, on the average, wires are shorter.

Actel is planning to introduce two SPGA families, the first of which will be the company's new reprogrammable SPGAs in the ES family, designed for efficient soft IP inclusion, and to be followed by embedded

SPGAs for combining ASIC cores with the ES programmable logic architecture. The ES architecture offers fine-grain logic modules that enable efficient and flexible mapping of the logic elements during synthesis and high-level design. In addition, the ES architecture is based on a fine-grained routing architecture. The ES devices come with or without the embedded soft blocks or cores and range in density from 50,000 to 400,000 gates.

The SPGA architecture uses SRAM technology, which is a departure from Actel's traditional antifuse-based ACT series devices. However, the SPGA design is switch technology independent, which would allow future chips to use antifuse or flash technology as the programming element. Table 8.2 lists the planned A65ESxxx series members and their typical characteristics, such as usable gates range, module blocks, embedded RAM bits and embedded RAM blocks, dedicated flip-flops, and user I/Os.

TABLE 8.2 Actel's Planned SPGA A65ESxxx Series Family Members and Their Typical Characteristics ***(from Ref. 7, with permission of Actel Corp.)***

	A65ES50	A65ES75	A65ES100	A65ES150	A65ES200	A65ES400
Typical gates (logic and SRAM)	50,000	75,000	100,000	150,000	200,000	400,000
Usable gates	33,000 to 84,000	50,000 to 123,000	66,000 to 164,000	100,000 to 250,000	131,000 to 328,000	263,000 to 655,000
ModuleBlocks	2,048	3,072	4,096	6,144	8,192	16,384
Embedded RAM bits	4,096	8,192	16,384	24,576	32,768	65,536
Embedded RAM blocks	2	4	8	12	16	32
Dedicated flip-flops	2,048	3,072	4,096	6,144	8,192	16,384
User I/O	200	240	320	400	480	640

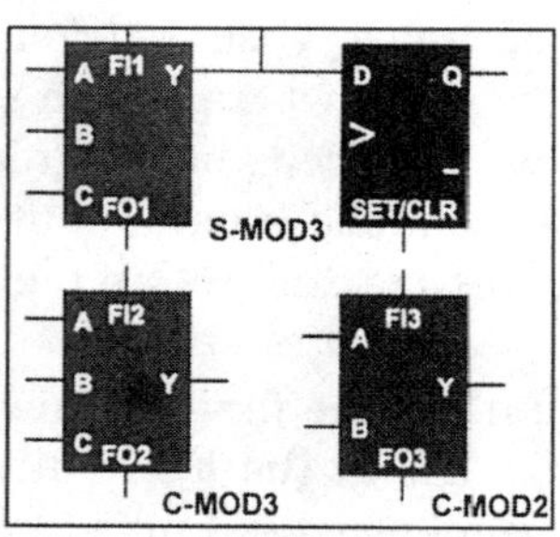

The first chip to be sampled will be A65ES100, a chip that packs 8.5 million transistors and can implement 100,000 gates, 16 kb of dual-ported RAMs (8 blocks of 2 kb each), and 320 I/O lines. The ES array overall structure is multitiered, starting at the lowest level with a logic cluster that contains the equivalent of 16 gates, which are arranged to form three basic modules (or function generators), as follows: a two-input function generator, a three-input function generator, and a combination of three-input function generator plus a D-type flip-flop.

The key to Actel's SPGA architectural efficiency is the use of a multitiered routing approach, which can be divided into two basic categories. The first of these is Local Connect, which provides inter- and intramodule block routing. The second category is the MultiDrive active routing technique, a hierarchical expressway system used to drive heavy loads and transition signals across long spans of logic. In the MultiDrive active routing scheme, separate transistors are used to implement logic and to drive the interconnects. By separating these functions, more transistors can be included per chip, which translates into smaller die size and more efficient designs. The ES architecture makes greater use of hierarchy than current programmable architectures. For example, Actel's A65ES100 has seven levels of routing hierarchy and a unique feature to this scheme is the constant, maximum routing delay associated with each level, providing the device with fanout-independent delays.

The ES architecture has three types of local connect routing: FastConnect, Rt1 (Route 1), and Rt4 (Route 4). FastConnect is a high-performance direct connect between modules in adjacent module blocks that provides a vertical connect path from the top to the bottom of the device. Rt1 and Rt4 are two-dimensional routing meshes spanning one or four ModuleBlocks, respectively. The Rt1 mesh provides linkage within and between adjacent blocks for low fanout connections. The Rt4 performs dual functions of providing mesh connections within and between the ModuleBlock of a 2×2 ModuleBlock array and provides on/off-ramp access into the hierarchical expressway routing. The expressway system consists of routing tracks that span distances in 2^N manner, beginning with Rt16, and repeat this approach until the entire device array has been spanned.

In addition to logic and routing resources at each level, the FPGAs contain up to 32 two-kb blocks of dual-ported SRAM located at the center of the chip. These SRAM blocks can be used as register files, data buffers, look-up tables, and other system functions. The reduced-size cell yields a typical gate density of about 240 usable gates/mm^2 for the SRAM-based SPGA programmable logic, which makes them about 50 percent denser than the antifuse-based FPGAs. The programmable

ES functions are specified for operation with a clock rate as high as 75 MHz system operation and with a 200 MHz datapath throughput.[8]

In addition to ES family, Actel is planning to offer the ES embedded SPGAs family, which will include an embedded cell-based array (CBA) architecture. This embedded SPGAs family will be targeted toward implementing complex system elements such as third-party IP cores or system level functions developed in house by customers. QuickLogic is also working on devices that can mate its one-time programmable logic technology with hardwired blocks.

Lucent has also announced plans to build hybrid chips that combine masked function blocks with SRAM-based reprogrammable logic. The first FPGA that Lucent is bringing out in the 3C series, the 3C55, is just a generic FPGA—an 18 × 18 array of programmable logic cells, equivalent to 55,000 gates. To create its first field-programmable system chip (FPSC), Lucent will eliminate four rows of logic from the chip, leaving behind 14 × 18 array of 40,000 programmable gates. The 4 × 18 array of cleared space will be used to implement 80,000 gates of standard cell logic, which will expand the 55,000 gate device into a 120,000-gate device. The hybrid chip's intended applications will be in soft core functions such as the PCI, USB, and fast ATM designs.

8.3 Analog FPGAs

Analog FPGAs were first introduced by a company called IMP, Inc., and were basically analog ICs that could be programmed in functionality, interconnection, and performance characteristics.[9] These devices were fabricated in electrically programmable analog circuit (EPAC) technology based on IMP's mixed-signal 1.2 μm CMOS process with on-chip SRAM and EEPROM memory. The devices could be programmed either off line or in real time, which provided the user means to make changes *on the fly* (i.e., in-system programming capability) while leaving the original EEPROM configuration unchanged. The EPAC devices utilized a certain architectural environment, referred to as the EPAC *framework,* which provided for programming features, debugging aides, reference modules, and so forth.

Figure 8.4*a* shows the key elements of this framework including the serial interface, the EEPROM memory module, a *utility* section, and the functional section containing the programmable analog modules.[10] The serial programming interface not only allows IC configuration but also processes a number of useful commands, including probing of internal analog or digital signals, read-back of the configuration data, power management control, and other functions.

In a microprocessor-based system, multiple EPAC devices can be cascaded and programmed in sequence or individually. No specific de-

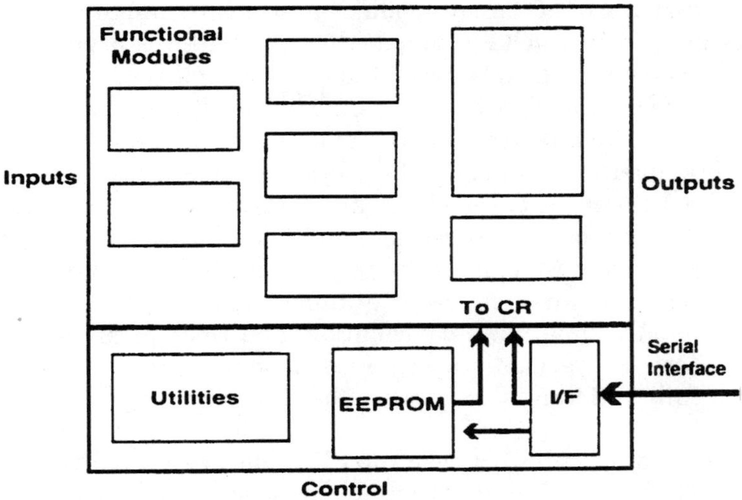

(a)

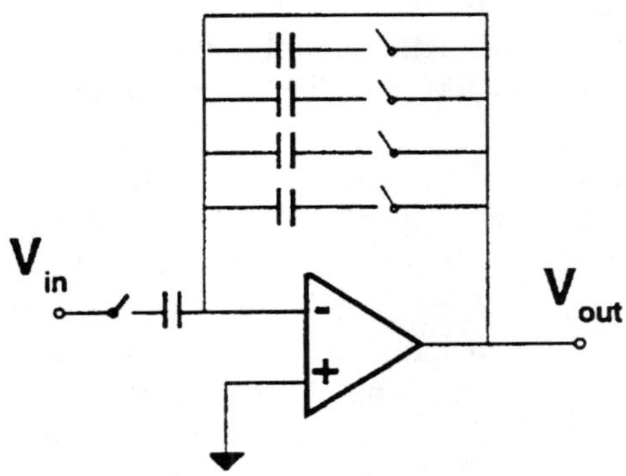

(b)

Figure 8.4 (a) EPAC framework and (b) conceptual schematic of switched capacitor (SC) circuit modules *(from Ref. 10 with permission of IMP, Inc.).*

vice programmer is required to interface with the chip. The EPAC framework allows configuration of analog modules using the on-chip EEPROM that acts as a *shadow memory* for the SRAM-based configuration register (CR). On powering up, the EEPROM contents are downloaded into a CR carrying the data to the appropriate modules. The register can be loaded directly from the serial interface so that the multiple configurations can be executed without disturbing the contents of the EEPROM. The contents of the CR and the EEPROM can be redefined any time—even while the chip is in the normal operation.

The IMP EPAC uses the switched-capacitor (SC) principle as the prime implementation technology for modules. Figure 8.4*b* shows the conceptual schematic of SC circuit modules in which the input capacitor samples an incoming voltage signal while the feedback capacitors determine the amount of gain between the input and output.

When the charge transfers from the input to feedback capacitors, the ratio between them determines the amount of voltage needed to achieve and complete the transfer. Therefore, it is the ratio of the capacitors and not their absolute values that determines the gain, and the capacitor ratios can be precisely manufactured so that gain matching of 0.1 percent is achieved. To increase flexibility and performance, all capacitors can be made programmable; multiphase clocking can enable offset and noise cancellation; and clock rates can be altered to allow for various speeds. Also, modules can be implemented in a fully differential manner, effectively doubling the feedback circuitry.

One of the key differences between the EPAC devices and FPGAs is the granularity of the building blocks. In comparison to FPGAs that contain a large number of relatively low-level cells, EPAC provides access to high-level cells called *modules*. The high-level cells implement functions such as D/A conversion, amplification, and comparison, as opposed to merely offering components like amplifiers, resistors, capacitors, and so on.

The EPAC family included three devices: IMP50E10, IMP50E20, and IMP50E30. Recently, the original developer of EPAC family, IMP Inc., made an announcement that it has discontinued further production of this device family.

Motorola Semiconductor (Phoenix, AZ) has introduced a field programmable analog array (FPAA) as a true analog equivalent to an FPGA.[11] This FPAA contains an array of 20 identical CMOS analog cells arranged in a 5 × 4 matrix. The cell is based on switched-capacitor operational amplifier (op-amp) that serves as the core programmable element. The cell also includes a comparator, capacitor arrays, switches, and SRAMs, as shown in block diagram Fig. 8.5*a*.

The first device introduction MPAA020 contains configurable analog blocks (CABs) that incorporate a switched capacitor CMOS op-amp, comparator, capacitor arrays, CMOS switches, and SRAM. The data stored in SRAM controls the switches that program various capacitance values (static and dynamic) in the input and feedback signal paths of the op-amp. The MPAA020 contains 41 op-amps, 100 programmable capacitors, and 6864 switches.[12] The arrays are structured in a grid that contains 20 CABs arranged in a 4 × 5 matrix and 13 I/Os with their own amplifiers that can be used for buffering and/or filtering (anti-aliasing or smoothing). An 8-bit programmable bandgap voltage reference is available to each CAB. Figure 8.5*b* shows the block diagram of a MPAA020 device architecture.

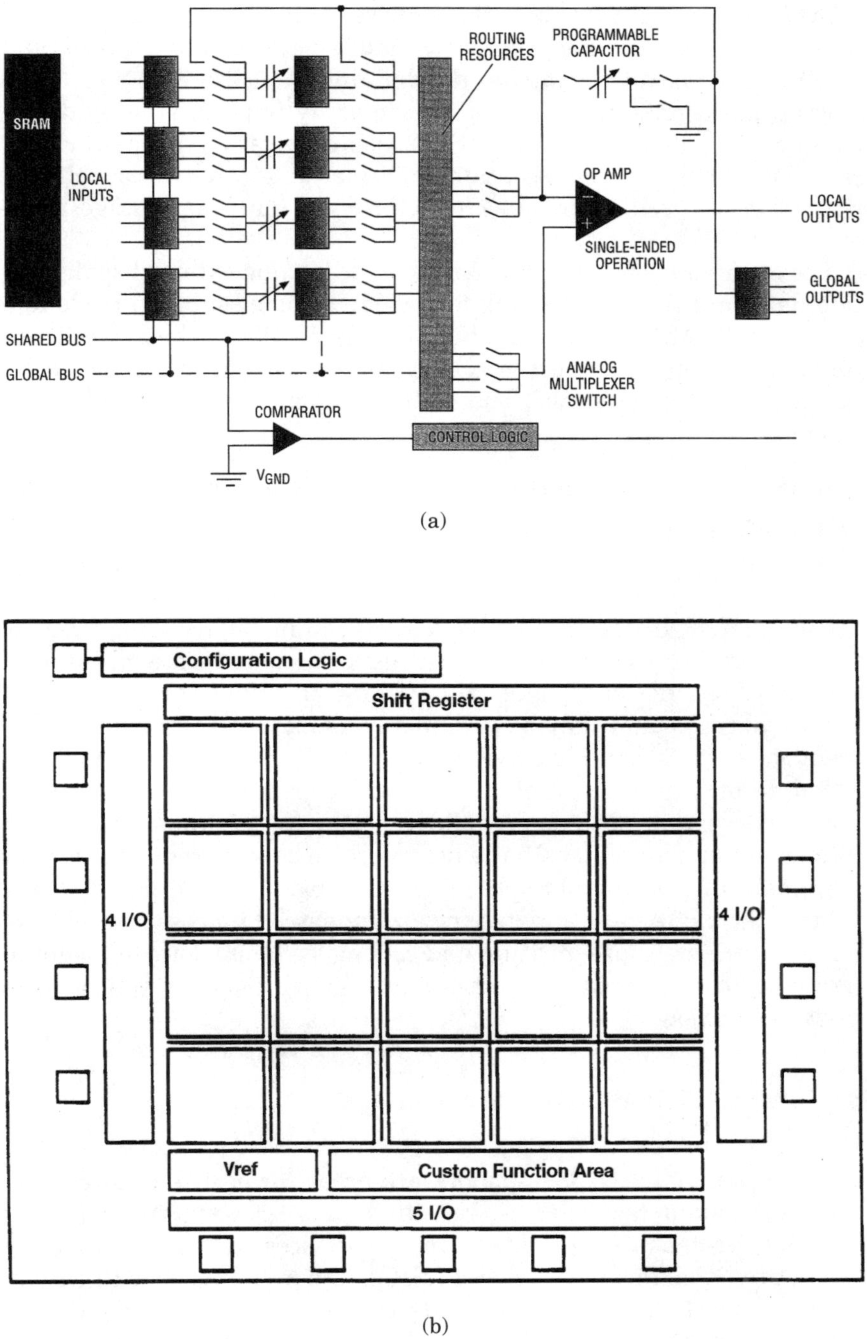

Figure 8.5 Motorola field programmable analog array (FPAA): (a) cell block diagram and (b) MPAA020 device architecture *(from Ref. 12 with permission of Motorola Corp.)*.

Two buses move data from the shift register to the CABs; the data control bus retrieves and moves the data from the shift register, and the transfer control bus latches the data into the local SRAM. The analog design configuration is within the array is performed by downloading 6k bits of data via RS232 communication from a PC or EPROM. The data stream contains information for individual cell configuration, the cell interconnections, internal bandgap voltage reference, and I/Os.

The configuration logic is responsible for loading external configuration data into the device using following two modes: micro mode and boot from ROM (BFR) mode. The micro mode allows the device to be configured from a microprocessor (or similar system) using a conventional peripheral interface. The BFR mode supports three submodes as follows:

1. Use of an internal address generator with a byte wide data format
2. Use of an external address generator with a serial data format
3. Use of an external address generator with a byte-wide data format

For MPAA020, the recommended maximum signal frequency is 200 kHz, with 1 MHz specified as the maximum clock rate for internal sampling. The newer designs will operate at higher frequencies and higher voltages, and they have as many as 32 CABs. MPAA development support is provided by EasyAnalog Design software tool, PC based, to provide manual place-and-route capability. The EasyAnalog macrocell library contains predefined macros (e.g., gain stage, biquad filter, summing amplifier) that a designer can choose from a library for placement in the array for subsequent routing.

Motorola FPAAs are targeted toward industrial process control (e.g., temperature, pressure, heating and cooling systems), low- to medium-frequency communication applications, and automotive and medical instrumentation.

8.4 Latest PLD Developments and Future Trends

The programmable logic suppliers are accelerating use of advanced, deep submicron technologies, resulting in a growing trend toward even higher densities and lower voltage devices that are being more competitive than even traditional ASICs. It is projected that PLDs will become one of the largest segments of the logic market, surpassing both standard logic and MPGAs. FPGAs manufactured using advanced 0.35 μm and 0.25 μm process technologies offering in excess of 100,000 logic gates are currently available. For example, the new Xilinx XC4085XL FPGA fabricated on a 0.35 μm, three-layer metal process has 16 million transistors, more than three times the number of transistors in Intel's Pentium Pro.

The trend toward higher-density programmable logic is requiring migration to lower supply voltages such as 3.3 V and 2.5 V. The scaling down to 0.25 μm and finer geometries implies the PLD product families as incrementally moving down the *voltage staircase,* as shown in Figure 8.6.[13] The I/O standards must also change as the supply voltages drop. The latest PLD families accept 5 V signal on all I/Os and can drive TTL levels into any 5 V device. The 3.3 V LVTTL standard is also being widely used. Future devices will have multivoltage I/Os capable of interfacing between variety of standards.

For product families that do not have sufficiently high performance demands to require the latest processes (3.3 V or lower), it would be more economical for PLD vendors to continue building products on a more mature (and less expensive) 5 V process.

The following are some examples of the latest PLD technologies under development:

- *Actel MX series FPGAs.* Actel's new high-density family of Integrator series FPGAs, called MX series, has chip complexities ranging from 2000 to 52,000 logic gates.[14] This family is based on a 0.45 μm triple-level metal, antifuse-based process, which provides enhanced routing resources as compared to the company's previous dual-level metal designs. The 40MX and 42MX families together have seven members. The 40MX family devices will operate in 5 V or 3.3 V only systems, whereas the 42MX devices will operate in 5 V, 3.3 V, and mixed 5V/3.3V systems. The 42MX FPGAs include system-level fea-

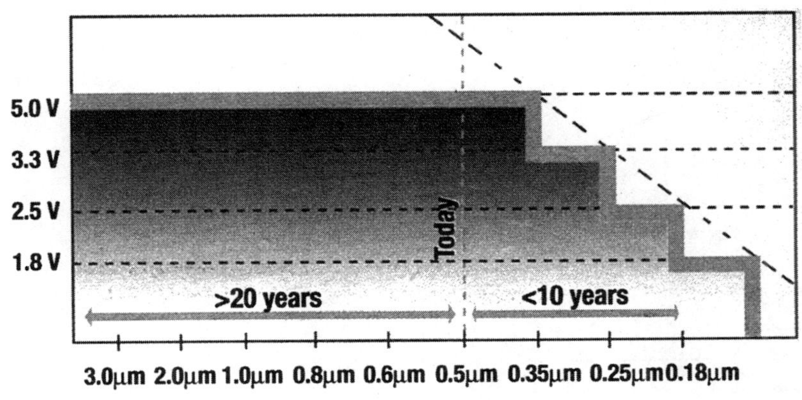

Figure 8.6 PLD families incrementally moving down the voltage staircase *(from Ref. 13).*

tures such as JTAG support, dual-port SRAM, fast wide-decode modules, and programmable PCI interface.

The device family array includes dedicated flip-flops, ranging from 147 on the 40MX02 to 1888 on the 42MX52. The two largest devices, the 42MX36 and 42MX52, include dedicated blocks of dual-port/single-port RAM. The chips are logic compatible with the company's ACT 1, XL, and DX families and include redesigned clock buffers and clock distribution networks to reduce the clock-to-Q (pad-to-pad) times to as little as 5.6 ns. A key improvement in the MX series is a Multiplex I/O cell that includes selectable, enhanced 3.3 V input switching.

The Integrator series 40MX and 42MX devices consist of logic modules, routing resources, clock networks, and I/O modules. The 40MX logic module is an eight input, one-output logic, as shown in Figure 8.7. The logic module can implement the four basic logic functions (NAND, AND, OR, and NOR) in gates of two, three, or four inputs. Each function may have many versions, with different combinations of active-low inputs. The logic module can also implement a variety of D-latches, AND-ORs, and OR-ANDs.

The Actel 42MX family contains three types of logic modules: combinatorial (C-modules), sequential (S-modules), and decode (D-

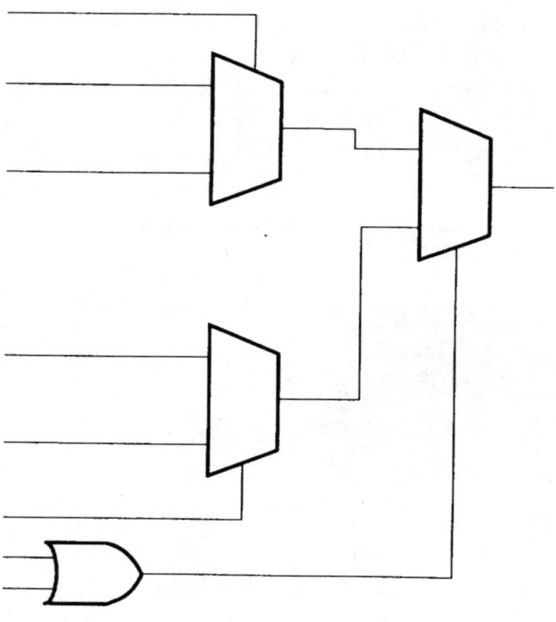

Figure 8.7 Actel MX40 FPGA family logic module *(from Ref. 14 with permission of Actel Corporation).*

modules). The S-module implements the same combinatorial logic function as the C-module while adding a sequential element that can be configured as either a D flip-flop or a transparent latch. Some of the 42MX devices contain a third type of logic module called D-module that contains wide decoding circuitry to provide a wide-input AND function similar to that found in product term architectures. The D-module allows 42MX devices to perform wide-decode functions at speeds comparable to the CPLDs and PAL devices. The D-module output is hard wired to an output pin, or it can be fed back into the array to be incorporated into other logic. Figure 8.8 shows 42MX family (8.8*a*) C-module, (8.8*b*) S-module implementation of up to 7-input function plus D-type flip-flop with clear, and (8.8*c*) D-module.

Several 42MX devices contain dual-port SRAM modules arranged in 256-bit blocks that can be configured as 32 × 8 or 64 × 4. These SRAM modules can be cascaded together to form memory spaces of user-definable width and depth. The SRAM modules are dual-port structures containing independent read and write ports and are optimized for high-speed buffered applications requiring FIFO and LIFO queues.

The MX series I/O modules provide the interface between the device pins and the logic array. The I/O modules are flexible structures, in that each output pin has a dedicated output-enable control. The I/O module can be used to latch input or output data, or both. All 42MX I/O modules contain a tristate buffer, with input and output latches that can be configured for input, output, or bidirectional operation. The MX architecture uses vertical and horizontal routing tracks to interconnect the various logic and I/O modules. The routing tracks are metal interconnects that may either be of continuous length or broken into pieces, called *segments,* of varying lengths. The segments can be joined together at the ends using antifuses.

The Integrator series MX family is supported by Actel's Designer Series development software, which can provide a seamless integration into any ASIC design flow. The Designer Series development tools offer automatic placement and routing, static timing analysis, user programming/debugging, and diagnostic probe capabilities.

- *Xilinx Virtex FPGA family.* Xilinx is developing a next-generation family of SRAM-based FPGAs to offer from less than 20k to over 500k gates in the initial release with over 90 percent gate utilization, and plans for future enhancements to provide up to 1 million gates and the ability to operate at system clock rates of over 100 MHz.[15] The initial member of this Virtex family of FPGAs will include devices from 1500 to more than 32,000 logic cells. The core

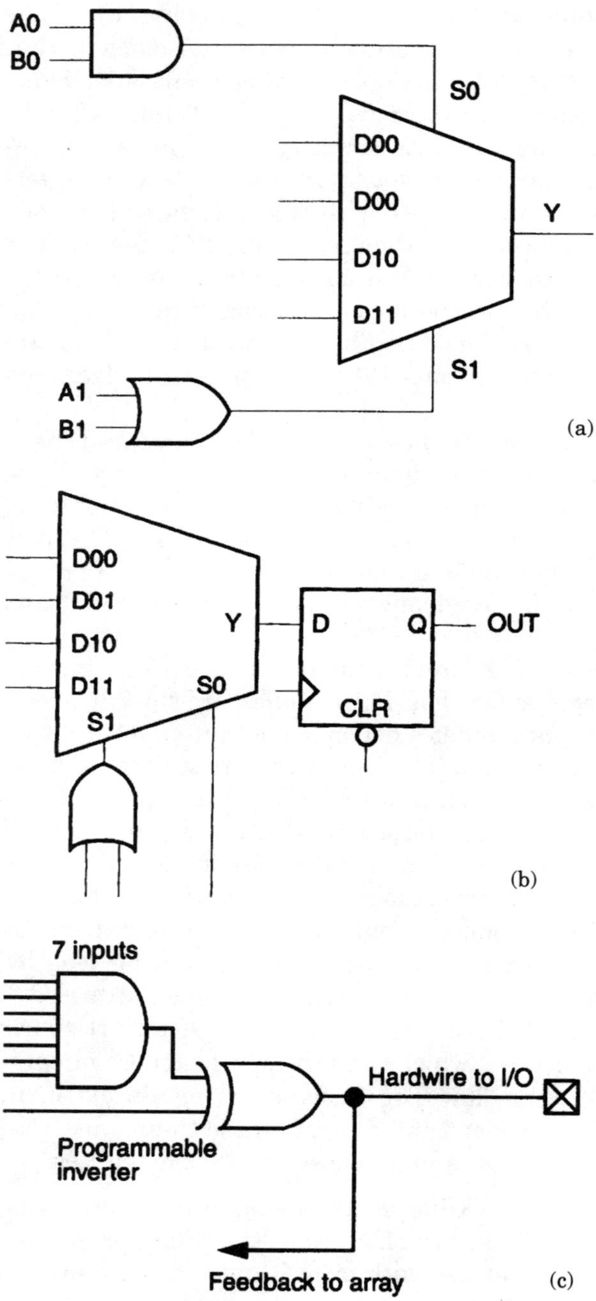

Figure 8.8 Actel MX42 FPGA family (a) C-module, (b) S-module implementation of up to seven-input function plus D-type flip-flop with clear, and (c) D-module *(from Ref. 14 with permission of Actel Corporation).*

logic in Virtex family is designed for 2.5 V operation and I/O buffers for 3.3 V operation (5 V tolerant). In addition to the abundant routing resources, this family includes a novel virtual diagonal interconnect (VDI) scheme, also referred to as vector-based interconnect, to allow accurate prediction of net delays. Figure 8.9*a* shows the Virtex family functional block diagram.

The improved logic cell in Virtex family contains a four-input LUT and a D-type flip-flop or a latch with common clock, clock-enable, and set/reset signals. Logic cells are paired together to form a slice, and two slices are grouped together to form a configurable logic block (CLB). All four LUTs in the CLB can connect to each other through a fast local interconnect that provides known delay between LUTs in a given CLB. Also, on-chip, phase lock loops (PLLs) are provided for high-speed clocking and to minimize signal skews to handle stringent timing requirements for applications such as a 66 MHz version of PCI. The CLB cells also include high-speed carry-propagation logic to minimize delays when multiple blocks of logic are cascaded. Dedicated carry logic is included to provide efficient implementation of multipliers that require shift and add operations.

Each LUT is based on a 16-bit SRAM that can either be used as logic function lookup or serve as a 16-word × 1-bit static memory that can be combined with other LUT RAMs to form registers, buffers, and so on. Additionally, two four-input LUTs can be combined to form a five-input LUT or a 4:1 multiplexer. The arrays also contain dedicated blocks of true dual-port static RAM (called *selectRAM*) that are fully compatible with the XC4000X family. These *selectRAM* blocks of 4k bits each and can be configured into several aspect ratios (word depth by word width) such as 4k × 1, 2k × 2, 512 × 8, and so forth. Figure 8.9*b* shows the Virtex family CLB.

The signals going in and out of a CLB are connected to other CLBs through a switch matrix, which has no directional dependence. Inputs can come from any of the four directions and, similarly, outputs can also propagate in any of four directions. The routing resources are divided into three general categories: global routing, I/O routing, and dedicated routing. Global routing resources that are two tiered (primary and secondary) are used to distribute signals with high fanout throughout the entire device. All CLB and I/O buffer clock pins can be driven by a global buffer. Additional routing resources on the chip periphery support the I/O buffers and form the interface between core logic and I/O buffers.

The core of FPGA is surrounded by I/O buffer cells, which contain registered inputs, registered outputs, and three-state enable control. The input-output block (IOB) is designed for speed and flexibility. There are registers (flip-flop or latch) on the input, output, and

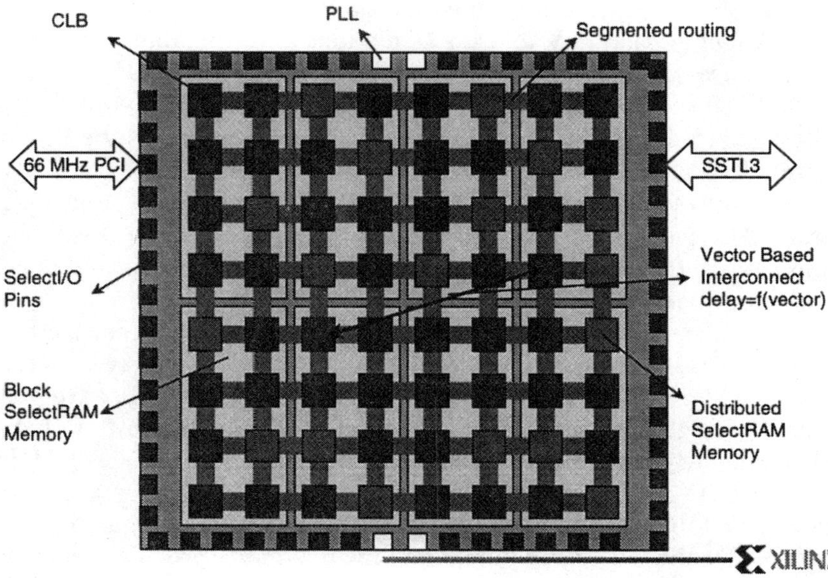

(a)

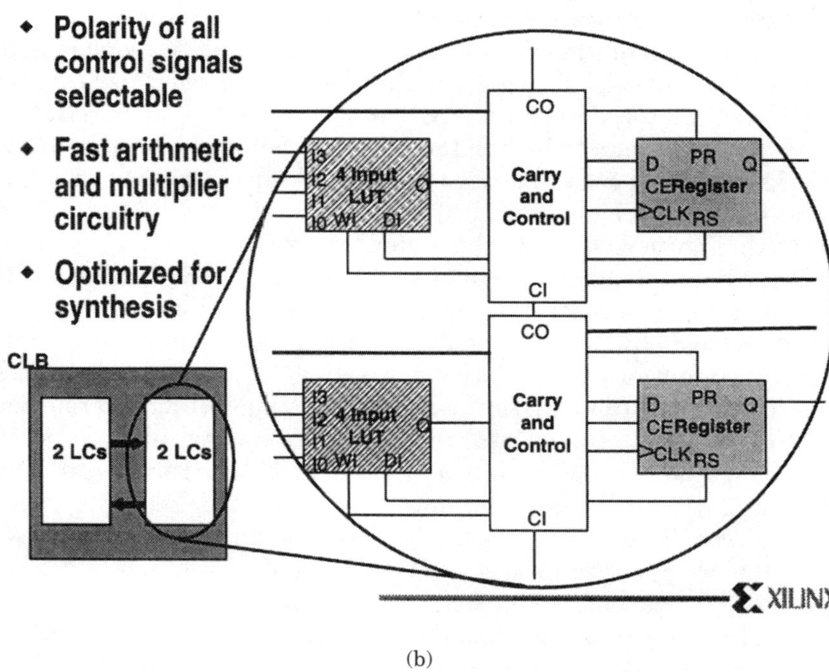

(b)

Figure 8.9 Virtex FPGA family showing (a) functional block diagram and (b) configurable logic block (CLB) *(from Ref. 16 with permission of ©Xilinx, Inc.).*

three-state enable pins. The IOB registers have a common clock and separate clock enable inputs for all three registers.

The logic designs in Virtex family will be supported by the Xilinx's Alliance series software. The architecture for Virtex FPGA family is developed to support IP core based designs. AllianceCORE™ is a cooperative effort between Xilinx and independent third-party core developers to produce a broad selection of industry standard core solutions. Currently, these third-party suppliers are shipping Xilinx-specific functions for applications such as USB, PCMCIA, and microprocessor peripherals.

The new high-density, in-system programmable devices offer significant advantages over the ASICs to board-level designers by providing the flexibility of changing product features by device reconfiguration. It means that a single PCB design can be made to work in multiple versions of a product offering different features by simply changing the FPGA configuration.

A growing trend toward high-level design languages VHDL and Verilog requires that synthesis tools have embedded knowledge of how a design can be best optimized for a given architecture with all its dedicated functions and capabilities. As an example, Viewlogic's new FPGA/CPLD synthesis software tool, Aurora, generates optimized results for each PLD type by providing appropriate VHDL and Verilog capabilities required. This is possible because of Viewlogic's agreement with PLD vendors, who will ship a subset of Viewlogic's system's solutions, including schematic and gate level simulation, synthesis, optional VHDL simulation tools, and their own layout tools to their customers.

Therefore, partnerships between the CAE vendors and the PLD vendors are absolutely necessary to ensure high-quality synthesis for the growing complexity of PLD family-based designs. The PLD vendors, who already have access to the core technologies for processors, memory, and special function ASICs, will be in a best position to offer a system-on-a-chip solution that will provide the cost effectiveness and high-performance solutions required by the PLD users.

References

1. Larry Walker, et al., "Focus Report: Cores and Megacells," *Integrated System Design,* Oct. 1996, pp. 34–54.
2. Jim Lipmann, "The hard facts about soft cores," *EDN,* Sept. 2, 1996, pp. 38–48.
3. Steve Svoboda, "IP tames today's DSP design challenges," *Electronic Design,* Nov. 18, 1996, pp. 99–108.
4. Brian Barrera, "Block based IP standards for chip design are merely the tip of the iceberg," *Electronic Design,* April 14, 1997, pp. 76–78.
5. *Altera Corporation publication News & Views,* Nov. 1996 and Feb. 1997 issues.

6. Debora Grass, "FPGAs: A matter of cores," *EDN,* March 27,1997, pp. 61–75.
7. Actel data sheets and latest product development briefs.
8. Dave Bursky, "Programmable arrays mix FPGA and ASIC blocks," *Electronic Design,* Oct. 14, 1996, pp. 69–74.
9. Hans W. Klein, "Circuit development using EPAC technology: An analog FPGA," *SPIE,* vol. 2607, pp. 135–144.
10. *Electrically Programmable Analog Circuit Design Handbook,* IMP Corp.
11. David Morrison, "Programmable IC delivers analog designs with a digital ease," *Electronic Products,* July 1997, pp. 15–16.
12. Motorola Semiconductor Technical Summary 1997: "Introducing Motorola's Field Programmable Analog Arrays."
13. Sandeep Vij, "The future of programmable logic," *ECN Technological Horizons,* June 1997, pp. 13–14.
14. Actel Integrator Series FPGA MX Family data sheets.
15. Dave Bursky, "High density FPGA family delivers megagate capacity," *Electronic Design,* Nov. 17, 1997, pp. 67–70.
16. Virtex FPGA family viewgraphs supplied courtesy of ©Xilinx, Inc.

Glossary

This is primarily a glossary of acronyms for programmable logic devices, technology, and reconfigurable logic terms used in this book. It is not intended to be an exhaustive listing of all acronyms used throughout the programmable logic devices industry. It does not list acronyms for multitude of designs and software development tools used for PLDs.

ABEL Advanced Boolean Expression Language (Data I/O trademark). A universal PLD design language and supporting software from Data I/O.

ACB Architecture Control Block. A term used by Altera for the EP600/EP900 configurable I/O macrocell.

ACT Application Configurable Technology (Actel trademark). A family of field programmable gate arrays (FPGAs) offered by Actel Corporation.

ADC Analog-to-digital conversion (or converter).

ADL Actel design language. A netlist format.

ADPCM Adaptive pulse code modulation.

AFM Atomic force microscopy.

AHDL A high-level design entry hardware description language (HDL) developed by Altera for its design compilers.

AL Anomalously large. A term used to categorize a type of solar flare phenomenon.

ALU Arithmetic logic unit.

AmPAL AMD Programmable Array Logic (AMD trademark).

AMPP Altera Megafunction Partner Program. An alliance set up by Altera with other suppliers to provide megafunctions for its FLEX series FPGAs.

APEEL ICT trademark for the assembler for PEEL PLD product line.

APR Automatic place and route.

ASIC Application specific integrated circuit. An integrated circuit designed or tailored by an end-user for a particular application.

ASPLD Application specific PLD. A PLD containing fixed-function circuitry targeted for a specific class of applications.

ATM Asynchronous transfer mode.

ATPG or **ATVG** Automated test pattern (or vector) generation. Computer generation of production test vectors with the goal of providing adequate fault coverage.

BCP Best commercial practices.

BER Bit error rate.

BGA Ball grid array. A type of packaging configuration that uses solder balls at the bonding pads instead of wire bonds.

BFM Bus function model.

BFR Boot from ROM. This terms refers to one of the basic modes for loading external configuration data into an SRAM (or LUT) based FPGAs.

BiCMOS Bipolar-CMOS mixed technology.

BIDI Bidirectional buffer. A part of routing resources used by Lucent Technologies ORCA FPGAs.

BLMC Buried logic macro cell.

BOP Billion operations per second. A term typically used to describe the performance of a computing system.

BST Boundary scan test.

CAD Computer aided design.

CAE Computer aided engineering.

CBA Cell-based array. An architecture licensed by Synopsys Corp.

CGS Code generation system.

CHAMP Configurable hardware algorithm mappable processor.

CIP Configurable interconnect point. A switching circuitry that is part of routing resources used by Lucent Technologies ORCA FPGAs.

CLB Configurable logic block. A grouping of logic elements used in FPGA architectures.

CMOS Complementary MOS. A MOS technology that makes use of both N-channel and P-channel transistors.

CODEC Coder/decoder.

CoolRunner™ Philips Semiconductors CPLD family based on fast zero power design techniques.

CPLD Complex PLD. A term used to describe a PLD consisting of several logic blocks, such that overall architecture of the blocks is similar to that of PALs.

CPU Central processing unit.

CR Configuration register.

CUPL Universal Compiler for Programmable Logic (Logical Devices trademark).

DIP Dual in-line package.

DMA Direct memory access.

DNA Deoxyribonucleic acid.

DPGA Dynamically programmable gate array.

DR Dynamic reconfiguration.

DRAM Dynamic random access memory.

DRC Design rule check.

DRHW Dynamically reconfigurable hardware.

DRL Dynamically reconfigurable logic.

DSP Digital signal processing (or processor).

DUT Device under test.

EAB Embedded array block. This is used in Altera FLEX 10K FPGA family to build memory blocks and megafunctions such as microprocessors, multipliers, and DSPs.

ECL Emitter coupled logic. An extremely high-speed family of bipolar logic and memory devices.

EDA Electronic design automation.

EDAC Error detection and correction.

EDIF Electronic design interchange format. A standard format used to transfer design data in the form of schematics and netlists.

EECMOS Electrically erasable CMOS technology.

EEPLD Electrically erasable PLD. A PLD that uses EEPROM memory cells to hold the programmable logic configuration.

EEPROM Electrically erasable programmable read-only memory.

EPLD Erasable PLD. A PLD that uses EPROM memory cells to hold the programmable logic configuration.

EPROM Erasable programmable read-only memory (usually refers to ultraviolet erasable technology, rather than electrically erasable).

ERA Electrically reconfigurable array (Plessey's name for its FPGA products).

ERC Electrical rule check.

ES Embedded SPGAs introduced by Actel Corporation.

ESD Electrostatic discharge.

FastFlash A term used to describe Xilinx XC9500 CPLD family.

FastCONNECT™ A switch matrix used in Xilinx XC9500 FastFlash CPLD family architecture.

FFB Fast function block. A type of function block used in Xilinx XC7300 EPLD family.

FFM Full functional model.

FFT Fast Fourier transform.

FIFO First-in-first-out. A semiconductor memory device buffer that stores the information in the order in which it receives and then releases that beginning with first element received and progressing sequentially.

FIR Finite impulse response.

FISP Foundry independent standard product.

FIT Failure in time.

FLASH A nonvolatile memory technology used in several PLD and CPLD families.

FLEX An FPGA architecture from Altera.

FPAA Field programmable analog array. An analog equivalent of an FPGA introduced by Motorola Semiconductor (Phoenix, AZ).

FPCB Field programmable circuit board.

FPGA Field programmable gate array. A high-density PLD architecture with multiple levels of logic and programmable interconnect.

FPIC Field programmable integrated component.

FPID Field programmable interconnect device.

FPLA Field programmable logic array. Another term used for simple PLDs with both programmable AND and programmable OR terms.

FPLD Field programmable logic devices. In general, this term used in the book refers to all programmable logic devices including CPLDs and FPGAs.

FPLS Field programmable logic sequencer. These are FPLAs that feature registers and are intended for sequencing and state machine applications.

FPS Field programmable system.

FPSC Field programmable system chip. A hybrid chip being introduced by Lucent Technologies that can be programmed in functionality, interconnection and performance characteristics.

FSM Finite state machine. A machine that can be in one of a finite number of states. The term is often used for logic circuits, which sequence through various states.

FZP™ Fast zero power. Philips Semiconductors' design technique for reducing power dissipation in their CoolRunner CPLD family.

GAL Generic Array Logic (Lattice Semiconductor trademark). Electrically erasable devices that are pin-to-pin compatible with popular PAL-type devices.

GERM General environment resource manager. This refers to control section (kernel) of the Virtual Computer system.

GFDR General force-directed relaxation algorithm. An algorithmic approach used by FPGA automatic placement tools.

GLB Generic logic block. A logic block used by Lattice Semiconductor for its CPLDs.

GRM General routing matrix. A term used for routing resources in Xilinx FPGAs.

GRP Global routing pool. A term used in Lattice ispLSI and pLSI CPLD families to describe interconnect resources for all internal logic.

HAL Hard Array Logic. HAL is a registered trademark of AMD for mask-programmed versions of PALs.

HCPLD High-capacity PLDs, including both CPLDs and FPGAs.

HDL Hardware description language. A textual design description format used for design entry and synthesis, or for circuit modeling and simulation.

HDS Hardware design system.

HIM Host interface module.

HLD High-level design.

IC Integrated circuit.

ICR In-circuit reconfigurable.

ICT International CMOS Technology. Manufacturer of the popular PEEL family electrically erasable PLDs.

IIR Infinite impulse response.

ILMC Input logic macro cell.

Integrator Series Actel FPGA product family name.

IOB Input-output block.

IOC I/O control block (e.g., in Altera MAX EPLD families).

IOLMC Input/output logic macro cell.

IP Intellectual property.

IRMW Infrared missile warning.

ISA Industry standard architecture.

ISDN Integrated services digital network.

ISM Instruction Set Model.

ISP In-system programmable. A term pioneered by Lattice Semiconductor used to describe programming/reprogramming of devices directly on the printed circuit board, from a PC or ATE. The term "isp" is a prefix for Lattice families and part numbers.

ISR In-system reprogramming.

JEDEC Joint Electronic Device Engineering Council. (1) A council that creates, approves, arbitrates, and/or oversees industry standards for electronic devices, and is a part of Electronic Industries Association (EIA). (2) In programmable logic, refers to a computer file containing information about the programming of a device. The file format is JEDEC-approved standard.

JTAG Joint Task Action Group. A JEDEC designated group for some specific activity such as creation (or revision) of a standard.

K (or Karnaugh) Map A graphic tool for minimizing sum-of-products or product-of-sums logic functions.

LAB Logic array block. Altera's term for groupings of 16 macrocells.

LAN Local area network.

LCA Logic cell array. A term used by Xilinx for its SRAM-based FPGAs.

LCC Leadless chip carrier.

LE Logic element. A term used by Altera.

LEO Low earth orbit.

LET Linear energy transfer. A measure of energy deposition rate with units of MeV/[cm^2/mg] or MeV/μm.

LIM Local interconnect matrix.

LINAC Linear accelerator.

LM Logic module. A term used by Actel.

LPCVD Low-pressure chemical vapor deposition.

LPF Lowpass filter.

LPGA Laser programmable gate arrays. A new technology variation on MPGAs, introduced by a company called Chip Express.

LPM Library of parameterized modules. A new standard emerging that allows the synthesis compiler to pass high-level module description (as opposed to the gate level) to the target device vendor's tools for optimization.

LS Low-power Schottky TTL family. Lower power version of the standard Schottky TTL family.

LSB Least significant bit.

LSI Large scale integration. A term used for describing relative complexity of integrated circuits.

LTPD Lot tolerance percent defective.

LUT Look-up table.

LVS Logic versus schematic.

LVT Low-voltage technology.

LVTTL Low-voltage transistor-transistor logic.

MAC Multiply and accumulate.

MACH Macro Array CMOS High speed. AMD/Vantis trademark for its family of CMOS CPLDs.

MACO Mask array conversion for ORCA FPGA family from Lucent Technologies.

MAX Multiple Array Matrix. Altera's trademark for the architecture used in its EPLD family of products.

Maxterm A sum in the canonical product-of-sums form. Each maxterm contains every input variable, in either true or complement form.

Minterm A product in the canonical sum-of-products form. Each minterm contains every input variable, either in true or complement form.

MCM Multichip module.

MOP Million operations per second. A term typically used to express performance of a computing system.

MORRPH Modular and reprogrammable real-time processing hardware.

MOS Metal-oxide-semiconductor transistor. One of the two basic types of transistors. In logic design, used for NMOS, PMOS, and CMOS families.

MPA Motorola Programmable Array. A Motorola FPGA family based on Pilkington architecture.

MPGA Masked programmed gate array.

MSB Most significant bit.

MSI Medium scale integration. A measure of complexity of integrated circuits relative to small or large scale integrated circuits.

MTBF Mean time between failures.

MTTF Mean time to failure.

MUX Multiplexer. A digital circuit that selects one or more inputs from a large number of inputs.

MX A new family of Integrator series FPGAs introduced by Actel Corp.

Netlist A shorthand description for a circuit schematic indicating logical elements connections, usually expressed in EDIF format.

NMOS N-channel MOS. A technology, which makes exclusive use of N-channel MOS transistors.

NRE Nonrecurring engineering.

NSSDC National Space Sciences Data Center.

OE Output enable.

OFFM Optimized full functional model.

OLMC Output logic macrocell. A logic block architecture used by Lattice Semiconductor for its ispLSI/pLSI CPLD and GAL families.

OMF Open Model Forum. An IP packaging/interface standard being developed by EDA industry.

OMI Open model interface.

ONO Oxide-nitride-oxide. A dielectric-based antifuse technology.

OR Ordinary event. A term used to categorize a type of solar flare phenomenon.

ORCA Optimized Reconfigurable Cell Array. Lucent Technologies trademark for its FPGA architecture.

ORP Output routing pool. An architectural term used by Lattice Semiconductor.

OTP One-time programmable. Refers to programmable devices that are (1) UV-erasable, but are not packaged in windowed packages, so that there is no way to erase and reprogram them, and (2) fuse-based devices that cannot be erased and, therefore, can be used only once.

PAL Programmable Array Logic. AMD trademark for PLD that implements logic via a programmable AND logic array driving a fixed OR logic array.

PALCE Programmable array logic CMOS electrically erasable (e.g., AMD PALCE family).

PALASM PAL assembler (AMD trademark).

PALE Programmable atomic logic element (PALE). A basic logic primitive for PLASMA chip used in Teramac Computing System.

PAM Programmable active memories.

pASIC A term used by Quicklogic for its FPGA architecture.

PBX Private branch exchange.

PCI Peripheral component interconnect. A term used for industry standard bus interconnect specification.

PDA Percent defective allowable.

PDEF Physical Design Exchange Format. A proprietary file format used by Synopsys Corp.

PDR Preliminary design review.

PDS PAL Design Specification. The name of design language used in PALASM II.

PEEL Programmable Electrically Erasable Logic. ICT trademark and name for its family of EEPLD devices.

PEM Plastic encapsulated module.

PFU Programmable function unit. The basic logic cell in ORCA FPGAs.

PFQP Plastic quad flat pack.

PGA Pin grid array (package).

PIA Programmable interconnect array. A term used in some Altera devices.

PIC Programmable integrated circuits. A generic term used to describe all types of ICs that can be programmed including PROMs, PLDs, FPGAs. In Lucent Technologies ORCA FPGAs, PIC refers to programmed input-output cells.

PIM Programmable interconnect matrix (e.g., in Cypress FLASH370I ISR CPLD family).

PIND Particle impact noise detection.

PIP Programmable interconnect point.

PLA Programmable logic array. An architecture using a programmable AND array driving a programmable OR array, with both AND/OR arrays as programmable.

PLASMA Programmable logic and switch matrix. A basic building block of the Teramac Custom Computing System.

PLC Programmable logic cell. A major building block of ORCA FPGAs.

PLD Programmable logic device. A generic term for logic device, whose function can be configured by the user.

PLICE Programmable Low-Impedance Circuit Element. Actel's trademark and name for its antifuse technology.

PLS Programmable logic sequencer. A term used by Philips Semiconductor for some of its FPLAs.

pLSI A name used by Lattice Semiconductor for its CPLD family.

PLUS A Philips-Signetics product family name.

PMD Programmable multilevel device.

PML Programmable macro logic. A Philips Semiconductors term for a PLD family.

PMOS P-channel MOS. A technology that makes exclusive use of P-channel MOS transistors.

PREP Programmable Logic Performance Corporation. A nonprofit industry consortium for developing standard benchmarks to compare speed performance and functional capacity of various PLD families.

PRL Paris Research Laboratory. Digital Equipment Corporation (DEC) facility that has been investigating the use of SRAM-based FPGAs for building reconfigurable hardware processors.

Product-of-Sums (POS) A representation of a logic function wherein the input signals are individually inverted (if necessary), then ORed together to form sums that are ANDed together. Any combinational logic function can be represented in product-of-sums form.

Product Term (pterm, p-term) An AND gate in a PLD that implements sum-of-products logic.

PSID™ Programmable switching interconnect devices. An enhancement of the FPID product family introduced by I-Cube Corp.

PTSA Product terms sharing array.

QCI Quality conformance inspection.

QFP Quad flat pack.

QML Qualified manufacturers list.

RAD-PAK An IC packaging concept provided by a company called Space Electronics Inc. (SEI) for shielding of semiconductor chips from total dose ionizing effects.

RAM Random access memory.

RH Radiation hardened.

RHA Radiation hardness assurance.

RIC Reconfigurable interface card (e.g., Giga Operations G900 RIC for image and signal processing).

ROM Read-only memory.

RTL Register transfer level.

SAA South Atlantic anomaly. This refers to apparent depression of the magnetic field over the cost of Brazil, where the Van Allen radiation belts dip low into earth's atmosphere.

SC Switched capacitor. This refers to switched capacitor principle as the prime implementation technology for EPAC modules, developed by a company called IMP, Inc.

SDA Serial distributed arithmetic.

SDF Standard delay format.

SEL Single-event latchup.

SEP Single-event phenomenon. This refers to ionization effects due to single high energy particles in the space environment striking the sensitive nodes of microelectronic devices and causing logic upsets or latchup.

SEU Single-event upset.

SLIC Supplemental logic and interconnect cells. A type of new logic block for ORCA 3C/3TA FPGAs that is roughly equivalent to a 22V10 PAL block.

SMD Standard military drawing.

SMT Surface mount technology.

SPC Statistical process control.

SPICE Simulation Program with Integrated Circuit Emphasis. A popular simulation software package.

SPGA System programmable gate arrays. They combine the architectural attributes of FPGAs and masked-programmed ASICs.

SRAM Static random-access memory.

SRC Supercomputing Research Center.

TAP Test access port. Refers to JTAG IEEE 1149.1 boundary scan architecture testability standard.

TDDB Time-dependent dielectric breakdown.

TID Total ionizing dose.

TLM Triple-layer metal. A term used to describe number of metallization layers in a submicron CMOS process.

TMR Triple modular redundancy.

TQFP Thin quad flat pack.

TTL Transistor-transistor logic.

UART Universal asynchronous receiver transmitter.

USB Universal serial bus.

UUT Unit under test.

ViaLink An amorphous (non-crystalline) silicon-based antifuse technology used by QuickLogic pASIC FPGA families.

VHDL VHSIC hardware description language.

VHSIC Very high speed integrated circuit.

Virtex A new family of high density FPGAs being introduced by Xilinx Corp.

VITAL VHDL Initiative toward ASIC Libraries. An IEEE standard.

VLSI Very large-scale integrated circuit. A measure of circuit complexity relative to large scale integrated circuit.

VQFP Very-thin quad flat pack.

VSI Virtual socket interface. An IP packaging/interface standard being developed by EDA industry.

X-BLOX Xilinx logic block architecture.

XNF Xilinx netlist format.

XPLF™ Extended programmable logic array. Philips Semiconductors' CoolRunner CPLD family architecture.

ZIA Zero power interconnect. Array used in Philips Semiconductors' CoolRunner CPLD family.

Appendix

Chapters 2 and 3 of this book provided an introduction to major vendors of programmable logic device, including CPLD and FPGA product families, architectures, and design and development tools. This appendix supplements that information by providing a detailed list of all PLD vendors and product families and subcategories, as well as design and software tool suppliers, including the software packages available from third-party suppliers. The appendix is divided into two sections. Section I is a listing of PLD suppliers and their major product families (or recent introductions). Section II is a listing of software design and development tools that are applicable to PLDs as well as gate arrays and custom ASICs.

Section I provides the following information:

- Company location, phone number, and e-mail/web page address
- Product family
- Technology
- Architecture
- Basic cell type
- Programmability
- Number of macros in cell library
- Maximum circuit density

This information is based on PLD vendors listed in Integrated System Design magazine, December 1997 issue. It should be noted that all PLD vendors are continuously improving their product families' processes and architectures for better yield, maximum circuit density, and higher performance. Therefore, PLD vendors' web pages are often valuable resources for new CPLD and FPGA families being introduced, or the older ones that are being discontinued; device data sheets and application notes; software design tool revision updates; third-party licensing agreements, and so on. Many vendors' web pages often include a frequently asked questions (FAQ) guide that provides quick review of problems concerning their product families, design applications, and software compatibility issues.

Section II provides a listing of design and development tools suppliers and software products for the PC-based Windows platform. Due to advancements made in Windows NT performance, many of the EDA vendors are looking at the Windows NT operating system using a Pentium Pro or Alpha processor as a feasible alternative to a UNIX-only platform. These vendors are beginning to port their software products to Windows NT. In the next few years, cost and performance considerations will be the major factors for migration from the UNIX-based platforms to the Windows NT environment. Section II provides a listing of each tool supplier and its location, including e-mail/web page address, product name and description, and the Window platforms supported. The following types of design tool suppliers and product categories are listed:

- FPGA and CPLD design tools
- Design entry tools
- Schematic entry tools
- Simulation tools
- Synthesis tools
- Integrated tool suites

This information is based on Windows EDA tool suppliers listed in Integrated System Design magazine, October 1997 issue. The selection of an appropriate tool depends on such factors as the user's design requirements, existing design environment, portability issues, cost, and so forth. The tool suppliers' web pages can provide additional details regarding product capabilities and, in some cases, free demo software downloads on a trial basis.

TABLE I Programmable Logic Device (PLD) Suppliers and Product Families[1]

Company and location	Product family	Technology	Architecture	Basic cell type	Programmability	Number of macros in cell library	Maximum circuit density
Actel Sunnyvale, Calif. www.actel.com	MX	CMOS	Interconnect architecture block	Mux	Antifuse	251 hard, 84 soft	30,000 gates + 3 kb memory
Atmel San Jose, Calif. pld@atmel.com www.atmel.com	Max 7000	CMOS	Multiple array matrix with continuous interconnect	EEPROM	External programmer	>300	600–5,000 usable gates
	Max 7000S	CMOS	Multiple array matrix with continuous interconnect	EEPROM	In-system programmable	>300	600–5,000 usable gates
	Flex 8000	CMOS	Flexible logic element with continuous interconnect	SRAM	In-system programmable	>300	2,500–16,000 usable gates
	Max 9000	CMOS	Multiple array matrix with continuous interconnect	EEPROM	In-system programmable	>300	6,000–12,000 usable gates
	Flex 10K	CMOS	Embedded with continuous interconnect	SRAM	In-system programmable	>300	10,000–158,000 usable gates
	Flex 6000	CMOS	OptiFlex, featuring continuous interconnect	SRAM	In-system programmable	>300	10,000–24,000 usable gates
Atmel San Jose, Calif. pld@atmel.com www.atmel.com	16V8	CMOS	Macrocells, product terms, interconnect	Combinatorial and sequential	Electrically erasable	Not applicable	250 gates
	20V8	CMOS	Macrocells, product terms, interconnect	Combinatorial and sequential	Electrically erasable	Not applicable	300 gates
	22V10	CMOS	Macrocells, product terms, interconnect	Combinatorial and sequential	Electrically erasable	Not applicable	500 gates
	V750	CMOS	Macrocells, product terms, interconnect	Combinatorial and sequential	Electrically erasable	Not applicable	750 gates
	V2500B	CMOS	Macrocells, product terms, interconnect	Combinatorial and sequential	Electrically erasable	Not applicable	2,500 gates
	F1500	CMOS	Macrocells, product terms, interconnect	Combinatorial and sequential	Electrically erasable	Not applicable	1,500 gates
	F1504AS	CMOS	Macrocells, product terms, interconnect	Combinatorial and sequential	Electrically erasable	Not applicable	2,500 gates

TABLE I Programmable Logic Device (PLD) Suppliers and Product Families[1] *(continued)*

Company and location	Product family	Technology	Architecture	Basic cell type	Programmability	Number of macros in cell library	Maximum circuit density
	F1508AS	CMOS	Macrocells, product terms, interconnect	Combinatorial and sequential	Electrically erasable	Not applicable	4,000 gates
Chip Express Santa Clara, Calif. (408) 988-2445 moreinfo@chipx.com www.chipexpress.com	QYH500 LPGA	CMOS	Gate array	Two-input NAND gate	Laser programmable	400	75,000 gates
	CX2000 LPGA	CMOS	Gate array	Module	Laser programmable	>1,500	160,000 gates + 128 kb memory
	CX2001 LPGA	CMOS	Gate array	Module	Laser programmable	>1,500	160,000 gates + 96 kb memory
	CX2002 LPGA	CMOS	Gate array	Module	Laser programmable	>1,500	160,000 gates + 128 kb memory
Cypress Semiconductor San Jose, Calif. (408) 232-4580 jxl@cypress.com www.cypress.com	Flash370i	CMOS	Programmable I/O blocks, logic blocks, interconnect	Combinational and sequential	Flash	>350	3,000 gates
	Ultra37000	CMOS	Programmable I/O blocks, logic blocks, interconnect	Combinational and sequential	Electrically erasable	>350	10,000 gates
	Max340	CMOS	Programmable I/O blocks, logic blocks, interconnect	Combinational and sequential	EPROM	>350	3,000 gates
DynaChip Sunnyvale, Calif. (408) 481-3100 support@dyna.com www.dyna.com	DL5000 Fast Field Programmable Gate Array (FFPGA)	BiCMOS	ECL/PECL/GTL I/O, active repeater interconnect, logic blocks	AND/OR, mux, arithmetic, sequential	SRAM	>200	10,000 gates
Gatefield Fremont, Calif. (800) 818-5052 (510) 429-5757 gfinfo@gatefield.com www.gatefield.com	GF250F	CMOS flash	Fine-grained, gate array-like sea of tiles, optimized for synthesis	3 inputs/1 output, combinational or sequential	In-system programmable	>1,000	25,000–100,000 gates
	GF260F	CMOS flash	Fine-grained, gate array-like, seal of tiles with configurable embedded memory blocks, optimized for synthesis	3 inputs/1 output, combinational or sequential	In-system programmable	>1,000	92,000 gates + 23 kb memory
ICT San Jose, Calif. (480) 434-0678 www.ictpld.com	PEEL array	CMOS	Programmable interconnect and I/O macrocells	Combinatorial and sequential	Electrically erasable	Not applicable	10,000 gates

Company and location	Product family	Technology	Architecture	Basic cell type	Programmability	Number of macros in cell library	Maximum circuit density
Lattice Semiconductor Hillsboro, Ore. (503) 681-0118 steve_stark@latticesemi.com www.latticesemi.com	ispLSI 1000/E family PLDs	EECMOS	Complex PLD. GLBs fully interconnected by global routing pool. Output routing pool connects GLBs to I/O pins.	Generic logic block PLA with product terms sharing feeds and program registers	In-system programmable	>300	2,000–8,000 gates + 96–288 registers
	ispLSI 2000 family PLDs	EECMOS	Complex PLD. GLBs fully interconnected by global routing pool. Output routing pool connects GLBs to I/O pins.	Generic logic block PLA with product terms sharing feeds and program registers	In-system programmable	>300	1,000–6,000 gates + 32–128 registers
	ispLSI 2000V family PLDs	EECMOS	Complex PLD. GLBs fully interconnected by global routing pool. Output routing pool connects GLBs to I/O pins.	Generic logic block PLA with product terms sharing feeds and program registers	In-system programmable	>300	1,000–6,000 gates + 32–128 registers
	ispLSI 3000 family PLDs	EECMOS	Complex PLD. GLBs fully interconnected by global routing pool. Output routing pool connects GLBs to I/O pins.	Twin generic logic blocks PLA with product terms sharing feeds and program registers, JTAG	In-system programmable	>300	7,000–14,000 gates + 192–512 registers
	ispLSI 6000 family PLDs	EECMOS	Complex PLD with twin GLBs, register/counter module, and memory modules available in FIFO and single- or dual-port SRAM configurations	Twin generic logic blocks PLA with product terms sharing feeds and program registers, JTAG	In-system programmable	192	25,000 gates + 4 kb memory/416 registers
	ispGAL22V10 and GAL22V10	EECMOS	22V10	Programmable AND and fixed OR array with registers	In-system programmable and electrically erasable	Not applicable	750 PLD gates
	GAL16V8 AND GAL20V8	EECMOS	GAL architecture	Programmable AND and fixed OR array with registers	Electrically erasable	Not applicable	300/400 PLD gates
	GAL18V10 and GAL26CV12	EECMOS	GAL architectures, subset, and superset	Programmable AND and fixed OR array with registers	Electrically erasable	Not applicable	550/850
	GAL20RA10	EECMOS	Asynchronous PLD	Programmable AND and fixed OR array with registers	Electrically erasable	Not applicable	700 PLD gates
	GAL6002 and GAL6001	EECMOS	FPLA	Programmable AND and programmable OR array with buried registers	Electrically erasable	Not applicable	1,200 PLD gates

TABLE I Programmable Logic Device (PLD) Suppliers and Product Families[1] *(continued)*

Company and location	Product family	Technology	Architecture	Basic cell type	Programmability	Number of macros in cell library	Maximum circuit density
	GAL16VP8 and GAL20VP8	EECMOS	16V8 and 20V8	Programmable AND and fixed OR array with registers	Electrically erasable	Not applicable	300/400 PLD gates
	GAL16LV8, GAL20LV8, GAL22LV10, and GAL26CLV12	EECMOS	GAL architecture	Programmable AND and fixed OR array with registers	Electrically erasable	Not applicable	300–900 PLD gates
	GAL16LV8ZD, GAL20LV8ZD, GAL22LV10ZD, and GAL20V8ZD	EECMOS	GAL architecture	Programmable AND and fixed OR array with registers	Electrically erasable	Not applicable	300–750 PLD gates
	GAL16V8Z, GAL16V8ZD, GAL20V8Z, and GAL20V8ZD	EECMOS	GAL architecture	Programmable AND and fixed OR array with registers	Electrically erasable	Not applicable	300–400 PLD gates
	GAL20XV10	EECMOS	XOR	Programmable AND and fixed OR array with registers	Electrically erasable	Not applicable	650 PLD gates
	ispGDS family generic digital switches	EECMOS	7 × 7 to 11 × 11 programmable switch matrix	Configurable I/O cell as either an input, an inverting or noninverting output, or a fixed TTL-high	In-system programmable	Not applicable	200 PLD gates
	ispGDX family generic digital crosspoint devices	EECMOS	80 × 80 to 160 × 160 programmable switch matrix	Combinatorial, latched, and registered inputs or outputs with individual tristate and polarity control	In-system programmable, programmable I/O cells	Not applicable	80–160 I/O pins, 80–160 registers
Lucent Technologies Allentown, Pa. (800) 372-2447 orcafpga@aloft.lucent.com www.lucent.com/micro/fpga	ATT3000 series	CMOS	Symmetrical cell array	Look-up table	SRAM	>275	1,800–4,500 gates
	OR2CxxA series	CMOS	Configurable-grain logic array	Look-up table	SRAM	>300	4,800 gates and up
	OR2TxxA series	CMOS	Configurable-grain logic array	Look-up table	SRAM	>300	4,800 gates and up
	OR3Cxx series	CMOS	Configurable-grain logic array	Look-up table/PAL	SRAM	>300	40,000 gates and up

Company and location	Product family	Technology	Architecture	Basic cell type	Programmability	Number of macros in cell library	Maximum circuit density
	OR3Txxx series	CMOS	Configurable-grain logic array	Look-up table/PAL	SRAM	>300	40,000 gates and up
Motorola Semiconductor Products Sector Chandler, Ariz. (602) 732-3742 sps.motorola.com/fpga	MPA1000	CMOS	SRAM-based, fine-grain architecture using hierarchical routing	Two-input NAND with programmable inverter. Each cell may also have a secondary function (flip-flop, latch, XOR, wired-OR)	SRAM-based, external EPROM and EEPROM	290	22,000 gates
Philips Semiconductors Albuquerque, N.M. (505) 822-7829 coolpld@abq.sc.philips.com www.coolpld.com	CoolRunner CPLDs	EECMOS	Extended PLA (XPLA) logic blocks with programmable AND and programmable OR array, interconnect array, I/O, asynchronous and synchronous clocking	Logic blocks with inputs up to 36 wide and 4 OEs, macrocell I/Os with configurable registers or combinatorial logic	Electrically erasable, some in-system programmable	32–128	1,000–4,000 gates
	GAL-type devices	EECMOS, EPROM, CMOS, BiCMOS	V architecture, combinatorial, and D register I/O	Macrocell I/O	Electrically erasable (EECMOS), one-time programmable (EPROM AND BiCMOS)	Not applicable	500–600 gates
	Programmable logic arrays	Bipolar	Programmable AND programmable OR array	Bidirectional I/Os or combinatorial output	One-time programmable	Not applicable	500–700 gates
	Sequencers	Bipolar and EPROM CMOS	Programmable AND and programmable OR array, buried and output registers (one part with asynchronous and synchronous clocks)	JK registered outputs, combinational and JK registered I/O, SR registered outputs with feedback paths, or buried or output D, T, or JK registers with separate feedback paths, combinational I/O with feedback	One-time programmable	Not applicable	500–700 gates
QuickLogic Sunnyvale, Calif. (408) 990-4000 (800) 842-FPGA info@quicklogic.com www.quicklogic.com	pASIC 3	CMOS	Variable-grained logic blocks optimized for Verilog/VHDL synthesis, programmable I/O cell with input registers, and JTAG boundary scan	Variable granularity, 16 simultaneous inputs, 5 concurrent outputs, 1 flip-flop, plus registers on all pins, dual-port RAM	Antifuse	>700	65,000 gates + 15 kb memory

TABLE I Programmable Logic Device (PLD) Suppliers and Product Families[1] *(continued)*

Company and location	Product family	Technology	Architecture	Basic cell type	Programmability	Number of macros in cell library	Maximum circuit density
	pASIC 2	CMOS	Variable-grained logic blocks optimized for Verilog/VHDL synthesis, programmable I/O cell with input registers, and JTAG boundary scan	30 inputs, 5 concurrent outputs, 1 flip-flop, plus registers on all pins	Antifuse	>700	16,000 gates
	pASIC 1	CMOS	Coarse-grained logic blocks optimized for Verilog/VHDL synthesis, individual OE control on each pin	14 simultaneous inputs, 5 concurrent outputs, 1 flip-flop	Antifuse	>700	14,000 gates
Vantis Sunnyvale, Calif. (888) VANTIS2 www.vantis.com	Mach 5 CPLD family	CMOS	Multiple PAL blocks with switch matrix interconnect	Electrically erasable	In-system programmable/ JTAG	0 (full library in development)	20,000 gates
	Mach 4	CMOS	SpeedLocked multiple PAL blocks with triple switch matrix interconnect	Electrically erasable	In-system programmable/ JTAG	0 (full library in development)	10,000 gates
	Mach 1 and 2	CMOS	SpeedLocked multiple PAL blocks with switch matrix interconnect	Electrically erasable	In-system programmable/ JTAG	0 (full library in development)	3,600 gates
WSI Fremont, Calif. dprull@wsitd.com www.wsipsd.com	PSD3XX	CMOS	PLD, I/O, programmable address decoder, EPROM, SRAM, programmable MCU interface	Registered or combinatorial	EPROM	No libraries	800 gates + 32 kb RAM, 128 kB EPROM
	PAD4XXA1	CMOS	CPLD, I/O, programmable address decoder, EPROM, SRAM, programmable MCU interface	Registered or combinatorial	EPROM	No libraries	1,500 gates + 32 kb RAM, 128 kB EPROM
	PSD4XXA2	CMOS	CPLD, I/O, programmable address decoder, EPROM, SRAM, programmable MCU interface	Registered or combinatorial	EPROM	No libraries	2,500 gates + 32 kb RAM, 128 kB EPROM

Company and location	Product family	Technology	Architecture	Basic cell type	Programmability	Number of macros in cell library	Maximum circuit density
	PSD5XX	CMOS	CPLD, I/O, programmable address decoder, EPROM, SRAM, programmable MCU interface, 4 programmable timer/counters, 8-level priority programmable interrupt controller	Registered or combinatorial	EPROM	No libraries	3,000 gates + 32 kb RAM, 128 kB EPROM
	PSD6XX	CMOS	CPLD, I/O, programmable address decoder, EPROM, SRAM, programmable MCU interface	Registered or combinatorial	EPROM	No libraries	1,500 gates + 4 kb RAM, 128 kB EPROM
	PSD7XX	CMOS	CPLD, I/O, programmable address decoder, EPROM, SRAM, programmable MCU interface, programmable power-on/reset, programmable pulse width, programmable brown-out detection, programmable watchdog timer	Registered or combinatorial	EPROM	No libraries	1,500 gates + 4 kb RAM, 128 kB EPROM
Xilinx San Jose, Calif.	XC4000E FPGA family	CMOS	Programmable logic blocks, I/O blocks, interconnect, plus user RAM, wide decoders, boundary-scan logic, fast-carry logic, 8 global nets, asynchronous/synchronous dual-port RAM option	Two 4-input function generators, one 3-input function generator, and two registers per cell; each 4-input function generator can be used as logic or user RAM	SRAM	482 total elements (macros and primitives)	3,000–25,000 gates + 32 kb memory
	XC4000XL/XV FPGA family	CMOS	Programmable logic blocks, I/O blocks, interconnect, plus user RAM, wide decoders, boundary-scan logic, fast-carry logic, 8 global nets, asynchronous/synchronous dual-port RAM option	Two 4-input function generators, one 3-input function generator, and two registers per cell; each 4-input function generator can be used as logic or user RAM	SRAM	500 total elements (macros and primitives)	28,000–125,000 gates + 148 kb RAM

TABLE I Programmable Logic Device (PLD) Suppliers and Product Families[1] *(continued)*

Company and location	Product family	Technology	Architecture	Basic cell type	Programmability	Number of macros in cell library	Maximum circuit density
	XC6200 reconfigurable coprocessor FPGA family	CMOS	SRAM-based FPGA, programmable logic block I/O block and interconnect; built-in processor interface; optimized for structured logic, datapath, reconfigurable coprocessing applications	Two-input logic function and flip-flop	Partially reconfigurable, SRAM-based	>300 total elements (macros and primitives)	9,000–100,000 gates + 256 kb memory
	XC9500 CPLD family	CMOS	Advanced 36V18 function blocks interconnected by fully populated switch matrix	Flash cell	In-system programmable, with 10,000 erase programming cycles	450 total elements (macros and primitives)	800–6,400 gates

Reference

1. Alexander Braun, "Focus Report: Programmable Logic," *ISD* magazine, Dec. 1997.

TABLE II PLD Software Tool Vendors for PC-Based Windows Platforms[1]

Company and location	Product name	Description	Windows platforms
FPGA and CPLD Design			
Atmel San Jose, Calif. (408) 441-0311 fpga@atmel.com www.atmel.com	FPGA/DSP Designer 4.0	Design suite for implementing 2,000- to 100,000-gate designs into AT6000 series coprocessor FPGAs.	3.1, 95, and NT
Cypress Semiconductors San Jose, Calif. (408) 943-2600 lis@cypress.com	Warp2	Compiles VHDL design input and then outputs a JEDEC file for the selected PLD or CPLD. For simulation, provides VHDL and Verilog models.	3.1, 95, and NT
	Warp2Sim	All the functions of Warp2 plus timing simulation provided with ViewLogic's ViewSim simulator.	3.1, 95, and NT
	Warp3	All the capabilities of Warp2Sim with the addition of schematic capture, mixed-mode design entry, and Speedwave source-level debugging.	3.1, 95, and NT
Lattice Semiconductor Hillsboro, Ore. (800) ISP-PLDS isp@latticesemi.com www.latticesemi.com	ispDS+ HDL Synthesis-Optimized Logic Fitter	Interfaces with all leading CAE vendor tools. Includes GUI, explore tool, pin editor, static timing analyzer, ispDownload.	3.1, 95, and NT
	ispVHDL Viewlogic System	Includes Viewlogic VHDL synthesis, capture, timing simulator, and ispDS+ Fitter, plus GUI, explore tool, pin editor, static timing analyzer, ispDownload.	3.1, 95, and NT
	ISP Synario System	Includes ABEL-HDL, capture, functional simulator, and ispDS+ Fitter, GUI, explore tool, pin editor, static timing analyzer, ispDownload	3.1, 95, and NT
	ispHDL Advanced System for Synplicity	Includes Synplify VHDL and Verilog synthesis and ispDS+ Fitter, GUI, explore tool, pin editor, static timing analyzer, ispDownload.	3.1, 95, and NT
	ispDS	Includes Lattice-HDL, functional simulator, and ispDS, GUI, interactive logic editor, ispDownload.	3.1, 95, and NT
Mentor Graphics Wilsonville, Ore. (503) 685-7000 www.mentorg.com	Galileo Extreme	Provides an environment for FPGA and CPLD design. Includes Logic Explorer for bilingual HDL synthesis and architecture-specific area and timing optimization.	95 and NT
Morphologic Nashua, N.H. (888) MORPH-IT info@morpholgic.com www.morphologic.com	Evolution	Multichip, multivendor tool for implementing DSP algorithms in FPGA-based systems. Includes graphical entry tool, simulator, multichip floorplanner, and rules-based requirements monitor.	95 and NT
Protel Technology Santa Clara, Calif. (800) 544-4186 salesusa@protel.com www.protel.com	Advanced PLD	Universal programmable logic design for Protel's EDA/Client platform. Frequent library updates available from Protel's Web site.	3.1, 95, and NT
QuickLogic Sunnyvale, Calif. (408) 990-4000 info@quicklogic.com www.quicklogic.com	QuickWorks	Integrated FPGA development environment with schematic capture, text entry, VHDL and Verilog synthesis, place and route, timing analysis, simulation, and programming for all QuickLogic FPGAs	3.1, 95, and NT
	QuickTools	Place and route, timing analysis, third-party CAE tool interfaces, and programming for all QuickLogic FPGAs	3.1, 95, and NT
VeriBest Boulder, Colo. (800) VERIBEST info@veribest.com www.veribest.com	VeriBest FPGA Synthesis	FPGA synthesis powered by Synopsys FPGA Express for VHDL/Verilog synthesis and optimization for FPGAs and CPLDs	NT
	VeriBest FPGA Design View	Integrates the FPGA/CPLD design flow for schematic, mixed schematic-HDL, and pure HDL flows, including programmable logic vendors' tools	NT

TABLE II PLD Software Tool Vendors for PC-Based Windows Platforms[1] *(continued)*

Company and location	Product name	Description	Windows platforms
	Veribest PLD	Provides logic synthesis/optimization for PLD design. Includes automatic partitioning and HDL simulation model generation.	NT
Viewlogic Systems Marlboro, Mass. (800) 873-8439 viewdirect@viewlogic.com www.viewlogic.com	Intelliflow	Design manager. Recognizes the types of source files and technology to automatically configure flows to step the designer through synthesis, simulation, place-and-route, and analysis. Supports both schematic and HDL (VHDL and Verilog) design methodologies.	95 and NT
	ViewPLD	Allows easy PLD design specification using ABEL with hierarchy support. Optimizes designs with automatic device fitting.	95 and NT
Xilinx San Jose, Calif. (408) 559-7778 fpga@xilinx.com www.xilinx.com	Alliance Series Standard	FPGA/CPLD design solutions leveraging "Open Systems" integration with EDA partners.	95 and NT
	Alliance Series Base	FPGA/CPLD design solutions leveraging "Open Systems" integration with EDA partners. Provides design entry, simulation, synthesis, and device implementation in a	95 and NT
	Foundation Series	tightly integrated design environment for a broad range of IC densities.	95 and NT
		Design Entry	
Escalade Santa Clara, Calif. (408) 654-1600 info@escalade.com www.escalade.com	DesignBook	Allows designers to capture functionality or design intent, synthesis constraints, and simulation test benches using a combination of language-independent paradigms. Its symbolic entry ensures portability between Verilog HDL and VHDL.	95 and NT
	Design Explorer	Offers Escalade DesignBook users automation for design space exploration. Parameter space configuration and constraining drives the launch and management of multiple jobs. Job queue management and incremental results extraction permit real-time progress tracking.	95 and NT
Mentor Graphics Wilsonville, Ore. (503) 685-7000 www.mentorg.com	FlowXpert	For project managers or technical leads. Tracks key project deliverables, provides progressive disclosure of project status, and automatically captured process metrics and history.	95 and NT
	Renior	Highly automated HDL graphical entry tool. Using state, block, flowchart, and truth editors, automatically generates synthesizable VHDL or Verilog code.	95 and NT
VeriBest Boulder, Colo. (800) VERIBEST info@veribest.com www.veribest.com	VeriBest Graphical High-Level Design	Graphical state machine design suite that generates synthesizable VHDL or Verilog. Integrated with Design Capture.	95 and NT
	VeriBest HDL Writer	VHDL/Verilog editor integrated with Design Capture, FPGA DesignView, VHDL Simulator, and FPGA Synthesis.	NT
	VeriBest Design Capture	Schematic and design entry environment for layout and simulation tools. Support for many netlist formats, including ASCII and DEIF. Tight Integration maintained for downstream VeriBest tools.	95 and NT
Viewlogic Systems Marlboro, Mass. (800) 873-8439 viewdirect@viewlogic.com www.viewlogic.com	StateCAD Workview Office Edition	Graphical editor for entering state diagrams. Produces error-free, synthesizable VHDL, Verilog, ABEL-HDL, and C from state diagrams.	95 and NT
		Schematic	
Ivex Design International Beaverton, Ore. (503) 531-3555 info@ivex.com www.ivex.com	WinDraft 2	Includes libraries, parts palette, TrueType fonts, ability to "copy" and "paste" to other applications. Full ASCII importing and exporting, and selectable line and bus thickness.	3.1, 95, and NT

TABLE II PLD Software Tool Vendors for PC-Based Windows Platforms[1] *(continued)*

Company and location	Product name	Description	Windows platforms
Mentor Graphics Wilsonville, Ore. (503) 685-7000 www.mentorg.com	Personal Architect Pro	Allows designers to create and edit Design Architect schematics from any PC. Works in traditional and satellite environments, whether in or out of the office.	3.1, 95, and NT
	Design Architect NT	Complete schematic environment for IC and PCB design. Integrated with simulation and board layout.	NT
MyCAD Sunnyvale, Calif. (408) 745-6785 sales@mycad.com www.mycad.com	MyLogic Station	Contains logic simulation. Supports multiple windows, views, pages, and functions; unlimited hierarchical design; and SPICE output. Includes EDIF netlist import/export interface.	95 and NT
OrCAD Beaverton, Ore. (800) 671-9505 info@orcad.com www.orcad.com	OrCAD Capture	Schematic entry tool for PCB and FPGA/CPLD design, including integrated schematic and editors, VBA-based macros, customizable BOM and DRC. Also includes library with 22,000+ parts and 30+ netlist formats, including VHDL and Verilog.	3.1, 95, and NT
Protel Technology Santa Clara, Calif. (800) 544-4186 salesusa@protel.com www.protel.com	Advanced Schematic 3	Fully hierarchical entry platform that features unparalleled automation, 42,000+ parts, user customization, and integration support for Protel and third-party tools	3.1, 95, and NT
Viewlogic Systems Marlboro, Mass. (800) 873-8439 viewdirect@viewlogic.com www.viewlogic.com	ViewDraw for Windows	Graphical tool for design capture and analysis. Includes schematic editor with OLE automation interface and graphical design manager. Linked to synthesis, simulation, timing analysis, and layout.	95 and NT
		Simulation	
Accolade Design Automation Duvall, Wash. (800) 470-2686 (425) 844-6479	PeakVHDL Pro	Includes 32-bit direct-compiled VHDL simulator with waveform viewer and source-level debugger, built-in source code editor, hierarchy browser, and VHDL Wizards	95 and NT
	Peak VHDL Pro+VITAL	Includes all features of PeakVHDL and adds support for IEEE 1076.4 (VITAL) as well as advanced VHDL editing features.	95 and NT
Analogy Beaverton, Ore. (503) 626-9700 info@analogy.com www.analogy.com	Saber Designer for NT	Mixed-signal, mixed-technology design and analysis tool. Includes Analogy's HDL-based Saber simulator, SaberSketch design capture, and SaberScope waveform viewing applications.	NT
Avanti Fremont, Calif. info@avanticorp.com www.avanticorp.com	Polaris	Verilog interpreted and compiled code simulator that offers IEEE 1364 compliance, full SDF back-annotation, and PLI integration.	95 and NT
Cadence Design Systems San Jose, Calif. (408) 943-1234 info@cadence.com www.cadence.com	Verilog-XL, Turbo NT	ASIC sign-off simulator	NT
CAE Plus Austin, Tex. (512) 338-0165 comm@cae-plus.com www.cae-plus.com	ASVP-LAB	Automates the creation of application-specific virtual prototypes and improves the debugging productivity of system IC software, hardware, and performance.	95 and NT
Deutsch Research Mountain View, Calif. info@deutsch.com www.deutsch.com	Dr. Spice	Circuit simulator for PCB as well as IC designs. Tight links to several schematic packages.	3.1, 95, and NT
Mentor Graphics Wilsonville, Ore. (503) 685-7000 www.mentorg.com	Quick HDL Lite	HDL simulation environment with such features as direct-compiled code architecture, easy-to-learn user interface, and debugging.	3.1, 95, and NT

TABLE II PLD Software Tool Vendors for PC-Based Windows Platforms[1] *(continued)*

Company and location	Product name	Description	Windows platforms
MyCAD Sunnyvale, Calif. (408) 745-6785 sales@mycad.com www.mycad.com	MySpice Station	Analog circuit simulator based on Berkeley SPICE 3.E1. Supports dc sweep, ac sweep, noise, transient, transfer function, pole-zero analysis. Supports BSIM2, BSIM3, MOS level 6, and lossy transmission line models. Includes postprocessor.	95 and NT
	MyVHDL Station	VHDL simulator tool that contains source-level debugger, waveform editor, and displayer. Accepts standard IEEE 1076-1997. Provides many execution and display windows.	95 and NT
Simulation Technologies St. Paul, Minn. (612) 631-1858 info@simtech.com www.simtech.com	VirSim	Interactive multiwindow graphical verification debugging environment.	95 and NT
	V-CPU	Provides a virtual prototype, built with models and simulators, of the target system early in the design process	95 and NT
Simucad Union City, Calif. (510) 487-9700 silos@simucad.com www.simucad.com	Silos III	Performs logic and fault simulation for ASIC and FPGA designs using Verilog HDL at the behavioral, gate, and switch levels.	95 and NT
Tanner EDA Pasadena, Calif. (626) 792-3000 sales@tanner.com www.tanner.com/eda	T-Spice Pro	Circuit simulator for analog and mixed-signal IC design. Includes BSIM3V2 and BSIM3V3 models. Offers table-based simulation for large designs (300,000+ devices).	95 and NT
Tatum Labs Ann Arbor, Mich. (313) 663-8810 tatumlabs.@provide.net www.tatumlabs.com	SpiceAge*4W	Analog/mixed-signal simulator. Provides basic and advanced analyses—dc, frequency, transient, Fourier, Monte Carlo, etc., and Bode, Smith, and cascade plots.	3.1, 95, and NT
VeriBest Boulder, Colo. (800) VERIBEST info@veribest.com www.veribest.com	VeriBest VHDL Simulator	VHDL development environment that supports VHDL '87 and VHDL '93 for ASIC, FPGA, PLD, PCB, and system simulation.	NT
	VeriBest Verilog Simulator	Verilog-XL clone simulator for ASIC, FPGA, PLD, PCB, and system simulation. Includes a comprehensive simulation and analysis environment.	NT
	VeriBest Wave Bench	Stimulus creation environment. Stimulus is created by drawing waveforms graphically or from a text format. Generates VHDL or Verilog.	NT
	VeriBest Analog	Analog simulation environment for IC, board, and system simulation. Behavioral modeling options, SPICE compatibility, tight integration with design entry tools.	NT
Viewlogic Systems Marlboro, Mass. (800) 873-8439 viewdirect@viewlogic.com www.viewlogic.com	ViewSim for Windows	Gate-level simulator and waveform viewer. Built-in graphical navigator to view the hierarchy. Offers extensive library support for systems and FPGA design.	95 and NT
	ViewAnalog for Windows	SPICE simulator for verifying analog and mixed-signal designs. Provides modeling libraries with 10,000+ components including rf, digital, and power supply libraries. Integrated digital simulator for analyzing circuits with digital gates.	95 and NT
	Speedwave for Windows	VHDL simulator with VITAL support. Includes hierarchical design browser, source-level debugger, and behavioral, gate-level, and RTL waveform viewer. Support for Smartmodels and SWIFT models.	95 and NT
	VCS for Windows	Verilog simulator. Includes hierarchical design browser, source-level debugger, and waveform viewer. Complete PLI support.	95 and NT
	Fusion for Windows	Single cosimulation environment for mixed designs containing gates, VHDL, Verilog, and SPICE models. Includes a common source-level debugger, user interface, and waveform viewer.	95 and NT

TABLE II PLD Software Tool Vendors for PC-Based Windows Platforms[1] *(continued)*

Company and location	Product name	Description	Windows platforms
		Synthesis	
Accolade Design Automation Duvall, Wash. (800) 470-2686 (425) 844-6479	PeakFPGA Synthesis	Provides FPGA synthesis for devices from all major FPGA vendors, including Actel, Altera, Lattice, Lucent, QuickLogic, and Xilinx	95 and NT
Aceo Technology Fremont, Calif. info@aceo.com www.aceo.com	Asyn, Gatran, SoftWire	Provides logic synthesis, intelligent design migration, and multi-FPGA partitioning. Includes timing analysis, hierarchy browser, constraint management, and schematic viewing.	95 and NT
Exemplar Logic Alameda, Calif. (510) 337-3700 info@exemplar.com www.exemplar.com	Galileo Extreme	Automated logic synthesis, simulation, and timing analysis tools for technology-independent design methods for FPGAs and CPLDs	95 and NT
	Leonardo	Interactive logic synthesis, simulation, optimization, and timing analysis tools for technology-independent design methods for FPGAs, CPLDs, and ASICs	95 and NT
Mentor Graphics Wilsonville, Ore. (503) 685-7000 www.mentorg.com	Leonardo	Platform-independent FPGA and ASIC synthesis environment. Supports bilingual HDL synthesis, along with architecture-specific area and timing optimization.	95 and NT
Synplicity Sunnyvale, Calif. (408) 617-6000 marketing@synplicity.co m www.synplicity.com	Synplify	Verilog and VHDL synthesis for high-performance FPGAs and CPLDs.	3.1, 95, and NT
	HDL Analyst	Add-on to Synplify that automatically creates RTL and gate-level schematics. Provides cross-probing with the HDL source.	3.1, 95, and NT
Tatum Labs Ann Arbor, Mich. (313) 663-8810 tatumlabs.@provide.net www.tatumlabs.com	SuperFilter	Passive, active, switched-capacitor, and digital filter synthesis. Unlimited orders in multiple mixed cascades. Generates topology, component values, coefficients based on filter type, poles, etc.	3.1, 95, and NT
Viewlogic Systems Marlboro, Mass. (800) 873-8439 viewdirect@viewlogic.com www.viewlogic.com	ViewSynthesis for Windows	VHDL and Verilog synthesis and optimization for FPGAs and ASICs. Contains technology-specific FPGA optimizations to obtain maximum use and speed.	95 and NT
		Tool Suites	
Accel Technologies San Diego, Calif. (800) 488-0680 sales@acceltech.com www.acceltech.com	Accel ED	Software for analog, digital, and mixed PCBs. Integrated schematic capture, PCB layout, and high-performance autorouting and autoplace applications.	95 and NT
Aldec Henderson, Nev. (702) 456-1222 info@aldec.com www.aldec.com	Active-CAD	An integrated solution with schematic, HDL entry tools integrated with Netlist simulation, VHDL simulation and synthesis. Project Manager includes direct interfaces to all popular FPGA/CPLD place-and-route tools.	3.1, 95, and NT
CAD-Migos Software Tools Redwood City, Calif. (415) 369-5853 info@cadmigos.com www.cadmigos.com	Spice-It! for Windows	An integrated electronic circuit design and mixed-mode SPICE simulation system.	95 and NT
Hewlett-Packard, HP EEsof Division Westlake Village, Calif. (818) 879-6200 www.hp.com/go/hpeesof	Advanced Design System	Integrates a range of communication system design technologies, including microwave, rf, DSP, and electromagnetic simulation tools.	95 and NT
	DSP Designer and DSP Synthesis	Includes technologies and models needed for DSP design and synthesis, including behavioral synthesis.	95 and NT

TABLE II PLD Software Tool Vendors for PC-Based Windows Platforms[1] *(continued)*

Company and location	Product name	Description	Windows platforms
	Series IV/PC	Schematic, layout, circuit simulators, and libraries of 90,000+ parts.	95 and NT
	Touchstone and Libra for Windows	Linear and nonlinear frequency-domain circuit simulators.	3.1, 95, and NT
OrCAD Beaverton, Ore. (800) 671-9505 info@orcad.com www.orcad.com	OrCAD Capture, Enterprise Edition	All features of OrCAD Capture plus synchronization of CIS/manufacturing-related data with schematic parts used in a design. Creates and validates designs complete with part numbers, footprints, package types, and other production data.	95 and NT
	OrCAD Express	Board-level and multivendor programmable device design environment. Offers mixed schematic and VHDL design entry, VHDL simulation, and synthesis. Integrated with place-and-route tools from Actel, Altera, Lattice, Lucent, Vantis, and Xilinx.	3.1, 95, and NT
	OrCAD Express, Enterprise Edition	All features of OrCAD Express plus synchronization of CIS/manufacturing-related data with schematic parts used in a design. Creates and validates designs complete with part numbers, footprints, package types, and other production data.	3.1, 95, and NT
Synario Design Automation Redmond, Wash. (888) SYNARIO edasales@synario.com www.synario.com	Synario Programmable IC Design Solution	Suite of tightly integrated EDA tools for FPGA, CPLD, and PLD design	95 and NT
	Synario Board Design Solution	Suite of tightly integrated EDA tools for system and board design.	95 and NT
Tanner EDA Pasadena, Calif. (626) 792-3000 sales@tanner.com www.tanner.com/eda	Tanner Tools Pro	Integrated solution, including schematic capture, circuit simulation, automatic place-and-route, full custom layout, DRC, netlist extraction, LVS, and standard-cell layout libraries	3.1, 95, and NT
	Tanner Tools, MEMS Pro	Package offering schematic drawing, electrical and mechanical simulation, layout drawing with cross- section viewing and MEMS element libraries. Includes Tanner Tools Pro for IC design.	95 and NT
	Tanner Tools MCM Pro	Integrated MCM design, including schematic capture, DIE file libraries, automatic place-and-route thermal analysis, layout, verification, and signal integrity analysis.	95 and NT
Viewlogic Systems Marlboro, Mass. (800) 873-8439 viewdirect@viewlogic.com www.viewlogic.com	Workview Office	Tool suite including schematic capture; synthesis; simulation; analysis of ASICs, FPGAs, PCBs, and systems.	95 and NT

Reference

1. Carolyn Mathas, "Focus Report: Windows EDA Tools," *ISD* magazine, Oct. 1997.

Index

Index